The Quantum Nature of Light

From photon states to quantum fluids of light

IOP Series in Quantum Technology

Series Editor: **Barry Garraway** (School of Mathematical and Physical Sciences, University of Sussex, UK), **Barry Sanders** (Institute for Quantum Science and Technology, University of Calgary, Canada) and **Lincoln Carr** (Quantum Engineering Program, Colorado School of Mines, USA)

About the series

The IOP Series in Quantum Technology is dedicated to bringing together the most up to date texts and reference books from across the emerging field of quantum science and its technological applications. Prepared by leading experts, the series is intended for graduate students and researchers either already working in or intending to enter the field. The series seeks (but is not restricted to) publications in the following topics:

- Quantum biology
- Quantum communication
- Quantum computation
- Quantum control
- Quantum cryptography
- Quantum engineering
- Quantum machine learning and intelligence
- Quantum materials
- Quantum metrology
- Quantum optics
- Quantum sensing
- Quantum simulation
- Quantum software, algorithms and code
- Quantum thermodynamics
- Hybrid quantum systems

A full list of titles published in this series can be found here: https://iopscience.iop.org/bookListInfo/iop-series-in-quantum-technology.

The Quantum Nature of Light

From photon states to quantum fluids of light

J T Mendonça

Instituto de Plasmas e Fusão Nuclear, Instituto Superior Técnico, Lisbon, Portugal

IOP Publishing, Bristol, UK

ISBN 978-0-7503-2786-2 (ebook)
ISBN 978-0-7503-2784-8 (print)
ISBN 978-0-7503-2787-9 (myPrint)
ISBN 978-0-7503-2785-5 (mobi)

DOI 10.1088/978-0-7503-2786-2

Version: 20220901

IOP ebooks

British Library Cataloguing-in-Publication Data: A catalogue record for this book is available from the British Library.

Published by IOP Publishing, wholly owned by The Institute of Physics, London

IOP Publishing, Temple Circus, Temple Way, Bristol, BS1 6HG, UK

US Office: IOP Publishing, Inc., 190 North Independence Mall West, Suite 601, Philadelphia, PA 19106, USA

Sôbolos rios que vão por Babilónia, me achei.

—Camões

Contents

Preface

This book results from my walks across the landscape of Physics, and was produced in nearly complete isolation, during the difficult pandemic years of 2020–2021. In previous years I had however visited many countries, and profited from long stays in different institutions, where I worked on a variety of problems, chosen by a mixture of chance and necessity. From this wandering around Physics and around the World, something stayed in my mind and eventually reached the pages of the present book.

I am thankful to a large number of people, from colleagues to students, but I will only refer explicitly to those who worked with me in the last 10 years. My longest and most repeated stays abroad in this period where in Brazil. I learned about the culture and life-style behind the distorted touristic image of Rio de Janeiro thanks to my friend António Serbeto, who also introduced me to the area of free-electron lasers and co-authored several papers on this and other topics. But I was first and mainly received in São Paulo, due to the friendly attention of Ricardo Galvão, a powerful theoretical mind with an experimental twist. Recently, he became famous in the World Press, when he courageously exposed the fake news delivered by the president of his country, concerning the Amazon deforestation. Fernando Haas, one of the best theoreticians of the new generation, made me visit Curitiba and later Rio-Grande do Sul, where not only did I work with him in interesting new physics, but discovered the best grills in the world. I was also invited to an incredible barbecue in São Carlos by Luis Marcassa, who introduced me to the problems of Rydberg atoms. In Brasilia, I worked with Victor Dodonov, an influential name in quantum optics. I became familiar with Bose–Einstein condensates thanks to my collaboration with Arnaldo Gammal, during long and not completely satisfying stays at USP.

In Europe, I profited from long stays in Greece, France, Spain and the UK, apart from a few shorter stays in Sweden. In Athens, I was generously received by Kyriakos Hizanidis, and I lived a few incredible months near the iconic Syntagma Square, while working with him on nonlinear problems. I will not forget the great wine and olive-oil, he brought from his piece of land in Patras. In the Island of Crete I met Ioannis Kourakis, a warm personality who made me visit the treacherous canyons near Xhania, where the insurrections against the Ottoman Empire started in the 19th century. I had first met Ioannis in Bochum, Germany, where he was a post-doc of my deceased old friend Padma Sukla, and in several other occasions. But I only managed to collaborate with him recently, during my visit to Abu Dhabi, where we worked on superfluid light, one of the topics in this book. I would like to thank Renato Fedele, who generously invited me to Naples for a couple of times. I could enjoy the warm hospitality and the culture of that wonderful city, while learning new aspects of accelerator beam physics. Finally, I was invited on several occasions to Umea, in the north of Sweden, for short periods of time, where I could profit from the hospitality, and from the solid and broad knowledge of Lennart Stenflo, Gert Brodin and Mattias Marklund.

I started my adventures in laser-cooling and atomic traps thanks to my repeated visits to Robin Kayser in Nice, who helped me to initiate the experimental activities in that area at IST Lisboa. In the meantime I visited the nearby city of Marseille, where I worked for a couple of months in other problems, not relevant to the present book, but had the opportunity to meet colleagues I had met several years earlier in Paris, such as Dominique Escande, Fabrice Doveil and Sadri Benkadda. I also spent a couple of months in Madrid, a city where I always feel at home, where I worked with Javier Honrubia on the physics of intense lasers. The same topic was part of my activities during a long stay in England, at the Rutherford Appleton Laboratory, where I discussed with Robert Bingham and Peter Norreys possible experiments on QED vacuum, using the novel Peta-Watt laser installations that were being launched at that time. My visits to Charles Wang in Aberdeen, to Gordon Robb in Glasgow and to Danielle Faccio in Edinburgh were also quite inspiring.

I am also thankful to a bunch of younger researchers, some of them who were my post-docs, such as Jonathan Davies and Shahid Ali, with whom I started my adventures in the land of light vortices and twisted light, sometimes also called structured light. I became more active in this area, thanks to a new collaboration with my young colleague Jorge Vieira. I also would like to thank Hugo Terças, one of my closest collaborators in recent years, with emphasis on photon condensation and ultra-cold matter. He also motivated me to return to the problem of axion–photon coupling and to the fascinating search for dark matter. I am particularly grateful to José António Rodrigues and to João Rodrigues, who despite the same family name are not relatives. They were both instrumental to the completion and successful operation of our MOT experiment (Magneto-Optical Trap). This was particularly difficult due to the absence of local tradition and knowledge in this area. I also want to thank Ruggero Giampaoli for wonderful experiments on photon bubbles, an important concept in astrophysics not previously observed in the laboratory. This led our MOT experiment to another level, and is relevant to collective photon processes, but for several reasons I could not include it in the present book.

Finally, I would like to thank my editors at the Institute of Physics, and in particular John Navas who was one of my editors of a previous book, published 20 years ago by IoP. He challenged me in the first place, and made me feel comfortable with my successive delays and failed deadlines to deliver the final manuscript. I hope he will not feel too disappointed.

Lisboa, January 2022

Author biography

J T Mendonça

José Tito Mendonça received the degree of Docteur ès-Sciences (PhD) by the University of Paris (Orsay). At present, he is a research professor at the Instituto Superior Técnico, the University of Lisboa. He is a former head of the Physics Department of IST, and leader of the Group of Lasers and Plasmas. He was the first director of the Euratom-IST Association. He worked at JET European Experiment in Abingdon (UK) as a scientific advisor, and at the Rutherford-Appleton Laboratory as an invited researcher.

He organised several groups and laboratories, in the areas of nuclear fusion, intense lasers and atomic laser-cooling. He published four books and more than four hundred research papers. He made contributions to a variety of topics, such as: photon acceleration, time-refraction, optical cascades, stochastic acceleration, quasi-particle turbulence, collective neutrino processes, axion-photon interactions, and twisted light. His current interests include: ultra-cold matter, superfluid light, photon condensation, laser accelerators, and quantum technologies.

IOP Publishing

The Quantum Nature of Light
From photon states to quantum fluids of light
J T Mendonça

Chapter 1

Introduction

1.1 Motivation

Light is at the heart of our understanding of Nature. Light brings continuously unsolicited information about the planets, the stars, and the cosmos. In parallel, we can use it to drill on the microcosmos, and to learn about the structure of atoms and molecules. Given its overwhelming presence in our lives, light is traditionally associated with wisdom and knowledge. Apollo, known for his wisdom and beauty, is sometimes called the Greek god of Light. In the sanctuary of Delphi, the Temple of Apollo plays a prominent role (see figure 1.1).

The wave-particle dualism dominated the historical debate on the nature of light, and was a precursor of the modern physics concepts of Quantum Mechanics. A quantum theory of light emerged after 1900, when Ernest Planck noticed that the atoms exchange a discrete amount of energy with radiation, and when in 1905 Albert Einstein assumed that these quanta of energy are indeed light particles, later called *photons*.

These famous contributions reintroduced the debate between the corpuscular and the undulatory nature of light, although placing it in a more modern scenario. Such a debate, transferred to the nature of electrons by Louis de Broglie in 1924, led in the years 20 of the last century to the theory of quantum mechanics, that revolutionised our way of describing the physical world. This tremendous scientific revolution has been described in several papers and books (see, for instance, [1–3]). The Quantum Theory of Light, which is an important piece of the modern Quantum Field Theory, can nevertheless be considered an independent discipline, with its own problems and methods.

This book gives a broad perspective of the quantum nature of light, as we see it today. It extends from the traditional area of Quantum Optics, into a more modern discussion on quantum fluids of light, and at the end, includes relativistic quantum electrodynamics. This work can eventually be relevant to students and researchers working in a large range of areas, from single photon processes to ultra-intense laser

doi:10.1088/978-0-7503-2786-2ch1

Figure 1.1. The author near the Temple of Apollo, Sanctuary of Delphi, Greece (Photo: T Pinheiro, August 2019).

physics. This includes people interested in atom manipulation with photons, ultra-fast lasers, Bose–Einstein condensation of photons, superfluid light, laboratory astrophysics, quantum computation, and the exploration of quantum vacuum with Peta-Watt lasers. We also discuss the axion–photon coupling, which is relevant to the modern search of dark matter.

The book is divided in three parts. The first part, called *Basic Photon States*, is more traditional, and is related with basic electromagnetic field quantisation, the characterisation of quantum photon states, and elementary photon–atom interactions. It also includes cavity quantum electrodynamic processes, and a global view on space-time boundaries, such as the Casimir pressure, time-refraction and time-crystals. Excluding the temporal effects, all these topics are usually covered by books on quantum optics.

The second part of this book is devoted to *Quantum Fluids of Light* and focuses on collective photon states. It includes, not only a traditional discussion on the basic properties of the laser, but also superradiance, and the more recent areas of Bose–Einstein condensation of photons, light vortices and superfluid light. The relations between laser instability and photon condensation are discussed. The laser is a dominant concept in modern physics, and its use is central to the understanding of the Quantum States of Light. It is discussed by many excellent books that have been published over the last half-century. Here, however, we include it in a novel perspective on collective quantum processes. Not only is the laser symmetrical with respect to superradiance, because the first is a result of stimulated emission and the second a consequence of spontaneous emission, but also because similarities and symmetries can be found between the laser instability and the photon condensation process. Furthermore, superfluid light can now be achieved without the need of condensation.

Finally, the third part *Quantum vacuum* is focused on a more complete description of the electromagnetic vacuum, and includes energetic processes that could eventually lead to the emission of electron–positron pairs. It is relevant to the modern exploration of vacuum using the new ultra-intense laser systems. This part also includes the search for eventually existing dark matter particles, which points to possible new physics.

Being diverse in their purpose, and using specific theoretical methods, these three different parts depend very strongly on the existence of lasers and cover the main manifestations of the quantum nature of light. In a sense, we can say that the invention of the laser opened the way to explore the different areas of quantum light, understand the nature of collective quantum phenomena, and establish the ultimate meaning of quantum vacuum.

The book is intended to make the bridge between these three somewhat distinct aspects of the quantum states of light: basic processes, collective effects and vacuum physics. This is achieved through the use of the same notation and basic concepts, such as Lagrangian densities, Hamiltonian operators and Wigner function representation. It will be mainly focused on quantum theory, but semi-classical models (where the atom states are quantised and the electromagnetic field is described as a classical field) will also be used for comparison whenever reasonable and possible.

1.2 Photons, waves and fields

Photons play a major role in this book, and can be seen as elementary excitations of the quantum field. For a historical account of the concept of *photon* see [4–6]. The concept of a quantum of electromagnetic energy emerged in 1905, when Einstein proposed his model of the photoelectric effect. It is known that Einstein took inspiration from the work of Max Planck on the black-body radiation. But, in contrast with Planck, who assumed that absorption and emission of radiation by atoms in quanta of energy is a property of the atoms, Einstein made a step forward and considered that these energy quanta were a property of radiation itself.

From an historical point of view, the photon was only accepted as a real particle, with energy and momentum, in the 1920s. A decisive contribution came from the discovery of the Compton effect in 1923, and the subsequent adoption of the name *photon*, proposed by Lewis in 1926. Photons are still seen today as somewhat elusive and mysterious particles, and many of their properties still need to be unveiled.

But photons are not alone in this game. In particular, they cannot accurately describe the electromagnetic field without the help of waves and fields. Or, more exactly, of electromagnetic waves and the electromagnetic field. It is well-known that this field is described by Maxwell's equations. These equations, first established in complete form by James Clark Maxwell in the 1860s, are an extraordinary example of resilience, because they survived the two major theoretical revolutions of the 20th century: Relativity and Quantum Mechanics. They survived relativity because these equations are Lorentz invariant. In that respect, they anticipate the theory of special relativity. They also survived quantum mechanics because they describe the wave properties of light, in the same way as Schrödinger or Dirac's equations describe the wave properties of matter particles. They even survived quantum field theory, because the electromagnetic field operators still satisfy Maxwell's equations. In their operator form, Maxwell's equations are indeed the Heisenberg equations for these field operators.

Photons are the modern version of the corpuscles of light, a concept that emerged from Antiquity and culminated in the work of Newton. The explanation of the rainbow was a great success of the old corpuscular theory of light. This is closely connected with the brilliant yet erroneous explanation given by Newton to the prism experiment. He assumed that the red corpuscles were more energetic than the blue ones, because they suffered a smaller deviation from the prism. In contrast, we know today that the red photons are less energetic than the blue photons.

The wave concept of light was proposed in the 17th century by Huygens, and consolidated in the early 19th century, due to the work of Young, Fresnel and others (see [7] for a detailed account). From the historical battle between Newton and Huygens ideas, between the corpuscular and the wave description of light, and the need to reconcile them in modern physics, we arrived at the concept of *wave-particle dualism*. This dualism can however be understood in the frame of a purely classical description of light, based on Maxwell's equation [8].

But, instead of the wave-particle dualism, the never solved contradiction of classical physics, also dominant in the early quantum mechanics, what we have today is a triadic relation between particles (photons), waves and fields. In this magical triad, the field can be seen as the synthesis that solves and reconciles the dialectic opposition between waves and particles. The photon-wave-field triad, never explicitly announced but ever present, is at the core of quantum field theory. Our present view of the quantum states of light is based on a blend, in equal proportions, of these three concepts.

1.3 A necessary note

The subject of this book is very broad and, even a single chapter requires a huge amount of knowledge. Our presentation is intended to give a unified view of very different areas, concepts and perspectives. However, some differences remain in detail and depth between different parts and different chapters. This is due to several reasons, one and most obvious being the familiarity of the author with the discussed subjects. For the most part, the book is written in a pedagogical way. Some exceptions, easily identified, are more research oriented and more difficult.

Given the impossible task of reading or citing all of the relevant publications, our choice was based on three main criteria: historical papers, selected review papers and relevant books on each topic. We also have included several of our own papers, where we have developed some specific topics, not usually covered in the literature. For this reason, the number of self-citations, particularly noticed in a couple of chapters, is completely out of proportion with the actual relevance of our own work.

References

[1] Kleppner D and Jackiw R 2000 One hundred years of quantum physics *Science* **289** 893–8
[2] Kragh H 1999 *Quantum Generations: A History of Physics in the Twentieth Century* (Princeton, NJ: Princeton University Press)
[3] Badino M and Navarro J (ed) 2017 *Research and Pedagogy: A History of Quantum Physics through Its Textbooks* Open Access (Berlin: Max Planck Institute for the History of Science)
[4] Pais A 1982 *Subtle is the Lord: The Science and the Life of Albert Einstein* (Oxford: Oxford University Press)
[5] Kragh H 2014 Photon: new light on an old name (arXiv:1401.0293v3); in The names of physics: plasma, fission, photon *Phys. J.* H **39** 263–81
[6] Kidd R, Ardini J and Anton A 1989 Evolution of the modern photon *Am. J. Phys.* **57** 27–35
[7] Whittaker E 1089 *A History of the Theories of Aether and Electricity* (New York: Dover)
[8] Mendonça J T 2008 Maxwell and the classical wave particle dualism *Phil. Trans. R. Soc.* A **366** 1771–80

Part I

Basic photon states

This first part gives a short account of some of the main concepts and results of Quantum Optics. It covers most of the topics considered in traditional books on quantum optics, such as field quantisation and the photon state vector, field coherence and the existence of nonclassical states, entanglement, and photon–atom interactions. It also includes vacuum effects, such as the Casimir force, and a discussion on the temporal processes associated with time-refraction and time-crystals.

IOP Publishing

The Quantum Nature of Light

From photon states to quantum fluids of light

J T Mendonça

Chapter 2

Field quantisation

In this introductory chapter we review the basic concepts of Quantum Mechanics and introduce the subject of electromagnetic field quantisation, which is central to the present work. Quantum mechanics was developed by a brilliant generation of physicists in the early 20th century. It has been described by several excellent textbooks, from historical treaties to more recent ones. Relevant examples are [1–5]. Here, we only need to review its main concepts such as the state vectors, wave-functions, operators and density matrix, in order to fix the notation. We also review its basic equations such as the Schrödinger and the Heisenberg equations. We introduce the quantum mechanical pictures, make a comparison between the Schrödinger and the Heisenberg picture, and review the basic position and momentum representations. The simple harmonic oscillator will be used to illustrate the quantum concepts and to make a preliminary step towards field quantisation.

It is known that the birth of quantum mechanics was involved in a difficult debate on the concept of physical reality, initiated by Niels Bohr and Albert Einstein in the early 30s of the last century. This debate lasted for several decades, and led to the introduction of alternative formulations of the theory that were revealed to be equivalent to each other. From such a debate, sometimes confusing and disappointing, emerged several important concepts that were incorporated in the theory, and included complementarity, the Bohm potential, the master equation, non-locality, hidden-variables, and entanglement. They have shown to be useful for theoretical purposes in different contexts and will be discussed, some in this first chapter and others in subsequent chapters. Representation of quantum states in the classical phase-space using Wigner–Weyl transformations and the subsequent use of a wave-kinetic equation of the Liouville type instead of the more standard Schrödinger equation, is also relevant and will be used throughout the book.

Quantisation of the electromagnetic field was first performed by Dirac [6] and is also explained in several treatises, such as [7–10]. Here we show that the electromagnetic field in vacuum can be described as a superposition of an infinite number

doi:10.1088/978-0-7503-2786-2ch2

of harmonic oscillators and introduce the photon annihilation and creation operators. Finally, we approach the delicate problem of the photon state. Starting from Maxwell's equations, we show that an equation formally analogous to the Schrödinger equation can be derived. For that purpose, we use the vector first proposed by Silverstein, as discussed by [11] and others. A connection with the Dirac equation is also established, but a more complete view of such a connection with only become apparent in the third part of this book.

For simplicity, we only discuss quantisation in vacuum, but extrapolation to the case of a non-dispersive optical medium is straightforward. However, the question of quantisation in a dispersive medium is more complicated and was approached by several researchers [12–17]. Of particular interest is the case of a plasma, which is a peculiar dispersive medium where, in contrast with typical optical media such as crystals or fibres, electric charges can freely move in the medium. This case allows us to introduce the concept of *photon mass*, first noted by Anderson [18], and to include a third polarisation state of the photon, which is longitudinal polarisation and can exist in media with free charged particles, such as semi-conductors, metals and gaseous discharges. This longitudinal photon state is called a *plasmon*.

2.1 Quantum mechanical background

We assume a basic knowledge of quantum mechanics and only recall here some of its main equations and concepts. This will be used to fix the notation. Additional details are given in appendix A. The state of a quantum system, and its temporal evolution, can be represented in many different ways that are summarised here. These descriptions are all equivalent and the variety of possible quantum representations is very useful for an appropriate description of the quantum states of light. Here, we introduce the Schrödinger equation, the Heisenberg picture, the von Newman equation and the Wigner function. For a simple view of the different formulations of quantum mechanics, see [19].

2.1.1 Schrödinger picture

The state of a quantum system can be characterised by a vector function $|\psi\rangle$, and the temporal evolution of this state is described by an equation of the form

$$i\hbar \frac{\partial}{\partial t} |\psi\rangle = H |\psi\rangle, \tag{2.1}$$

where $\hbar$ is the Planck's constant divided by 2π, and H is the *Hamiltonian operator*. This operator is hermitian, $H = H^\dagger$, and its explicit form will be dictated by the configuration of the system.

For a non-relativistic particle with mass m, moving with momentum p along the coordinate q it can be written as

$$H = \frac{p^2}{2m} + V(q, t), \tag{2.2}$$

where the first term represents the kinetic energy and $V(q, t)$ is the potential. In this case, the wave equation is usually called the *Schrödinger equation*. The general form of the wave equation stays valid for relativistic particles with semi-integer spin, as in the case of an electron or a positron, but with a different Hamiltonian. It also remains valid for the case of photons, which are particles with no rest mass and with spin-1, as will be shown later in this chapter. The Schrödinger equation (2.1) is the central piece of quantum mechanics, and was first formulated in 1926 [20, 21]. Quantisation requires that we postulate the following commutator relations involving the position and momentum variables, as

$$[q, p] \equiv qp - pq = i\hbar, \quad [q, q] = 0, \quad [p, p] = 0. \tag{2.3}$$

From here we can see that the limit $\hbar \to 0$ is the classical limit, where position and momentum commute with each other.

The quantity $\langle \psi \mid \psi \rangle$ is real and positive, and can be normalised, such that $\langle \psi \mid \psi \rangle = 1$. Any *observable* A, such as the position, the momentum, or the Hamiltonian, is represented by a hermitian operator, such that $A = A^\dagger$. For a given state of the system, its expectation value is real, and defined by

$$\langle A \rangle = \langle \psi \mid A \mid \psi \rangle. \tag{2.4}$$

Given two different state vectors $\mid \psi \rangle$ and $\mid \phi \rangle$, and a hermitian operator A, we can write

$$\langle \psi \mid A^\dagger \mid \phi \rangle = \langle \phi \mid A \mid \psi \rangle^*. \tag{2.5}$$

2.1.2 Representations

In quantum mechanics we can use different representations. They correspond to different choices of reference frames to represent the state vectors. These reference frames should be made of an ensemble $\mid u_i \rangle$ of state vectors satisfying the orthonormality condition $\langle u_i \mid u_j \rangle = \delta_{ij}$. This allows us to write

$$\mid \psi \rangle = \sum_i c_i \mid u_i \rangle, \quad c_i = \langle u_i \mid \psi \rangle, \tag{2.6}$$

where the complex numbers c_i are the coefficients defining the state vector on that frame of reference. This representation should be complete, in the sense that an operator identity I can be defined as

$$\sum_i \mid u_i \rangle \langle u_i \mid = I. \tag{2.7}$$

This can be generalised to a continuous reference frame $\mid u_\alpha \rangle$, where α is not a discrete but a real variable, satisfying the new orthonormality condition $\langle u_\alpha \mid u_{\alpha'} \rangle = \delta(\alpha - \alpha')$. In this case we have

$$| \psi\rangle = \int c(\alpha)| u_\alpha\rangle d\alpha, \quad c(\alpha) = \langle u_\alpha | \psi\rangle, \tag{2.8}$$

and the identity operator I is defined as

$$\int | u_\alpha\rangle\langle u_\alpha | \, d\alpha = I. \tag{2.9}$$

The internal product between two state vectors $| \psi\rangle$ and $| \phi\rangle$ can then be defined as

$$\langle\psi | \phi\rangle = \langle\psi | I | \phi\rangle = \sum_i \langle\psi | u_i\rangle\langle u_i | \phi\rangle = \sum_i c_i^* b_i, \tag{2.10}$$

where c_i and b_i are the coefficients of $| \psi\rangle$ and $| \phi\rangle$ respectively. A particular example of such a discrete representation is given in the next section, dealing with the simple harmonic oscillator, where the vectors $| u_i\rangle$ are energy state vectors. For a continuous reference frame, we have

$$\langle\psi | \phi\rangle = \langle\psi | I | \phi\rangle = \int \langle\psi | u_\alpha\rangle\langle u_\alpha | \phi\rangle d\alpha = \int c_\alpha^* b_\alpha \, d\alpha. \tag{2.11}$$

Important examples of a continuous reference frame are the position and momentum representations, where the real variable α is replaced by the real position q, or by the momentum p. In this case, the basic vectors $| u_\alpha\rangle$ can be written as $\left| u_q \right\rangle$ and $\left| u_p \right\rangle$, respectively, and the coefficients c_α are nothing but the wavefunction $\psi(q, t)$ and its Fourier transform $\psi(p, t)$. We get

$$\psi(q) = \langle q | \psi\rangle, \tag{2.12}$$

and

$$\psi(p) = \langle p | \psi\rangle = \int \psi(q) e^{-iqp/\hbar} \frac{dp}{2\pi\hbar}. \tag{2.13}$$

Generalisation to three dimensions is straightforward. In the position representation, the operator p is represented by $-i\hbar\partial/\partial q$, and the Schrödinger equation (2.1)–(2.2) takes its usual form

$$i\hbar\frac{\partial}{\partial t}\psi(q) = \left[-\frac{\hbar^2}{2m}\frac{\partial^2}{\partial q^2} + V(q, t)\right]\psi(q). \tag{2.14}$$

The wavefunction $\psi(q)$ is particularly useful to describe the state of non-relativistic massive particles, and its Fourier transform $\psi(p)$ is well adapted to describe the state of massless particles such as the photon, as shown later. Due to their non-commutativity, defined in equations (2.3), the position and momentum variables satisfy an *uncertainty relation* such that

$$\Delta q \, \Delta p \geqslant \frac{\hbar}{2}. \tag{2.15}$$

The uncertainty of an operator A can be defined by the deviations with respect to its expectation value $\langle A \rangle$, as

$$\Delta A = [\langle A^2 \rangle - \langle A \rangle^2]^{1/2}. \tag{2.16}$$

It can be shown that two pair of non-commuting observables A and B satisfy the inequality

$$\Delta A \, \Delta B \geqslant \frac{1}{2} \, |[A, B]|. \tag{2.17}$$

The position–momentum uncertainty is a particular case of this inequality.

2.1.3 Heisenberg picture

In the above description, usually called the *Schrödinger picture*, the operators are time independent, and the state vector $| \psi(t) \rangle$ is time dependent. In an alternative, the *Heisenberg picture* can be used where the operators evolve with time and the state vector stays constant. This new picture historically preceded that of Schrödinger by a few months, and was first explained in a series of papers by Heisenberg, Born and Jordan [22–24]. Let us denote these new quantities as q_H, p_H and $| \psi_H \rangle$. State vectors of the two pictures can be related by

$$| \psi_S(t) \rangle = U(t, t_0) | \psi_H(t_0) \rangle, \tag{2.18}$$

where $U(t, t_0)$ is the *evolution operator* (see appendix A). The Schrödinger state vector evolves in time, while the Heisenberg state vector is fixed. They only coincide at $t = t_0$. In this new picture, the temporal evolution of an operator A is determined by

$$\frac{d}{dt} A = \frac{1}{i\hbar} [A, H]. \tag{2.19}$$

The operator A is constant if it commutes with the Hamiltonian H. In the case of the position and momentum operators q and p, the Heisenberg equations take the form

$$\frac{dq}{dt} = \frac{\partial H}{\partial p}, \quad \frac{dp}{dt} = -\frac{\partial H}{\partial q}. \tag{2.20}$$

This is the quantum equivalent of the canonical equations of Classical Mechanics.

Another useful description of a quantum system is provided by the density operator ρ, which is defined as

$$\rho = | \psi \rangle \langle \psi |. \tag{2.21}$$

This definition shows that the density operator is hermitian, $\rho^\dagger = \rho$. If the state vector is represented in a discrete basis $| u_i \rangle$ (see appendix A), the identity operator will be represented on that basis by

$$\rho = \sum_{ij} c_i c_j^* \, | u_i \rangle \langle u_j |. \tag{2.22}$$

We can therefore say that ρ is represented on that basis by a matrix, usually called the *density matrix*, with elements

$$\rho_{ij} = \langle u_i \mid \rho \mid u_j \rangle = c_i c_j^*. \tag{2.23}$$

For a normalised state vector, the trace of the density matrix (the sum of its diagonal elements) is equal to 1

$$Tr.(\rho) = \sum_i \rho_{ii} = \sum_i \mid c_i \mid^2 = 1. \tag{2.24}$$

If we take the time derivative of equation (2.21), and use the Schrödinger equation (2.1), we obtain

$$\frac{\partial \rho}{\partial t} = -\frac{1}{i\hbar}[\rho, H]. \tag{2.25}$$

This is sometimes called the *von Neumann equation*. The density operator is particularly useful to study systems that are mixed states, and cannot be described by a single state vector $\mid \psi \rangle$. The three different approaches provided by the Schrödinger, the Heisenberg and the von Newman equations are the most relevant descriptions of a quantum system.

2.1.4 Wigner function

Another useful description is provided by the *Wigner function*. This was introduced in 1932 [25], with the aim of representing a quantum system in classical phase-space. This can be useful to compare quantum and classical regimes and to identify the genuine quantum effects.

The approach inaugurated by Wigner was completed by Moyal in 1946 [27], who established an evolution equation, sometimes called the *Wigner–Moyal equation*, or the *wave-kinetic equation*. Again, this is equivalent to the Schrödinger equation, and describes the temporal evolution of a quantum system in classical phase-space. This new representation was initially ignored in textbooks on quantum theory. But it became popular in Statistical Mechanics since the 80s of the last century, and has been slowly incorporated in the mainstream of quantum mechanics. It is particularly useful in quantum optics, and to describe the collective quantum states of light in the mean-field approximation. The Wigner approach has been discussed in several review papers along the years, such as [28–31], and more recently [32, 33]. It is also described in many book chapters.

We start with the definition of the Wigner function, the discussion of its meaning, and its relation with the *Weyl transformation* [26]. For a quantum system, described by the wavevector $\mid \psi \rangle$, we can define the corresponding Wigner function as

$$W(\mathbf{r}, \mathbf{p}, t) = \int \langle \mathbf{r} + \mathbf{s}/2 \mid \psi \rangle \langle \psi \mid \mathbf{r} - \mathbf{s}/2 \rangle \exp(i\mathbf{p} \cdot \mathbf{s}/\hbar) d\mathbf{s}. \tag{2.26}$$

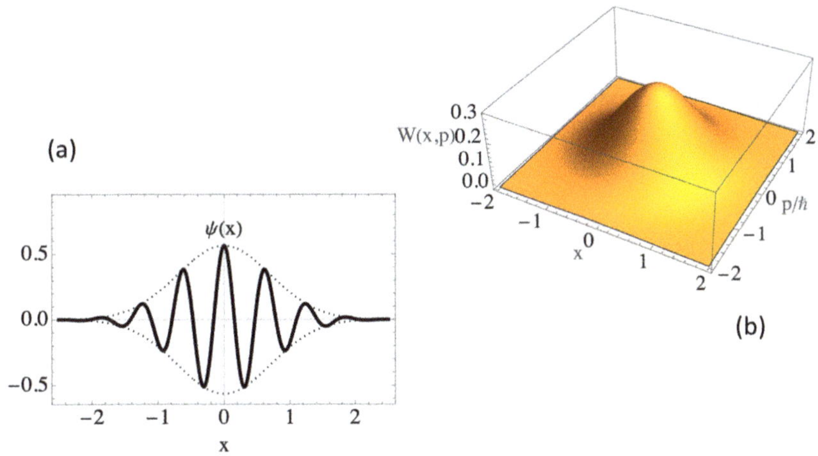

Figure 2.1. (a) Wavefunction $\psi(x) = \psi_0 \exp(-x^2/2)$, and (b) the corresponding Wigner function $W(x, p)$.

This allows us to represent the system in classical phase-space $(\mathbf{r}, \mathbf{p})$, which is 6-dimensional, instead of the simple representation in $\mathbf{r}$ as allowed by the wavefunction $\psi(\mathbf{r}) = \langle \psi \mid \mathbf{r} \rangle$. The additional momentum variable is introduced by the above definition, using a space Fourier transform. Such a transition, from the $\mathbf{r}$-representation of a quantum system using wavefunctions, to a representation in classical phase-space using the Wigner function, is illustrated in figure 2.1, for the simple case of a one-dimensional system evolving along x.

Furthermore, normalisation of the Wigner function is established by integrating over the classical phase-space, using

$$\langle \psi \mid \psi \rangle = \int \frac{d\mathbf{p}}{(2\pi\hbar)^3} \int d\mathbf{r}\, W(\mathbf{r}, \mathbf{p}, t) = 1. \tag{2.27}$$

This shows that, if we integrate the Wigner function over the momentum variable $\mathbf{p}$, we should get the quantum probability for finding the system at a given position $\mathbf{r}$, at time t, as

$$P(\mathbf{r}, t) \equiv \mid \psi(\mathbf{r}, t)^* \psi(\mathbf{r}, t)|^2 = \int W(\mathbf{r}, \mathbf{p}, t) d\mathbf{r}. \tag{2.28}$$

This quantity, by definition, is positive definite. In contrast, the Wigner function is real but not necessarily positive, and for that reason it cannot be called a probability. The Wigner function is usually called *quasi-probability*. Such a difference with respect to the classical concept of probability in phase-space is very important, because it helps to identify the deviations of a quantum system from a classical behaviour. This is an important qualitative aspect of the Wigner representation.

It should also be noticed that the Wigner function can be reduced to a particular case of the *Weyl transformation*. It was introduced in 1931, with the purpose of defining the quantum operator $\hat{A}$ associated with a given classical dynamical function $a(\mathbf{r}, \mathbf{p})$. This transformation can be defined as

$$a(\mathbf{r}, \mathbf{p}) = \int \langle \mathbf{r} + \mathbf{s}/2 \mid \hat{A} \mid \mathbf{r} - \mathbf{s}/2 \rangle \exp(i\mathbf{p} \cdot \mathbf{s}/\hbar) d\mathbf{s}. \tag{2.29}$$

Comparing this with equation (2.26) we can see that the Wigner function is nothing but a Weyl transformation, where the operator $\hat{A}$ is replaced by the density operator $\rho = \mid \psi \rangle \langle \psi \mid$. The mean value of an operator, $\hat{A}$ can be calculated using the Wigner function, as

$$\langle \hat{A} \rangle = \int d\mathbf{r} \int \frac{d\mathbf{p}}{(2\pi\hbar)^3} a(\mathbf{r}, \mathbf{p}) W(\mathbf{r}, \mathbf{p}, t). \tag{2.30}$$

These expression shows that the mean value of an arbitrary quantum operator can be defined by a purely classical expression, where the Wigner function W plays the role of a classical probability distribution. Except that, in this case, W is not a classical probability, but a quasi-probability that can take negative values.

2.2 Harmonic oscillator

The harmonic oscillator is one of the basic models of quantum mechanics. Not only can it be used to describe many different problems in atomic, molecular and condensed matter physics, but it also plays a central role in the description of electromagnetic field quantisation. Furthermore, due to its simplicity, it can be used to illustrate the basic quantum formalism introduced above. Due to its relevance to field quantisation, we describe the harmonic oscillator with some detail.

2.2.1 Energy levels

We consider a particle with unit mass $m = 1$, moving along the coordinate q under the influence of a quadratic and time-independent potential $V(q) \propto q^2$. The Hamiltonian operator is

$$H = \frac{1}{2}(p^2 + \omega^2 q^2), \tag{2.31}$$

where the constant ω determines the strength of the restoring force. We know that this is a hermitian operator, $H^\dagger = H$, and that q and p satisfy the commutation relation $[q, p] = i\hbar$. We can easily solve the Schrödinger equation (2.1) and get

$$\mid \psi(t) \rangle = U(t, 0) \mid \psi(0) \rangle, \quad U(t, 0) = \exp\left(-\frac{i}{\hbar} Ht\right). \tag{2.32}$$

In an alternative, we can use the Heisenberg picture and start from the equations of motion (2.20), which are

$$\frac{d}{dt} q = \frac{\partial H}{\partial p} = p, \quad \frac{d}{dt} p = -\frac{\partial H}{\partial q} = -\omega^2 q. \tag{2.33}$$

The solution is

$$q(t) = q_0 \cos(\omega t) + \frac{p_0}{\omega} \sin(\omega t), \quad p(t) = p_0 \cos(\omega t) - \omega q_0 \sin(\omega t), \tag{2.34}$$

where (q_0, p_0) give the initial conditions for $t = 0$. They are, in fact, the position and momentum operators used in the Schrödinger picture. It is now useful to introduce a pair of new operators a and $a^\dagger$, defined as

$$a = \frac{1}{\sqrt{2\hbar\omega}}(\omega q + ip), \quad a^\dagger = \frac{1}{\sqrt{2\hbar\omega}}(\omega q - ip). \tag{2.35}$$

Equivalently, we can write

$$q = \sqrt{\frac{\hbar}{2\omega}}(a + a^\dagger), \quad p = -i\sqrt{\frac{\hbar\omega}{2}}(a - a^\dagger). \tag{2.36}$$

Using these definitions, we can then calculate

$$a^\dagger a = \frac{1}{2\hbar\omega}(2H + i\omega[q, p]), \quad aa^\dagger = \frac{1}{2\hbar\omega}(2H - i\omega[q, p]). \tag{2.37}$$

Adding these two expressions, we obtain

$$H = \frac{1}{2}\hbar\omega(a^\dagger a + aa^\dagger). \tag{2.38}$$

And, by subtraction, we get the commutator

$$[a, a^\dagger] = (aa^\dagger - a^\dagger a) = 1. \tag{2.39}$$

From this we derive a new expression for the Hamiltonian

$$H = \hbar\omega\left(a^\dagger a + \frac{1}{2}\right). \tag{2.40}$$

At this point, it is useful to introduce a new operator

$$N = a^\dagger a, \tag{2.41}$$

which is hermitian, $N^\dagger = N$, like the Hamiltonian. It can easily be found that it satisfies the commutation relations

$$[a, N] = a, \quad [a^\dagger, N] = -a^\dagger. \tag{2.42}$$

Or, equivalently

$$Na = a(N - 1), \quad Na^\dagger = a^\dagger(N + 1). \tag{2.43}$$

If we use the Heisenberg picture, we can write the equations of motion for the operators a and $a^\dagger$ as

$$\frac{da}{dt} = \frac{1}{i\hbar}[a, H] = i\omega a, \quad \frac{da^\dagger}{dt} = \frac{1}{i\hbar}[a^\dagger, H] = -i\omega a^\dagger, \tag{2.44}$$

with the solutions

$$a(t) = a(0)\exp(i\omega t), \quad a^\dagger(t) = a^\dagger(0)\exp(-i\omega t). \tag{2.45}$$

Let us now consider the energy eigenstates E, satisfying the equation $H \mid \psi\rangle = E \mid \psi\rangle$. The eigenstate $\mid n\rangle$ of a given energy eigenvalue E_n is then determined by

$$H \mid n\rangle \equiv \hbar\omega\left(a^\dagger a + \frac{1}{2}\right) \mid n\rangle = E_n \mid n\rangle. \tag{2.46}$$

If we apply the operator a from the left to this equation, and use the commutator $[a, a^\dagger] = 1$, we get

$$Ha \mid n\rangle = (E_n - \hbar\omega)a \mid n\rangle. \tag{2.47}$$

This means that the state vector $(a \mid n\rangle)$ is the eigenstate of a new energy level $E_{n-1} = E_n - \hbar\omega$, apart from some normalisation factor c_n. We can then write

$$a \mid n\rangle = c_n \mid n - 1\rangle. \tag{2.48}$$

This allows us to state that

$$H \mid n - 1\rangle = E_{n-1} \mid n - 1\rangle. \tag{2.49}$$

Repeating this procedure n times, we eventually arrive at the energy state $\mid 0\rangle$, corresponding to the lowest energy value of the oscillator, E_0. This leads to

$$Ha \mid 0\rangle = (E_0 - \hbar\omega)a \mid 0\rangle. \tag{2.50}$$

Noting that an energy lower than E_0 cannot exist, we have to assume that

$$a \mid 0\rangle = 0. \tag{2.51}$$

This new state $\mid 0\rangle$ is then called the *vacuum state*. Going back to equation (2.46) we conclude that

$$H \mid 0\rangle = \frac{1}{2}\hbar\omega \mid 0\rangle = E_0 \mid 0\rangle. \tag{2.52}$$

This allows us to conclude that the energy states of the harmonic oscillator are

$$E_n = \hbar\omega\left(n + \frac{1}{2}\right), \tag{2.53}$$

where n is an integer. From equation (2.46) we also conclude that

$$N \mid n\rangle \equiv a^\dagger a \mid n\rangle = n \mid n\rangle. \tag{2.54}$$

The normalisation constant c_n introduced in equation (2.48) can now be determined using

$$\langle n \mid a^\dagger a \mid n\rangle = \langle n - 1 \mid c_n^* c_n \mid n - 1\rangle. \tag{2.55}$$

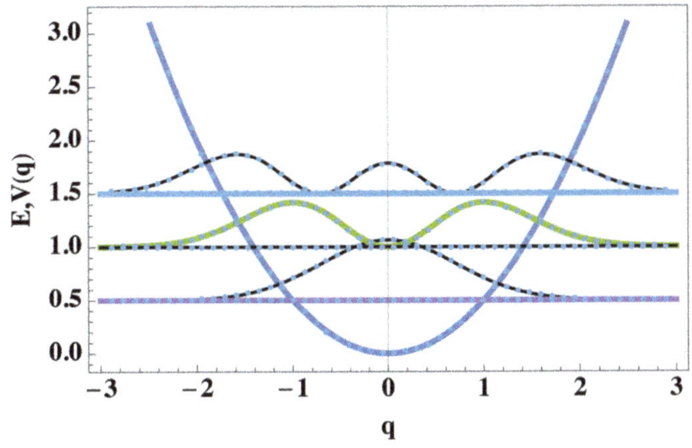

Figure 2.2. Harmonic oscillator: the parabolic potential $V(q)$, as well as the probability distribution $|\psi_n(q)|^2$ for the three lowest energy levels E_n, $n = 0, 1, 2$.

The normalisation of both $|\, n\rangle$ and $|\, n - 1\rangle$ implies that $|c_n|^2 = n$. We then have

$$a\,|\, n\rangle = \sqrt{n}\,|\, n - 1\rangle, \quad a^\dagger\,|\, n\rangle = \sqrt{n + 1}\,|\, n + 1\rangle. \tag{2.56}$$

This discussion shows that, by applying the operator $a^\dagger$ successively n times to the vacuum state $|\, 0\rangle$ we can generate the state $|\, n\rangle$. Including the normalisation constant, we get

$$|\, n\rangle = \frac{1}{\sqrt{n!}}(a^\dagger)^n|\, 0\rangle. \tag{2.57}$$

We finally notice that these energy eigenvectors satisfy the orthogonality and completeness conditions

$$\langle n\,|\, n'\rangle = \delta_{nn'}, \quad \sum_0^\infty |\, n\rangle\langle n\,| = I, \tag{2.58}$$

where I is the identity operator. This means that an arbitrary state vector $|\, \psi\rangle$ of the harmonic oscillator can be represented as a superposition of energy eigenstates $|\, n\rangle$ (figure 2.2).

2.2.2 Wavefunctions

Let us now consider the position representation. In this case the energy states will be represented by wavefunctions $\psi_n(q) = \langle q\,|\, n\rangle$. Let us consider the ground state $n = 0$. The corresponding wavefunction $\psi_0(q)$ can easily be determined by replacing the definition of a, as given by equation (2.35) in equation (2.51). Noting that p is replaced by $(-i\hbar\partial/\partial q)$ in this representation, we get

$$\left(\omega q + \hbar\frac{\partial}{\partial q}\right)\psi_0(q) = 0. \tag{2.59}$$

The solution of this equation is normalised as

$$\int_{-\infty}^{\infty} |\psi_0(q)|^2\, dq = 1, \tag{2.60}$$

which leads to

$$\psi_0(q) \equiv \langle q \mid 0 \rangle = \left(\frac{\omega}{\pi\hbar}\right)^{1/4} \exp\left(-\frac{\omega q^2}{2\hbar}\right). \tag{2.61}$$

This determines the ground state wavefunction. For the nearest energy level, we can use equation (2.56) and write

$$\psi_1(q) \equiv \langle q \mid 1 \rangle = \langle q \mid a^\dagger \mid 0 \rangle, \tag{2.62}$$

or

$$\psi_1(q) = \frac{1}{2\hbar\omega}\left(\omega q - \hbar\frac{\partial}{\partial q}\right)\psi_0(q). \tag{2.63}$$

The result is

$$\psi_0(q) = \left[\left(\frac{2\omega}{\hbar}\right)^2 \frac{\omega}{\pi\hbar}\right]^{1/4} q \exp\left(-\frac{\omega q^2}{2\hbar}\right). \tag{2.64}$$

Similarly, for all the other modes, we get

$$\psi_n(q) = \frac{1}{\sqrt{n!}}\langle q \,|(a^\dagger)^n|\, 0 \rangle = \frac{1}{n!}\frac{1}{2\hbar\omega}\left(\omega q - \hbar\frac{\partial}{\partial q}\right)^n \psi_0(q). \tag{2.65}$$

This then leads to

$$\psi_n(q) = \left(\frac{\alpha}{\sqrt{\pi}2^n n!}\right)^{1/2} H_n(\alpha q)\exp\left(-\frac{\alpha^2 q^2}{2}\right), \tag{2.66}$$

where $\alpha = \sqrt{\omega/\hbar}$, and $H_n(\alpha q)$ are the Hermite polynomials of order n. Using this expression, we can calculate the Wigner functions for the quantum states of the harmonic oscillator. Using equation (2.26), it can be shown that the corresponding Wigner functions are

$$W_n = (q, p) = 2(-1)^n L_n\left[2\frac{p^2}{\hbar^2\alpha^2} + \alpha^2 q^2\right]\exp\left(-\frac{p^2}{\hbar^2\alpha^2} - q^2\alpha^2\right), \tag{2.67}$$

where $L_n(x)$ are the *Laguerre polynomials*, as defined by Rodrigues's formula [34]

$$L_n(x) = \frac{e^x}{n!}\frac{d^n}{dx^n}(x^n e^{-x}). \tag{2.68}$$

To illustrate this result, let us consider two quantum states, the ground state $n = 0$, and the first excited state $n = 1$, with wavefunctions

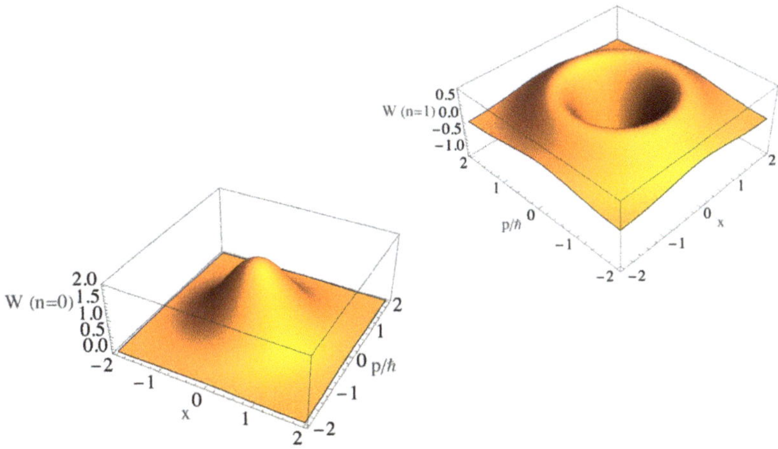

Figure 2.3. Wigner function $W(x, p)$, for ground state $n = 0$, and for the first excited state $n = 1$ of the harmonic oscillator. In these plots, we have used $\alpha = 1$.

$$\psi_0(q) = \frac{\sqrt{\alpha}}{\pi^{1/4}} \exp\left(-\frac{\alpha^2 q^2}{2}\right), \quad \psi_1(q) = \frac{\sqrt{2}\,\alpha^{3/2} q}{\pi^{1/4}} \exp\left(-\frac{\alpha^2 q^2}{2}\right). \tag{2.69}$$

Using equation (2.21), the corresponding Wigner functions can be written explicitly as

$$W_0(q, p) = 2\exp\left(-\frac{p^2}{\hbar^2\alpha^2} - q^2\alpha^2\right), \tag{2.70}$$

and

$$W_1(q, p) = 2\left[-1 + 2\frac{p^2}{\hbar^2\alpha^2} + 2q^2\alpha^2\right]\exp\left(-\frac{p^2}{\alpha^2\hbar^2} - q^2\alpha^2\right). \tag{2.71}$$

These functions are plotted in figure 2.3. We can see that the Wigner function for the excited state takes negative values, which is a clear quantum signature. It should however be noted that, although the Wigner function for the ground state is positive, it nevertheless corresponds to a state with a non-zero mean energy value, because $\langle H_0 \rangle = \hbar\omega/2$, in contrast with the zero energy value for a classical harmonic oscillator at rest. This means that, if a negative quasi-probability W is clearly a quantum signature, a positive quasi-probability is not necessarily a signature of the classical regime.

2.3 Electromagnetic field quantisation

2.3.1 Maxwell's equations

Let us start with Maxwell's equations, in their classical version. We can write them as

$$\nabla \times \mathbf{E} = -\frac{\partial \mathbf{B}}{\partial t}, \tag{2.72}$$

$$\nabla \times \mathbf{B} = \frac{1}{c^2}\frac{\partial \mathbf{E}}{\partial t} + \mu_0 \mathbf{J}, \tag{2.73}$$

$$\nabla \cdot \mathbf{E} = \frac{\rho}{\epsilon_0}, \tag{2.74}$$

$$\nabla \cdot \mathbf{B} = 0, \tag{2.75}$$

where $\mathbf{E}$ and $\mathbf{B}$ are the electric and induction fields produced by charges and currents, ρ and $\mathbf{J}$, distributed in vacuum. These charges and currents, on their turn, have to satisfy the continuity equation, which represents the total conservation of charge, as

$$\frac{\partial \rho}{\partial t} + \nabla \cdot \mathbf{J} = 0. \tag{2.76}$$

This can indeed be derived from equations (2.73) and (2.74). We have used the vacuum permittivity ϵ_0, and vacuum permeability μ_0, which specify the electric and magnetic properties of vacuum, assumed here as empty space, devoid of charges and currents. The speed of light in vacuum is a constant defined by the relation $c = 1/\sqrt{\epsilon_0 \mu_0}$. The fields can also be determined by the scalar and vector potentials, V and $\mathbf{A}$, using the definitions

$$\mathbf{E} = -\nabla V - \frac{\partial \mathbf{A}}{\partial t}, \qquad \mathbf{B} = \nabla \times \mathbf{A}. \tag{2.77}$$

It is known that Maxwell's equations are gauge invariant, which means that the same field can be described by a diversity of potentials. It is sometimes convenient to choose the *Coulomb gauge*, defined by the condition

$$\nabla \cdot \mathbf{A} = 0. \tag{2.78}$$

Using Maxwell's equations, we then derive the potential equations

$$\nabla^2 V = -\frac{1}{\epsilon_0}\rho, \tag{2.79}$$

$$\left(\nabla^2 - \frac{1}{c^2}\frac{\partial^2}{\partial t^2}\right)\mathbf{A} = -\mu_0 \mathbf{J} + \nabla\left(\frac{1}{c^2}\frac{\partial V}{\partial t}\right). \tag{2.80}$$

We can see that, in this gauge, the potential V simply determines the electrostatic field associated with the charge distribution ρ. At this point, it is useful to define the transverse and the longitudinal currents

$$\mathbf{J} = \mathbf{J}_T + \mathbf{J}_L, \tag{2.81}$$

satisfying the conditions

$$\nabla \cdot \mathbf{J}_T = 0, \qquad \nabla \times \mathbf{J}_L = 0. \tag{2.82}$$

This decomposition is generally valid, because it is known from vector analysis (*Helmholtz theorem*) that a vector field can be defined as a sum of two fields, one divergence free, and the other irrotational. From the continuity equation (2.76), we can easily get

$$\nabla \cdot \mathbf{J}_L = -\frac{\partial \rho}{\partial t}. \tag{2.83}$$

Taking the time derivative of the Poisson's equation (2.79), we can then conclude that

$$\mathbf{J}_L = \epsilon_0 \nabla \frac{\partial V}{\partial t}. \tag{2.84}$$

Replacing this in equation (2.80), we obtain

$$\left(\nabla^2 - \frac{1}{c^2}\frac{\partial^2}{\partial t^2}\right)\mathbf{A} = -\mu_0 \mathbf{J}_T. \tag{2.85}$$

This shows that, in the Coulomb gauge, the vector potential is only excited by the transverse current $\mathbf{J}_T$. Let us consider solutions of the potential equations in the absence of sources

$$\rho = 0, \qquad \mathbf{J}_T = 0. \tag{2.86}$$

In this case, we have $V = 0$, and the vector potential $\mathbf{A}$ is determined by a simple wave equation

$$\left(\nabla^2 - \frac{1}{c^2}\frac{\partial^2}{\partial t^2}\right)\mathbf{A} = 0. \tag{2.87}$$

We consider the fields existing inside an empty cubic cavity, with perfectly conducting walls of size L. In this case, we can expand the vector potential in a Fourier series of the form

$$\mathbf{A}(\mathbf{r}, t) = \sum_{\mathbf{k}} \mathbf{A}_{\mathbf{k}}(t) e^{i\mathbf{k}\cdot\mathbf{r}} + c.\,c., \tag{2.88}$$

where the components of $\mathbf{k}$, are compatible with the assumed boundary conditions, namely

$$k_j = \frac{2\pi}{L}\nu_j, \qquad (j = x, y, z), \tag{2.89}$$

with $\nu_j = \pm 1, \pm 2$, and so on. By expanding this hypothetical cavity to infinity, and taking the limit $L \to \infty$, we can transform the sum of a discrete number of modes into an integral over a continuum of modes, defined in free space with boundaries at infinity. The sum in equation (2.88) is then replaced by an integral, as

$$\sum_k \to 2\left(\frac{L}{2\pi}\right)^3 \int d\mathbf{k}. \tag{2.90}$$

Here, the additional factor of 2 accounts for the two independent polarisation states. It is interesting to note that the number of modes inside the infinitesimal interval $d\mathbf{k}$ is given by

$$d\nu \equiv d\nu_x d\nu_y d\nu_z = 2\left(\frac{L}{2\pi}\right)^3 dk_x dk_y dk_z. \tag{2.91}$$

In the unit volume $L^3 = 1$, the number of modes will then be equal to

$$d\nu = 2\frac{d\mathbf{k}}{(2\pi)^3}. \tag{2.92}$$

Returning to the Coulomb gauge condition (2.78), we can see that it implies that these modes are associated with transverse fields, satisfying

$$\mathbf{k} \cdot \mathbf{A}_\mathbf{k}(t) = \mathbf{k} \cdot \mathbf{A}_\mathbf{k}^*(t) = 0. \tag{2.93}$$

Replacing the mode expansion (2.88) in the wave equation (2.87), we get

$$\left(\frac{\partial^2}{\partial t^2} + \omega_k^2\right)\mathbf{A}_\mathbf{k}(t) = 0, \tag{2.94}$$

where the mode frequency is $\omega_k = kc$. The solution is

$$\mathbf{A}_\mathbf{k}(t) = \mathbf{A}_\mathbf{k} \exp(-i\omega_k t). \tag{2.95}$$

The mode expansion (2.88) then takes the form

$$\mathbf{A}(\mathbf{r}, t) = \sum_\mathbf{k} \mathbf{A}_\mathbf{k} \exp(i\mathbf{k} \cdot \mathbf{r} - i\omega_k t) + c.\ c. \tag{2.96}$$

The corresponding electric and magnetic fields are

$$\mathbf{E}(\mathbf{r}, t) = i\sum_k \omega_k \mathbf{A}_\mathbf{k} \exp(i\mathbf{k} \cdot \mathbf{r} - i\omega_k t) + c.\ c., \tag{2.97}$$

and

$$\mathbf{B}(\mathbf{r}, t) = i\sum_k (\mathbf{k} \times \mathbf{A}_\mathbf{k}) \exp(i\mathbf{k} \cdot \mathbf{r} - i\omega_k t) + c.\ c. \tag{2.98}$$

We can also introduce the *unit polarisation vector*, $\mathbf{e}_\mathbf{k}$, as defined by

$$\mathbf{e}_\mathbf{k} = \frac{\mathbf{A}_\mathbf{k}}{|A_\mathbf{k}|}, \quad \mathbf{k} \cdot \mathbf{e}_\mathbf{k} = 0. \tag{2.99}$$

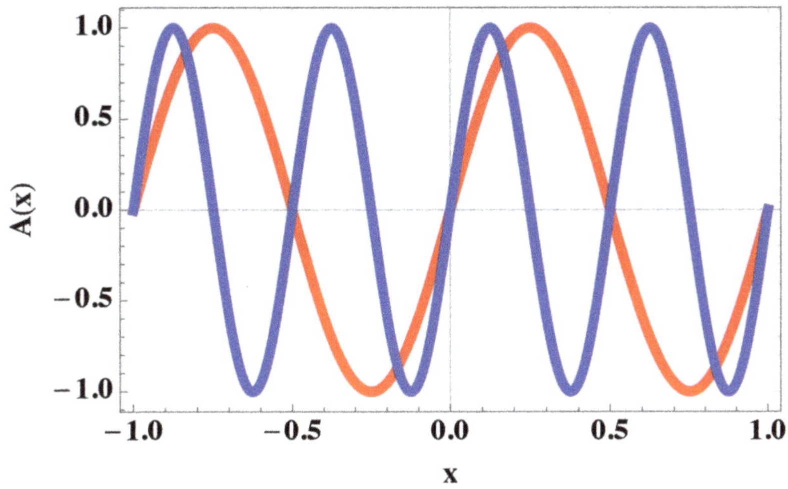

Figure 2.4. Representation of two modes of the electromagnetic field inside of a cavity with length L, in the x direction. The modes behave similarly in the other two directions, y and z.

Let us now consider the energy of the electromagnetic field inside the cavity, as given by

$$H_{\text{tot}} = \frac{1}{2} \int_V \left(\epsilon_0 E^2 + \frac{1}{\mu_0} B^2 \right) d\mathbf{r}. \tag{2.100}$$

Using the above mode expansion, and defining the electromagnetic energy per mode $H_{\mathbf{k}}$, we can easily obtain

$$H_{\text{tot}} = \sum_{\mathbf{k}} H_{\mathbf{k}}, \quad H_{\mathbf{k}} = 2\epsilon_0 V \omega_k^2 (\mathbf{A}_{\mathbf{k}} \cdot \mathbf{A}_{\mathbf{k}}^*), \tag{2.101}$$

where $V = L^3$ is the cavity volume (figure 2.4).

2.3.2 Field operators

In order to quantise the field and establish the field operators, it is useful to introduce a pair of canonical variables, Q_k and P_k, such that

$$\mathbf{A}_{\mathbf{k}} = \frac{1}{2\omega_k \sqrt{\epsilon_0 V}} (\omega_k Q_{\mathbf{k}} + i P_{\mathbf{k}}) \mathbf{e}_{\mathbf{k}}, \quad \mathbf{A}_{\mathbf{k}}^* = \frac{1}{2\omega_k \sqrt{\epsilon_0 V}} (\omega_k Q_{\mathbf{k}} - i P_{\mathbf{k}}) \mathbf{e}_{\mathbf{k}}^*, \tag{2.102}$$

where $\mathbf{e}_k$ is the unit polarisation vector, and the quantities Q_k and P_k are assumed real. Replacing this in the energy per mode $H_{\mathbf{k}}$, we get

$$H_{\mathbf{k}} = \frac{1}{2} \left(P_{\mathbf{k}}^2 + \omega_k^2 Q_{\mathbf{k}}^2 \right). \tag{2.103}$$

This is formally identical to the energy of a classical harmonic oscillator. We can then say that the electromagnetic field inside an empty cavity is formally identical to an infinite sum of harmonic oscillators. The electromagnetic field can then be quantised in the same way as we did for the harmonic oscillator, replacing the position and momentum-like quantities $Q_{\mathbf{k}}$ and $P_{\mathbf{k}}$ by the two operators, $\hat{Q}_{\mathbf{k}}$ and $\hat{P}_{\mathbf{k}}$, satisfying the commutation relations

$$\left[\hat{Q}_{\mathbf{k}}, \hat{Q}_{\mathbf{k}'}\right] = 0, \quad [\hat{P}_{\mathbf{k}}, \hat{P}_{\mathbf{k}'}] = 0, \quad \left[\hat{Q}_{\mathbf{k}}, \hat{P}_{\mathbf{k}'}\right] = i\hbar\delta(\mathbf{k} - \mathbf{k}'). \tag{2.104}$$

For a discrete sum of modes, we should use $\delta(\mathbf{k} - \mathbf{k}') = \delta_{\mathbf{kk}'}$. From here we can define annihilation and creation operators per mode, $\hat{a}_{\mathbf{k}}$ and $\hat{a}_{\mathbf{k}}^{\dagger}$, such that

$$\hat{a}_{\mathbf{k}} = \frac{1}{\sqrt{2\hbar\omega_k}}\left(\omega_k\hat{Q}_{\mathbf{k}} + i\hat{P}_{\mathbf{k}}\right), \quad \hat{a}_{\mathbf{k}}^{\dagger} = \frac{1}{\sqrt{2\hbar\omega_k}}\left(\omega_k\hat{Q}_{\mathbf{k}} - i\hat{P}_{\mathbf{k}}\right). \tag{2.105}$$

They are the operator equivalent to the above mode amplitude relations (2.102). These new operators satisfy the commutation relations

$$[\hat{a}_{\mathbf{k}}, \hat{a}_{\mathbf{k}'}] = 0, \quad [\hat{a}_{\mathbf{k}}^{\dagger}, \hat{a}_{\mathbf{k}'}^{\dagger}] = 0, \quad [\hat{a}_{\mathbf{k}}, \hat{a}_{\mathbf{k}'}^{\dagger}] = \delta(\mathbf{k} - \mathbf{k}'). \tag{2.106}$$

Using equation (2.101), we can now establish the total Hamilton operator $\hat{H}$, as

$$\hat{H} = \sum_{\mathbf{k}} \hbar\omega_k\left(\hat{N}_{\mathbf{k}} + \frac{1}{2}\right), \quad \hat{N}_{\mathbf{k}} \equiv \hat{a}_{\mathbf{k}}^{\dagger}\hat{a}_{\mathbf{k}}, \tag{2.107}$$

where $\hat{N}_{\mathbf{k}}$ is the number operator per mode. Each of these modes are independent and can be associated with number states $| n_{\mathbf{k}}\rangle$, similar to those introduced for the harmonic oscillator. Furthermore, the annihilation and creation operators satisfy

$$a_{\mathbf{k}} | n_{\mathbf{k}}\rangle = \sqrt{n_{\mathbf{k}}} \, | n_{\mathbf{k}} - 1\rangle, \quad a_{\mathbf{k}}^{\dagger} | n_{\mathbf{k}}\rangle = \sqrt{n_{\mathbf{k}} + 1} \, | n_{\mathbf{k}} + 1\rangle. \tag{2.108}$$

The state of the total electromagnetic field is then a product of states of all the modes, as

$$\left| n_{\mathbf{k}_1}, n_{\mathbf{k}_2}, \ldots, n_{\mathbf{k}_j}, \ldots\right\rangle = | n_{\mathbf{k}_1}\rangle \, | n_{\mathbf{k}_2}\rangle \ldots \left| n_{\mathbf{k}_j}\right\rangle \ldots \tag{2.109}$$

or, in a more compact form

$$|\{n_{\mathbf{k}}\}\rangle = \prod_k \, | n_k\rangle. \tag{2.110}$$

This is an *eigenstate* of the total energy operator (2.107), such that

$$\hat{H} \, |\{n_{\mathbf{k}}\}\rangle = E \, |\{n_{\mathbf{k}}\}\rangle. \tag{2.111}$$

Here, the energy eigenvalue is

$$E = \sum_{\mathbf{k}} \hbar\omega_k\left(n_{\mathbf{k}} + \frac{1}{2}\right), \tag{2.112}$$

where $n_{\mathbf{k}}$ is the number of photons per mode. This defines what is usually called a *Fock state*, or *number state*, of the electromagnetic radiation.

Finally, the field operators are obtained from equations (2.102) and (2.105), as

$$\hat{A}_{\mathbf{k}} = \sqrt{\frac{\hbar}{2\epsilon_0\omega_k V}}\hat{a}_{\mathbf{k}}, \quad \hat{A}_{\mathbf{k}}^{\dagger} = \sqrt{\frac{\hbar}{2\epsilon_0\omega_k V}}\hat{a}_{\mathbf{k}}^{\dagger}. \tag{2.113}$$

The total vector potential operator becomes

$$\hat{\mathbf{A}}(\mathbf{r}, t) = \sum_{\mathbf{k}} \sqrt{\frac{\hbar}{2\epsilon_0\omega_k V}}\mathbf{e}_{\mathbf{k}}\hat{a}_{\mathbf{k}}e^{i(\mathbf{k}\cdot\mathbf{r}-\omega_k t)} + h.\ c., \tag{2.114}$$

where $\mathbf{e}_{\mathbf{k}}$ is the unit polarisation vector. The corresponding electric and magnetic field operators $\hat{\mathbf{E}}(\mathbf{r}, t)$ and $\hat{\mathbf{B}}(\mathbf{r}, t)$ will be given by

$$\hat{\mathbf{E}}(\mathbf{r}, t) = i\sum_{\mathbf{k}} \sqrt{\frac{\hbar\omega_k}{2\epsilon_0 V}}\mathbf{e}_{\mathbf{k}}\hat{a}_{\mathbf{k}}e^{i(\mathbf{k}\cdot\mathbf{r}-\omega_k t)} + h.\ c., \tag{2.115}$$

and

$$\hat{\mathbf{B}}(\mathbf{r}, t) = i\sum_{\mathbf{k}} \sqrt{\frac{\hbar}{2\epsilon_0\omega_k V}}(\mathbf{k}\times\mathbf{e}_{\mathbf{k}})a_{\mathbf{k}}e^{i(\mathbf{k}\cdot\mathbf{r}-\omega_k t)} + h.\ c. \tag{2.116}$$

It is also useful to consider the two independent polarisation states for each field mode. We then need to replace the sum over $\mathbf{k}$, by a sum over $\mathbf{k}$ and σ, where σ takes two values representing the two orthogonal polarisation states. The mode operators $a_{\mathbf{k}}$ and $a_{\mathbf{k}}^{\dagger}$ are then replaced by $a_{\mathbf{k},\sigma}$ and $a_{\mathbf{k},\sigma}^{\dagger}$, and the commutation relations (2.106), by

$$\left[a_{\mathbf{k},\sigma}, a_{\mathbf{k}',\sigma'}\right] = 0, \quad \left[a_{\mathbf{k},\sigma}^{\dagger}, a_{\mathbf{k}',\sigma'}^{\dagger}\right] = 0, \quad \left[a_{\mathbf{k},\sigma}, a_{\mathbf{k}',\sigma'}^{\dagger}\right] = \delta(\mathbf{k}-\mathbf{k}')\delta_{\sigma,\sigma'}. \tag{2.117}$$

In chapter 3, the main properties of the quantum electromagnetic field, the different field representations and the differences between classical and quantum radiation states will be explored.

2.4 Canonical quantisation

Quantisation of the electromagnetic field in vacuum was performed using the analogy of each field mode with the harmonic oscillator. The field can then be seen as the superposition of an infinite number of quantum harmonic oscillators. A more formal but simple approach to field quantisation is considered here, similar to that used in [35]. It is known that all the basic equations in physics can be derived from a *variational principle*. We first describe the variation procedure leading to Maxwell's equations and then describe the canonical quantisation of the electromagnetic field.

2.4.1 Variational principle

We start by recalling the basic principles of classical mechanics. We assume a given physical system that can be described by a configuration variable $q \equiv (q_1, \ldots, q_N)$, which can be the position of each of the N elements of the system, and its time derivative $\dot{q}$. We can define the *Lagrangian* of the system as a function $L(q(t), \dot{q}(t))$, and the action integral

$$S = \int_{t_1}^{t_2} dtL(q(t), \dot{q}(t)). \tag{2.118}$$

The *Principle of Least Action* states that the action is a minimum along the trajectory $q(t)$ between the two instants t_1 and t_2, and can be represented by

$$\delta S = 0, \tag{2.119}$$

where δS results from taking a variation δq on the trajectory of the system. This implies that

$$\frac{\delta S}{\delta q(t)} \equiv \frac{\partial L}{\partial q} - \frac{d}{dt}\frac{\partial L}{\partial \dot{q}} = 0. \tag{2.120}$$

This is known as the *Lagrange equation*. We can then define the canonical momentum p, as the derivative of the Lagrangian with respect to $\dot{q}$, as

$$p = \frac{\partial L}{\partial \dot{q}}. \tag{2.121}$$

We can invert this equation and use it to express the velocity $\dot{q}$ in terms of the pairs of variables q and p, as $\dot{q} \equiv \dot{q}(q, p)$. We are now ready to define the *Hamiltonian* of the system, by using the so-called *Legendre transformation*, as

$$H(q, p) = p\dot{q} - L(q, \dot{q}(q, p)). \tag{2.122}$$

In general, the functions L and H depend explicitly on time t. This transformation allows us to write

$$L(q, \dot{q})dt = pdq - H(q, p)dt. \tag{2.123}$$

Inserting this in the above definition of the action, we get

$$S = \int_{t_1}^{t_2} pdq - H(q, p)dt. \tag{2.124}$$

The variational principle (2.119) is then satisfied if

$$\dot{q} = \frac{\partial H}{\partial p}, \quad \dot{p} = -\frac{\partial H}{\partial q}. \tag{2.125}$$

These are the *Hamilton equations* or the *canonical equations* of motion. Canonical quantisation is obtained by transforming the canonical variables (q, p) into operators $(\hat{q}, \hat{p})$, and by imposing (equal time) commutation relations of the form

$$[\hat{q}, \hat{q}] = 0, \quad [\hat{p}, \hat{p}] = 0, \quad [\hat{q}, \hat{p}] = i\hbar. \tag{2.126}$$

2.4.2 Lagrangian density

Let us now describe the electromagnetic field using the above Lagrangian approach. For that purpose, we assume the *Lagrangian density*, or density per unit volume $\mathcal{L}(V, \dot{V}, \mathbf{A}, \dot{\mathbf{A}})$, such that the total Lagrangian of the electromagnetic field is defined by

$$L = \int_V \mathcal{L}(A^\mu, \dot{A}^\mu) d\mathbf{r}, \tag{2.127}$$

where we have introduced the 4-vector $A^\mu = (V/c, \mathbf{A})$ and the index μ takes the values $\mu = 0, 1, 2, 3$. The equations of motion of the field satisfy the Lagrange equations, similar to equation (2.120), which take the form

$$\frac{\partial \mathcal{L}}{\partial A^\mu} - \sum_\nu \partial_\nu \frac{\partial \mathcal{L}}{\partial(\partial_\nu A^\mu)} = 0, \tag{2.128}$$

with $\partial_\nu \equiv \partial/\partial x^\nu$. We now choose the function $\mathcal{L}$ in such a way that Maxwell's equations are satisfied. For that purpose, we use

$$\mathcal{L} = \frac{1}{2}\epsilon_0 E^2 - \frac{1}{2\mu_0} B^2 = \frac{1}{2}\epsilon_0 \left(\frac{\partial \mathbf{A}}{\partial t} + \frac{1}{c}\nabla V\right)^2 - \frac{1}{2\mu_0}(\nabla \times \mathbf{A})^2. \tag{2.129}$$

Replacing this in equation (2.128), we indeed get $\nabla \cdot \mathbf{E} = 0$, for $\mu = 0$, and

$$\nabla \times \mathbf{B} = \frac{1}{c^2}\frac{\partial \mathbf{E}}{\partial t}, \tag{2.130}$$

for $\mu = 1, 2, 2$. Let us now establish the canonical momentum. In analogy with equation (2.121), it is determined by

$$\Pi^\mu = \frac{\partial \mathcal{L}}{\partial \dot{\mathbf{A}}^\mu}. \tag{2.131}$$

Or, more explicitly

$$\Pi^0 = c\frac{\partial \mathcal{L}}{\partial \dot{V}} = 0, \quad \Pi^i = \frac{\partial \mathcal{L}}{\partial \dot{\mathbf{A}}^i} = -\epsilon_0 E^i, \tag{2.132}$$

for $i = 1, 2, 3$. In vectorial notation, we write $\mathbf{\Pi} = -\epsilon_0 \mathbf{E}$. The corresponding Hamiltonian density $\mathcal{H}$ can now be established, according to equation (2.122), as

$$\mathcal{H} = \sum_\mu \dot{\mathbf{A}}^\mu \Pi^\mu - \mathcal{L} = \frac{1}{2}\epsilon_0 E^2 + \frac{1}{2\mu_0} B^2 + \frac{\epsilon_0}{c}\mathbf{E} \cdot \nabla V. \tag{2.133}$$

As for the total Hamiltonian we get, after integrating the last term by parts

$$H_{\text{tot}} = \int_V \mathcal{H}(\mathbf{A}, \mathbf{\Pi}) d\mathbf{r} - \frac{\epsilon_0}{c}\int V(\nabla \cdot \mathbf{E}) d\mathbf{r}. \tag{2.134}$$

The canonical equations for the electromagnetic field can also be obtained in analogy with equation (2.125), leading to

$$\dot{\Pi}^i = \frac{\partial \mathcal{H}}{\partial A^i} \tag{2.135}$$

for $i = 1, 2, 3$. It can easily be recognised that

$$\frac{\partial \mathcal{H}}{\partial A^i} = \frac{1}{\mu_0}(\nabla \times \mathbf{B})^i. \tag{2.136}$$

Using the vectorial notation for the canonical momentum, $\mathbf{\Pi} = -\epsilon_0 \mathbf{E}$, we can easily recover equation (2.130). Based on the above Lagrangian description of the electromagnetic field, we can quantise the field assuming that the field operators satisfy (equal time) commutation relations, similar to equations (2.126), of the form

$$\left[\hat{A}^\mu, \hat{A}^\nu\right] = 0, \quad \left[\hat{\Pi}^\mu, \hat{\Pi}^\nu\right] = 0, \quad \left[\hat{A}^\mu(\mathbf{r}, t), \hat{\Pi}^\nu(\mathbf{r}', t)\right] = i\hbar\delta_{\mu\nu}\delta(\mathbf{r} - \mathbf{r}'). \tag{2.137}$$

This quantisation procedure can be described in a more consistent covariant formulation, to be considered later.

2.5 Photon wavefunction

2.5.1 Riemann–Silverstein vector

Let us return to Maxwell's equations (2.72)–(2.74) assuming for simplicity an empty space, in the absence of charges and currents, ($\rho = 0$, $\mathbf{J} = 0$). It is useful to introduce the auxiliary fields

$$\mathbf{E}' = \sqrt{\epsilon_0}\mathbf{E}, \quad \mathbf{B}' = \frac{i}{\sqrt{\mu_0}}\mathbf{B}. \tag{2.138}$$

We can write them as

$$c(\nabla \times \mathbf{E}') = i\frac{\partial \mathbf{B}'}{\partial t}, \quad \nabla \cdot \mathbf{E}' = 0, \tag{2.139}$$

$$c(\nabla \times \mathbf{B}') = i\frac{\partial \mathbf{E}'}{\partial t}, \quad \nabla \cdot \mathbf{B}' = 0. \tag{2.140}$$

We can now use the following electromagnetic vector, first introduced by Silverstein in 1901, and sometimes called the *Riemann–Silberstein vector*, defined as

$$\mathbf{F} = \frac{1}{\sqrt{2}}(\mathbf{E}' + \mathbf{B}') \equiv \sqrt{\frac{\epsilon_0}{2}}(\mathbf{E} + ic\mathbf{B}). \tag{2.141}$$

This allows us to formulate Maxwell's equations in vacuum in a more compact form, as

$$i\frac{\partial \mathbf{F}}{\partial t} = c(\nabla \times \mathbf{F}), \quad \nabla \cdot \mathbf{F} = 0. \tag{2.142}$$

It is now important to notice that, for an arbitrary vector field $\mathbf{a}$, defined in the three-dimensional space, the following identity is verified

$$\mathbf{s} \cdot \nabla \mathbf{a} = i(\nabla \times \mathbf{a}), \tag{2.143}$$

where $\mathbf{s}$ is a 3×3 matrix vector, with components

$$s_x = \begin{bmatrix} 0 & 0 & 0 \\ 0 & 0 & -i \\ 0 & i & 0 \end{bmatrix}, \quad s_y = \begin{bmatrix} 0 & 0 & i \\ 0 & 0 & 0 \\ -i & 0 & 0 \end{bmatrix}, \quad s_z = \begin{bmatrix} 0 & -i & 0 \\ i & 0 & 0 \\ 0 & 0 & 0 \end{bmatrix}. \tag{2.144}$$

These are the three-dimensional analogues of the *Pauli matrices*, and can be used to describe the polarisation state of a *spin-1* particle as a photon. We are then allowed to write the first of equations (2.142) in the form

$$\frac{\partial \mathbf{F}}{\partial t} = -\mathbf{s}\nabla \cdot \mathbf{F}. \tag{2.145}$$

It is also useful to consider the electromagnetic field energy in terms of this new vector field. Using the above definition of H_{tot}, we can easily get

$$H_{\text{tot}} = \int_V \mathbf{F}(\mathbf{r}, t) \cdot \mathbf{F}^*(\mathbf{r}, t) d\mathbf{r}. \tag{2.146}$$

Let us now consider field quantisation, where the electric and magnetic fields $\mathbf{E}$ and $\mathbf{B}$ are replaced by the corresponding quantum operators. In this case, equation (2.145) allows us to define a new field operator $\mathbf{F}$, governed by the equation

$$i\hbar \frac{\partial \mathbf{F}}{\partial t} = c\frac{\hbar}{i}\mathbf{s}\nabla \cdot \mathbf{F}. \tag{2.147}$$

Noting that $\mathbf{p} = -i\hbar\nabla$ is the momentum operator in the $\mathbf{r}$-representation, we introduce the Hamiltonian operator

$$\hat{H} = c(\mathbf{s} \cdot \mathbf{p}) \equiv c\frac{\hbar}{i}(\mathbf{s} \cdot \nabla), \tag{2.148}$$

which allows us to write equation (2.147) in the form of a standard Schrödinger equation, as

$$i\hbar \frac{\partial \mathbf{F}}{\partial t} = \hat{H}\mathbf{F}, \tag{2.149}$$

where $\mathbf{F}$ plays the role of the photon state vector operator. It is also useful to write a more explicit expression for this operator, in terms of the annihilation and creation operators $\hat{a}_{\mathbf{k}}$ and $\hat{a}_{\mathbf{k}}^\dagger$. Equation (2.141) allows us to write

$$\mathbf{F} = \frac{i}{2} \sum_{\mathbf{k}} \sqrt{\hbar\omega_k}\, V \epsilon_{\mathbf{k}} \hat{a}_{\mathbf{k}} e^{i(\mathbf{k}\cdot\mathbf{r} - \omega_r t)} + h.\ c., \tag{2.150}$$

with the new (non-normalised) polarisation vector

$$\boldsymbol{\epsilon}_{\mathbf{k}} = \mathbf{e}_{\mathbf{k}} + i\frac{c}{\omega_k}(\mathbf{k} \times \mathbf{e}_{\mathbf{k}}), \tag{2.151}$$

where $\mathbf{e}_{\mathbf{k}}$ is the unit polarisation vector determining the direction of the electric field.

2.5.2 Spinor field

Instead of the photon operator $\mathbf{F}$, it is also possible to define the photon state vector in terms of a spinor operator of the form

$$\Psi = \begin{bmatrix} \varphi \\ \chi \end{bmatrix}, \tag{2.152}$$

where φ and χ are 1×3 vector columns defined by

$$\varphi = \mathbf{E}' \equiv \sqrt{\epsilon_0}\mathbf{E}, \quad \chi = \mathbf{B}' \equiv \frac{i}{\sqrt{\mu_0}}\mathbf{B}. \tag{2.153}$$

Using equations (2.139)–(2.140) and the identity (2.143), we can then write

$$i\frac{\partial \varphi}{\partial t} = -ic\mathbf{s} \cdot \nabla\chi, \quad i\frac{\partial \chi}{\partial t} = -ic\mathbf{s} \cdot \nabla\varphi. \tag{2.154}$$

We can also make use of the momentum operator $\mathbf{p} = i\hbar\nabla$ to write these two equations in a matricial form, as

$$i\hbar\frac{\partial}{\partial t}\begin{bmatrix} \varphi \\ \chi \end{bmatrix} = \begin{bmatrix} 0 & c\mathbf{s} \cdot \mathbf{p} \\ c\mathbf{s} \cdot \mathbf{p} & 0 \end{bmatrix}\begin{bmatrix} \varphi \\ \chi \end{bmatrix}. \tag{2.155}$$

This equation is formally identical to the *Dirac equation*, to be discussed in chapter 11, but with the bi-spinors replaced by the three-spinors φ and χ, as defined by equation (2.153), and the 2×2 Pauli matrix vector σ replaced by the 3×3 matrix vector $\mathbf{s}$, with components defined by equation (2.144). This dimensional difference results from the fact that the Dirac equation describes particles with spin-1/2, like electrons and positrons, defined in the two-dimensional space of spins, while here we are dealing with photons, with spin-1, and a three-dimensional spin (or polarisation) space. Finally, noting that $\mathbf{s}^\dagger = -\mathbf{s}$, we can derive from equations (2.154) a continuity equation of the form

$$\frac{\partial}{\partial t}\Psi^\dagger\Psi + \nabla \cdot \mathbf{j} = 0, \tag{2.156}$$

where $\mathbf{j}$ is the current of probability defined by

$$\mathbf{j} = \Psi^\dagger\hat{\mathbf{v}}\Psi, \quad \hat{\mathbf{v}} = c\begin{bmatrix} 0 & -\mathbf{s} \\ \mathbf{s} & 0 \end{bmatrix}. \tag{2.157}$$

This continuity equation simply shows that (in the absence of any interaction with matter) the photon probability $\Psi^\dagger\Psi$ is conserved.

2.6 Quantisation in a medium

Field quantisation in a material medium is not a trivial theoretical problem, and has been discussed by many authors along the years. See, the following relevant papers [13–17]. Two simple examples are given here, and more general approaches will be considered later in this book. The simplest example is that of a non-dispersive medium. For waves with wavelength much larger than the inter-particle distance, we can describe the medium as a continuous medium, characterised by appropriate permittivity ϵ and permeability μ. The above description therefore stays valid, if we replace the vacuum constants ϵ_0 and μ_0, by new constants ϵ and μ, and the velocity of light in vacuum c by the velocity of light in the medium, $v = 1/\sqrt{\epsilon\mu}$. This means that the field operators (2.114) to (2.116) will become

$$\hat{\mathbf{A}}_{\mathbf{k}} = \sum_{\mathbf{k}} C_{\mathbf{k}} \mathbf{e}_{\mathbf{k}} \hat{a}_{\mathbf{k}} e^{i(\mathbf{k}\cdot\mathbf{r} - \omega_k t)} + h.\ c., \tag{2.158}$$

$$\hat{\mathbf{E}}(\mathbf{r},\, t) = i \sum_{\mathbf{k}} \omega_{\mathbf{k}} C_{\mathbf{k}} \mathbf{e}_{\mathbf{k}} \hat{a}_{\mathbf{k}} e^{i(\mathbf{k}\cdot\mathbf{r} - \omega_k t)} + h.\ c., \tag{2.159}$$

and

$$\hat{\mathbf{B}}(\mathbf{r},\, t) = i \sum_{\mathbf{k}} C_{\mathbf{k}} (\mathbf{k} \times \mathbf{e}_{\mathbf{k}}) a_{\mathbf{k}} e^{i(\mathbf{k}\cdot\mathbf{r} - \omega_k t)} + h.\ c., \tag{2.160}$$

where the quantity $C_{\mathbf{k}}$ is defined by

$$C_{\mathbf{k}} = \sqrt{\frac{\hbar}{2\epsilon\omega_k V}}. \tag{2.161}$$

This is, of course, valid in the absence of wave dispersion. As a simple but relevant example of a dispersive medium, we consider field quantisation in isotropic plasmas. The medium is electrically neutral, but contains free electrons and positive ions, with total charge neutrality. As previously discussed, the electromagnetic potentials are described in the Coulomb gauge by

$$\left(\nabla^2 - \frac{1}{c^2}\frac{\partial^2}{\partial t^2}\right)\mathbf{A} = -\mu_0 \mathbf{J}, \quad \nabla^2 V = -\frac{\rho}{\epsilon_0}, \tag{2.162}$$

where $\mathbf{J}$ is the transverse current. For high frequency fields, the ions can be assumed immobile, and provide a neutralising electric background. The charge and current densities, ρ and $\mathbf{J}$ are determined by the electron motion. We can define them as

$$\rho = e(n_0 - n), \quad \mathbf{J} = -en\mathbf{v}, \tag{2.163}$$

where e is the absolute value of the electron charge, and n_0 the equilibrium plasma density. For Z ionised ions, the ion density is simply $n_i = n_0/Z$. In order to calculate the electron mean density n, and the electron mean velocity $\mathbf{v}$, we use the continuity and momentum equations of the electron fluid

$$\frac{\partial n}{\partial t} + \nabla \cdot (n\mathbf{v}) = 0, \qquad \frac{\partial \mathbf{v}}{\partial t} + \mathbf{v} \cdot \nabla \mathbf{v} = -\frac{e}{m_e}(\mathbf{E} + \mathbf{v} \times \mathbf{B}) - \frac{\nabla P_e}{n m_e}, \tag{2.164}$$

where m_e is the electron mass and P_e is the classical pressure of the electron gas. Alternatively, a more exact description of the electron gas, based on a kinetic equation, could be used. We linearise these equations with respect to the perturbed density $\tilde{n} = (n - n_0)$, in order to study the elementary perturbations in the medium. We also assume an electron equation of state such that $\nabla P = m_e v_{\text{the}}^2 \nabla n$, where $v_{\text{the}} = \sqrt{3T_e/m_e}$ is the electron thermal velocity, and T_e the temperature of the electron gas. This allows us to transform equations (2.162) into two formally identical equations for the two potentials V and $\mathbf{A}$, of the form [36]

$$\left(\nabla^2 - \frac{1}{c^2}\frac{\partial^2}{\partial t^2}\right)\mathbf{A} = \frac{\omega_p^2}{c^2}, \qquad \left(\nabla^2 - \frac{1}{v_{\text{the}}^2}\frac{\partial^2}{\partial t^2}\right)V = \frac{\omega_p^2}{v_{\text{the}}^2}, \tag{2.165}$$

where $\omega_p = \sqrt{e^2 n_0/\epsilon_0 m_e}$ is the electron plasma frequency. This can easily be written in covariant form, as

$$\left(\nabla^2 - \frac{1}{c_\mu^2}\frac{\partial^2}{\partial t^2}\right)A^\mu = \frac{\omega_p^2}{c_\mu^2}, \tag{2.166}$$

where we have used the four-potential $A^\mu = (V/c, \mathbf{A})$. We have also used $c_\mu = v_{\text{the}}$ for $\mu = 0$, and $c_\mu = c$ for $\mu = 1, 2, 3$. This can be used as a starting point for the quantisation procedure. For that purpose, we make the space Fourier decomposition and, for each mode $\mathbf{k}$, we define three independent polarisation states σ, where two of them represent the transverse photons, as in vacuum. These transverse photon states are described by the vector potential $\mathbf{A}_{\sigma\mathbf{k}}$ with $\sigma = 1, 2$, and are characterised by unit polarisation vectors $\mathbf{e}_{\sigma\mathbf{k}} = \mathbf{A}_{\sigma\mathbf{k}}/|\mathbf{A}_{\sigma\mathbf{k}}|$, such that $(\mathbf{e}_{\sigma\mathbf{k}} \cdot \mathbf{k}) = 0$. They states are orthogonal to each other, meaning that $(\mathbf{e}_{\sigma\mathbf{k}} \cdot \mathbf{e}_{\sigma'\mathbf{k}}) = \delta_{\sigma\sigma'}$.

In addition to these two transverse states, we have a third state $\sigma = 3$, which is a longitudinal photon state, characterised by $(\mathbf{e}_{\sigma\mathbf{k}} \cdot \mathbf{k}) = k$. This is the so-called *plasmon* state, which did not exist in vacuum and is due to the existence of free electrons in the medium. The existence of these three states clearly reveals the nature of the spin-1 photon. Using equation (2.166) we can easily show that they all satisfy the oscillator equation

$$\left(\frac{d^2}{dt^2} + \omega_{\sigma\mathbf{k}}^2\right)A_{\sigma\mathbf{k}} = 0, \tag{2.167}$$

where we have for the longitudinal mode, $A_{\sigma=3,\mathbf{k}} = V_\mathbf{k}/c$. The mode frequencies satisfy the relation

$$\omega_{\sigma\mathbf{k}} = \sqrt{k^2 c_\sigma^2 + \omega_p^2}, \tag{2.168}$$

with $c_{\sigma=1,2} = c$, and $c_{\sigma=3} = v_{\text{the}}$. See figure 2.5 for an illustration. This is indeed a dispersion relation of *massive bosons*, with mass defined by

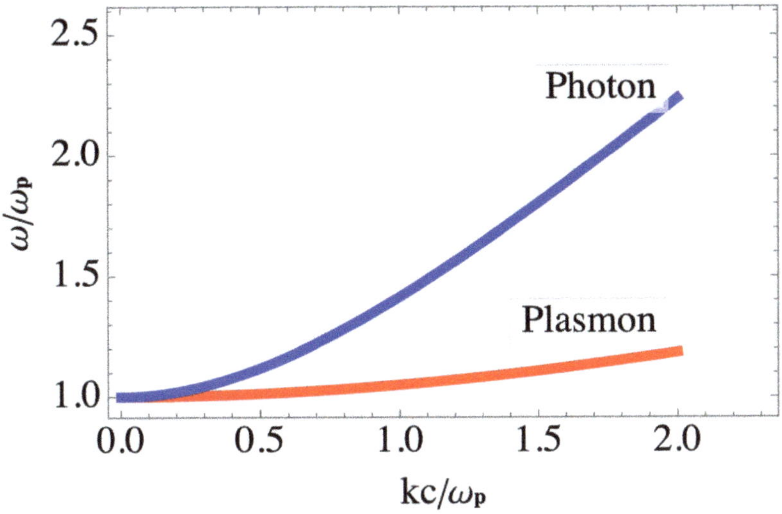

Figure 2.5. Dispersion relation for the massive photons in a plasma: normalised frequency ω/ω_p as a function of the normalised moment kc/ω_p, for $v_{\text{the}}/c = 0.3$. For transverse photons, the phase velocity ω/k tends asymptotically to c, and for plasmons to v_{the}.

$$m_\sigma = \frac{\hbar\omega_p}{c_\sigma^2}. \tag{2.169}$$

Given the difference between the speed of light c and the electron thermal velocity $v_{\text{the}} \ll c$, the mass of transverse photons is much larger than that of plasmons. A difference that is eventually reduced in a relativistic plasma, where however the electron fluid equations (2.164) are no longer valid. Quantisation of the harmonic oscillators described by equation (2.167) can then be performed in the usual way. However, the energy operator needs a new definition because, in a dispersive medium, the average energy of waves with frequency ω is determined by [37]

$$H = \frac{1}{2}\int\left[\left(\frac{\omega\epsilon(\omega)}{\partial\omega}\right)|\mathbf{E}|^2 + \frac{1}{\mu(\omega)}\left(\frac{\omega\mu(\omega)}{\partial\omega}\right)|\mathbf{B}|^2\right]d\mathbf{r}, \tag{2.170}$$

where $\epsilon(\omega)$ and $\mu(\omega)$ are the dielectric and magnetic functions of the medium. They therefore depend on the photon modes ($\sigma\mathbf{k}$). In a plasma, we usually neglect its diamagnetic properties, and use $\mu(\omega) = \mu_0$. Expanding this expression into modes, we obtain

$$H_{\mathbf{k}} = \sum_{\sigma=1,2,3} C_{\sigma\mathbf{k}} \mid \mathbf{A}_{\sigma\mathbf{k}} \mid^2, \tag{2.171}$$

where

$$C_{\sigma\mathbf{k}} = \left(\frac{\omega\epsilon(\omega)}{\partial\omega}\right)_{\sigma\mathbf{k}} + \frac{k^2}{2\mu_0}(1 - \delta_{3,\sigma}), \tag{2.172}$$

and the dielectric function is

$$\epsilon_{\sigma \mathbf{k}}(\omega) = \epsilon_0 \left[1 - \frac{\omega_p^2}{\omega^2} - \frac{k^2 v_{\text{the}}^2}{\omega^2} \delta_{3,\sigma} \right]. \tag{2.173}$$

Obviously, the last term in these two expressions only exists for $\sigma = 3$. Upon quantisation of the oscillator modes (2.167), we can then retrieve the Hamiltonian operator in the form

$$H_{\mathbf{k}} = \sum_{\sigma=1,2,3} \hbar \omega_{\sigma \mathbf{k}} \left(a_{\sigma \mathbf{k}}^{\dagger} a_{\sigma \mathbf{k}} + \frac{1}{2} \right). \tag{2.174}$$

This is formally identical to the case of vacuum, except that there is an additional polarisation state associated with plasmons. These three states will correspond to the two transverse photons with spin $s = \pm 1$, and the longitudinal photon or plasmon with $s = 0$. Extension of this method to a quantum plasma, where the electron gas is also quantised is straightforward [38].

References

[1] Dirac P A M 1958 *The Principles of Quantum Mechanics* 4th edn (Oxford: Clarendon)

[2] Schiff L I 1968 *Quantum Mechanics* (Toronto: McGraw-Hill)

[3] Landau L D and Lifshitz E M 1977 *Quantum Mechanics: Non-Relativistic Theory* 3rd edn (New York: Pergamon)

[4] Messiah A 2017 *Quantum Mechanics* (New York: Dover)

[5] Cohen-Tannoudji C, Diu B and Laloë F 1992 *Quantum Mechanics* vols I and II (New York: Wiley)

[6] Dirac P A M 1927 The quantum theory of the emission and absorption of radiation *Proc. R. Soc.* A **114** 243–65

[7] Heitler 1954 *Quantum Theory of Radiation* (Oxford: Oxford University Press)

[8] Louisell W H 1973 *Quantum Statistical Properties of Light* (New York: Wiley)

[9] Loudon R 1973 *The Quantum Theory of Light* (Oxford: Oxford University Press)

[10] Itzykson C and Zuber J-B 1980 *Quantum Field Theory* (Toronto: McGraw-Hill)

[11] Bialynicki-Birula I and Bialynicki-Birula Z 2013 The role of the Riemann-Silverstein vector in classical and quantum theories of electromagnetism *J. Phys. A: Math. Theor* **46** 053001

[12] Jauch J M and Watson K M 1948 Phenomenological quantum electrodynamics, Part II. Interaction of the field with charges *Phys. Rev.* **74** 1485

[13] Alekseev A I and Nikitin Yu P 1966 Quantisation of the electromagnetic field in a dispersive medium *Sov. Phys. JETP* **23** 608–14

[14] Drummond P D 1990 Electromagnetic quantization in dispersive inhomogeneous nonlinear dielectrics *Phys. Rev.* A **42** 6845–57

[15] Milonni P W 1995 Field quantization and radiative process in dispersive dielectric media *J. Mod. Opt.* **42** 1991–2004

[16] Dalton B J, Guerra E S and Knight P L 1996 Field quantization in dielectric media and the generalized multipolar Hamiltonian *Phys. Rev.* A **54** 2292–313

[17] Bennett R B, Barlow T M and Belge A 2016 A physically motivated quantization of the electromagnetic field *Eur. J. Phys.* **37** 014001

[18] Anderson P W 1973 Plasmons, gauge invariance, and mass *Phys. Rev.* **139** 439
[19] Styer D F *et al* 2002 Nine formulations of quantum mechanics *Am. J. Phys.* **70** 288–97
[20] Schrödinger E 1926 Quantisierung als Eigenwertproblem (Vierte Mitteilung) *Ann. Phys., Lpz.* **81** 109–39
[21] Schrödinger E 1978 *Collected Papers on Wave Mechanics* (New York: Chelsea)
[22] Heisenberg W 1925 Über die quantentheoretische Umdeutung kinematischer und mechanischer Beziehungen *Z. Phys.* **33** 879–93
[23] Born M and Jordan P 1925 Zur Quantenmechanik *Z. Phys.* **34** 858–88
[24] Born M, Heisenberg W and Jordan P 1926 Zur Quantenmechanik II *Z. Phys.* **35** 557–615
[25] Wigner E 1932 On the quantum correction for thermodynamic equilibrium *Phys. Rev.* **40** 749–59
[26] Weyl H 1932 *Theory of Groups and Quantum Mechanics* (London: Methuen and Company Ltd.)
[27] Moyal J E 1949 Quantum mechanics as a statistical theory *Math. Proc. Camb. Phil. Soc.* **45** 99–124
[28] Brittin W E and Chappell W R 1962 Second quantization in phase space *Rev. Mod. Phys.* **34** 620–7
[29] Hillery M, O'Connell R F, Scully M O and Wigner E P 1984 Distribution functions in physics: fundamentals *Phys. Rep.* **106** 121–67
[30] Tatarski V I 1983 The Wigner representation of quantum mechanics *Sov. Phys. - Usp.* **26** 311–27
[31] Lee H-W 1995 Theory and application of the quantum phase-space distribution functions *Phys. Rep.* **259** 147–211
[32] Case W B 2008 Wigner functions and Weyl transformations for pedestrians *Am. J. Phys.* **76** 937–46
[33] Weinbub J and Ferry D K 2018 Recent advances in Wigner function approaches *Appl. Phys. Rev.* **5** 041104
[34] Arfken G B and Weber H J 2005 *Mathematical Methods for Physicists* 6th edn (Amsterdam: Elsevier)
[35] Hillery M 2009 An introduction to the quantum theory of nonlinear optics *Acta Phys. Slov.* **59** 1–80
[36] Mendonça J T, Martins A M and Guerreiro A 2000 Field quantisation in a plasma: photon mass and charge *Phys. Rev.* E **62** 2989–91
[37] Landau L D and Lifshitz E M 1963 *Electrodynamics of Continuous Media* (Oxford: Pergamon)
[38] Mendonça J T 2021 Field quantization in quantum plasmas: photon and plasmon charge and mass *Contrib. Plasma Phys.* **62** e202100097

Chapter 3

Coherence

Here we discuss the coherence properties of the electromagnetic field, and in particular, where those properties deviate from their classical ones. This includes coherent states, as well as squeezed states. Furthermore, we discuss field and intensity correlations, which can be used to characterise an arbitrary quantum field state. The Bell inequalities are also discussed. All these topics are usually covered in many existing books on quantum optics, such as [1–5] and are presented here in a compact form.

We start with the properties of coherent states. The definition of these states was first given by Glauber in 1963 [6] (see also the review [7]), and plays a central role in the quantum theory of radiation. However, the existence of non-spreading wave-packet solutions of the harmonic oscillator, which contains the seed for the concept of coherent states, had already been noticed by Schrödinger in 1926 [8], at the dawn of quantum theory. The reasons for the relevance of coherent states are multiple. First, they can be seen as the quantum states closest to classical states. In particular, they can be excited by a classical radiation source. Second, they represent the minimum uncertainty states. And finally, they can be described as a displaced vacuum. Coherent states are, in a sense, the complete opposite of squeezed states, because the later cannot be reduced to a classical field.

Squeezed states are the simplest example of a non-classical quantum field, and were first considered by Kennard in 1927 [9]. They were formalised later by Husimi [10], and others [11, 12]. Several good reviews on squeezed states and non-classical states of light have been written along the years [13–17].

On the other hand, generalisations of Wigner's and Weyl's ideas on the classical representation of quantum states were proposed, allowing to study the evolution of generic field states in classical phase-space. This is also explained in the present chapter. We discuss the P-representation, the Q-representation, the W-representation (which is a reformulation of the Wigner function) and a generalised G-representation, which includes the previous three.

doi:10.1088/978-0-7503-2786-2ch3

This chapter also contains a discussion of quantum correlations, which are essential tools to analyse optical experiments and to prove the existence of quantum states of light. In particular, we define field correlations and illustrate this concept with the famous *Young experiment*. We also define intensity correlations, illustrated with the almost equally famous *Hanbury-Brown and Twiss experiment*. We conclude with a brief discussion of photon entanglement and Bell's theorem, which are the basic ingredients of the modern concept of teleportation, and of the more recent quantum computation developments.

3.1 Coherent states

3.1.1 Definition

We consider a single field state, with wavevector **k** and annihilation and creation operators, $\hat{a}_{\mathbf{k}}$ and $\hat{a}_{\mathbf{k}}^{\dagger}$. A *coherent state* can be defined as the eigenstate of the annihilation operator $\hat{a}_{\mathbf{k}}$. Dropping the subscript **k** for simplicity, we can write the corresponding eigenvalue equation as

$$\hat{a} \mid \alpha\rangle = \alpha \mid \alpha\rangle, \tag{3.1}$$

where α is a complex eigenvalue, and $\mid \alpha\rangle$ the state vector associated with this eigenvalue. Noting that a is not Hermitian, we cannot say that these eigenvalues are real, or that these eigenvectors form a complete orthogonal set. However we will see that, although non-orthogonal, they form an overcomplete set, which can be used to represent a generic field state.

First, we represent the coherent state $\mid \alpha\rangle$ in terms of the Fock or number state representation $\{\mid n\rangle\}$ discussed in the previous chapter, using

$$\mid \alpha\rangle = \sum_{n=0}^{\infty} C_n(\alpha)|n\rangle, \qquad C_n(\alpha) = \langle n|\alpha\rangle, \tag{3.2}$$

where $C_n(\alpha)$ are the expansion coefficients. Applying the operator $\hat{a}$ to this expression, we get

$$\hat{a} \mid \alpha\rangle = \sum_n C_n(\alpha)\sqrt{n} \mid n-1\rangle. \tag{3.3}$$

Replacing this in equation (3.1), we obtain a relation between consecutive coefficients, as

$$C_n(\alpha) = \frac{\alpha}{\sqrt{n}} C_{n-1}(\alpha). \tag{3.4}$$

For $n = 1$, 2 and 3, this gives

$$C_1 = \alpha C_0, \quad C_2 = \frac{\alpha}{\sqrt{2}} C_1 = \frac{\alpha^2}{\sqrt{2}} C_0, \quad C_3 = \frac{\alpha}{\sqrt{3}} C_2 = \frac{\alpha^3}{\sqrt{3!}} C_0. \tag{3.5}$$

For an arbitrary n, we then get

$$C_n(\alpha) = \frac{\alpha^n}{\sqrt{n!}} C_0(\alpha). \tag{3.6}$$

On the other hand, normalisation of the coherent state vector $|\,\alpha\rangle$, implies that

$$\langle\alpha\mid\alpha\rangle = \sum_n |C_n(\alpha)|^2 = 1. \tag{3.7}$$

From here, we then get

$$C_0(\alpha) = \exp\left(-\frac{|\alpha|^2}{2}\right). \tag{3.8}$$

Replacing this in equations (3.2), we obtain

$$|\,\alpha\rangle = \exp\left(-\frac{|\alpha|^2}{2}\right)\sum_{n=0}^{\infty} \frac{\alpha^n}{\sqrt{n!}} \,|\, n\rangle. \tag{3.9}$$

This expression defines a coherent state in terms of number states. We can see from here that the probability for finding n photons in a coherent state $|\,\alpha\rangle$ is given by

$$P_n(\alpha) \equiv |\,\langle n\,|\,\alpha\rangle|^2 = \exp\left(-|\alpha|^2\right)\frac{|\,\alpha\,|^{2n}}{n!}. \tag{3.10}$$

Such a discrete probability distribution was introduced for the first time by the French mathematician Siméon Poisson in the 19th century [18], and is called for that reason a *Poisson distribution.* It satisfies the following recurrence relation

$$P_n(\alpha) = \frac{|\,\alpha\,|^2}{n} P_{n-1}(\alpha). \tag{3.11}$$

The maximum probability is attained for $n \simeq |\,\alpha\,|^2$. This corresponds to the expectation value of the photon number operator, as given by

$$\langle\hat{N}\rangle \equiv \langle\alpha\mid\hat{a}^\dagger\hat{a}\mid\alpha\rangle = |\alpha|^2. \tag{3.12}$$

Notice however that, for $|\alpha|^2 \leqslant 1$, the maximum of the probability distribution occurs at $n = 0$. From the definition of $P_n(\alpha)$, we can also conclude that the mean photon number of a coherent state is

$$\langle\hat{N}\rangle = \sum_n P_n(\alpha)\, n. \tag{3.13}$$

The Poisson distribution is illustrated in figure 3.1, for two different values of α.

3.1.2 Overcompleteness

We have seen that the coherent state vector defined by equation (3.9) is normalised, $\langle\alpha\mid\alpha\rangle = 1$, but different coherent states are not orthogonal. Let us consider two

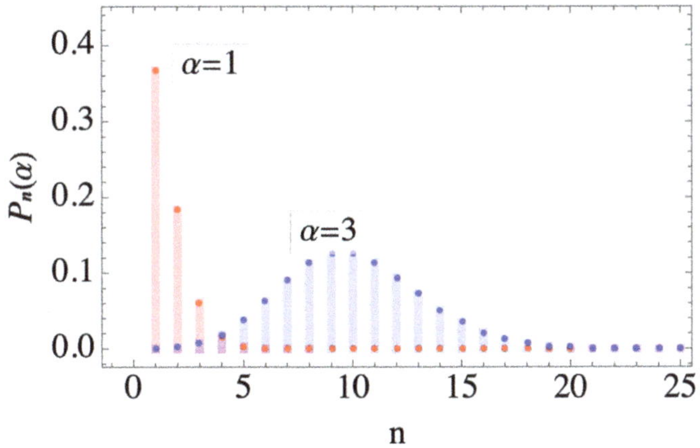

Figure 3.1. Poisson distribution, defined by equation (3.10), for two different values of the parameter: $\alpha = 1$ (in red) and $\alpha = 3$ (in blue).

different states $|\alpha\rangle$ and $|\beta\rangle$, where α and β are complex numbers. Taking the scalar product

$$\langle \alpha \mid \beta \rangle = \exp\left(-\frac{|\alpha|^2}{2} - \frac{|\beta|^2}{2}\right) \sum_{n=0}^{\infty} \frac{(\alpha^*)^n \beta^n}{n!} = \exp\left(-\frac{|\alpha|^2}{2} - \frac{|\beta|^2}{2} + \alpha^* \beta\right), \qquad (3.14)$$

we obtain

$$|\langle \alpha | \beta \rangle|^2 = \exp(-|\alpha - \beta|^2). \qquad (3.15)$$

This shows that different coherent states are non-orthogonal and that they only approach orthogonality when the complex numbers α and β are very different from each other, such that $| \alpha - \beta | \gg 1$. However, the coherent states form a *complete set* of states, which can be used to represent a general field state. This means that they can be used to define the identity operator I, using

$$\frac{1}{\pi} \int | \alpha \rangle \langle \alpha \mid d\alpha = I, \qquad (3.16)$$

where $d\alpha$ represents a double integration over the complex plane. In order to verify this expression, we start from the definition (3.9), which allows us to write

$$\int |\alpha\rangle\langle\alpha| \, d\alpha = \sum_{n,m} \frac{|n\rangle\langle m|}{\sqrt{n!m!}} \int e^{-|\alpha|^2} \alpha^n (\alpha^*)^m d\alpha. \qquad (3.17)$$

Using polar coordinates (r, θ), such that $\alpha = r \exp(i\theta)$, and $d\alpha = r dr d\theta$, we get

$$\int e^{-|\alpha|^2} \alpha^n (\alpha^*)^m d\alpha = \int_0^{\infty} dr e^{-r^2} r^{n+m+1} \int_0^{2\pi} d\theta \exp[i(n - m)\theta]. \qquad (3.18)$$

Integration in θ simply gives

$$\int_0^{2\pi} d\theta \exp[i(n - m)\theta] = 2\pi\delta_{nm}, \tag{3.19}$$

which allows us to write equation (3.17) as

$$\frac{1}{\pi}\int | \alpha\rangle\langle\alpha | d\alpha = \sum_n \frac{| n\rangle\langle n |}{n!}\int_0^{\infty} d\eta e^{-\eta}\eta^n, \tag{3.20}$$

where $\eta = r^2$. The integral over η is equal to $n!$ (see [19]), and we are reduced to

$$\frac{1}{\pi}\int | \alpha\rangle\langle\alpha | d\alpha = \sum_{n=0}^{\infty} | n\rangle\langle n | = I, \tag{3.21}$$

where, in the last step, we have used the completeness relation for the number states, equation (2.58). This demonstrates the above completeness relation for the coherent states, equation (3.16). But, because each state vector of this set can be represented in their own basis, as shown by equation (3.14), we call the set of coherent states, not just complete, but overcomplete.

3.1.3 Uncertainties

Let us now consider the uncertainty relations associated with a coherent state. First, we notice that the mean value of the square of the number operator, $\hat{N}^2 = \hat{a}^\dagger\hat{a}\hat{a}^\dagger\hat{a} = \hat{N} + (\hat{a}^{\dagger 2})(\hat{a}^2)$, is given by

$$\left\langle \hat{N}^2 \right\rangle = \langle\alpha | \hat{N}^2 | \alpha\rangle = |\alpha|^2 + |\alpha|^4. \tag{3.22}$$

The photon mean number deviation Δn with respect to the average will be

$$\Delta n \equiv \sqrt{\left\langle \hat{N}^2 \right\rangle - \langle \hat{N}\rangle^2} = \sqrt{|\alpha|^2 + |\alpha|^4 - |\alpha|^4} = |\alpha|. \tag{3.23}$$

The relative uncertainty of the mean number for a coherent state will then be

$$\frac{\Delta n}{\langle n\rangle} \equiv \frac{\Delta n}{\langle\alpha | \hat{N} | \alpha\rangle} = \frac{|\alpha|}{|\alpha|^2} = \frac{1}{|\alpha|}. \tag{3.24}$$

This shows that the photon number uncertainty is proportional to the inverse square root of the mean photon number. At this point, it is useful to introduce the quadrature operators $\hat{q}$ and $\hat{p}$, defined in the previous chapter, and calculate their mean values. We have

$$\langle\hat{q}\rangle = \sqrt{\frac{\hbar}{2\omega}}\langle\alpha |(\hat{a} + \hat{a}^\dagger)| \alpha\rangle = \sqrt{\frac{\hbar}{2\omega}}(\alpha + \alpha^*), \tag{3.25}$$

and

$$\langle\hat{p}\rangle = -i\sqrt{\frac{\hbar\omega}{2}}\langle\alpha |(\hat{a} - \hat{a}^\dagger)| \alpha\rangle = -i\sqrt{\frac{\hbar\omega}{2}}(\alpha - \alpha^*). \tag{3.26}$$

Similarly, we get

$$\langle \hat{q}^2 \rangle = \frac{\hbar}{2\omega}\langle \alpha \,|(\hat{a}^2 + \hat{a}^{\dagger 2} + \hat{a}\hat{a}^\dagger + \hat{a}^\dagger \hat{a})|\, \alpha \rangle = \frac{\hbar}{2\omega}(\alpha^2 + \alpha^{*2} + 2|\alpha|^2 + 1), \tag{3.27}$$

and

$$\langle \hat{p}^2 \rangle = -\frac{\hbar\omega}{2}(\alpha^2 + \alpha^{*2} - 2|\alpha|^2 - 1). \tag{3.28}$$

This allows us to calculate the respective variances, as

$$(\Delta q)^2 \equiv \langle \hat{q}^2 \rangle - \langle \hat{q} \rangle^2 = \frac{\hbar}{2\omega}, \quad (\Delta p)^2 \equiv \langle \hat{p}^2 \rangle - \langle \hat{p} \rangle^2 = \frac{\hbar\omega}{2}. \tag{3.29}$$

We can finally state the uncertainty relation for the coherent state as

$$\Delta q \cdot \Delta p = \frac{\hbar}{2}. \tag{3.30}$$

This shows that the coherent state satisfies the minimum value compatible with the uncertainty principle. It is a state of minimum uncertainty. It is also useful to discuss the uncertainties associated with the field operators. Using the electric field operator $\hat{E}$, we write

$$\langle \hat{E} \rangle = i\sqrt{\frac{\hbar\omega}{2\epsilon_0 V}}\left\{\langle \alpha \mid \hat{a} \mid \alpha \rangle e^{i(\mathbf{k}\cdot\mathbf{r}-\omega t)} - \langle \alpha \mid \hat{a}^\dagger \mid \alpha \rangle e^{-i(\mathbf{k}\cdot\mathbf{r}-\omega t)}\right\}. \tag{3.31}$$

In polar coordinates, such that $\alpha = r\exp(i\theta)$, we can easily get

$$\langle \hat{E} \rangle = -\sqrt{\frac{2\hbar\omega}{\epsilon_0 V}}\, |\alpha|\sin(\mathbf{k}\cdot\mathbf{r} - \omega t + \theta). \tag{3.32}$$

Similarly, we also could find that

$$\left\langle \hat{E}^2 \right\rangle = \langle \alpha \mid \hat{E}^2 \mid \alpha \rangle = \frac{\hbar\omega}{2\epsilon_0 V}\left\{1 + 4|\alpha|^2 \sin^2(\mathbf{k}\cdot\mathbf{r} - \omega t + \theta)\right\}. \tag{3.33}$$

The uncertainty of the electric field amplitude is then determined by

$$\Delta E = \sqrt{\left\langle \hat{E}^2 \right\rangle - \langle \hat{E} \rangle^2} = \left(\frac{\hbar\omega}{2\epsilon_0 V}\right)^{1/2}. \tag{3.34}$$

We can see that the field uncertainty is independent of the intensity, which for a coherent state is represented by $|\alpha|^2$. This shows that the relative field uncertainty will behave, for large $|\alpha| \gg 1$, as

$$\frac{\Delta E}{\langle \hat{E} \rangle} \propto \frac{1}{|\alpha|}, \tag{3.35}$$

thus tending to a classical field in the high intensity limit, where quantum uncertainties are nearly absent. Notice that this result strongly resembles equation (3.24), but we should not forget that the mean number of photons is proportional to $\left\langle \hat{E}^2 \right\rangle$. This means that the phase is also playing a role in the calculation of the field uncertainties.

3.1.4 Displaced vacuum

We now show that a coherent state can be described as a displaced vacuum. For that purpose, we use equation (2.57) to represent the number states in terms of the vacuum state $|\,0\rangle$, and equation (3.9) to represent a coherent state in terms of number states. We get

$$|\,\alpha\rangle = e^{-|\alpha|^2/2} \sum_n \frac{\alpha^n \hat{a}^{\dagger n}}{n!} \,|\,0\rangle = e^{-|\alpha|^2/2} e^{\alpha \hat{a}^\dagger} \,|\,0\rangle. \tag{3.36}$$

This can also be written as

$$|\,\alpha\rangle = \hat{D}(\alpha)|\,0\rangle, \tag{3.37}$$

where $\hat{D}(\alpha)$ is the *displacement operator* defined by

$$\hat{D}(\alpha) = e^{-|\alpha|^2/2} e^{\alpha \hat{a}^\dagger}. \tag{3.38}$$

Alternatively, this operator can be written in a more symmetric form, as

$$\hat{D}(\alpha) = \exp(\alpha \hat{a}^\dagger - \alpha^* \hat{a}). \tag{3.39}$$

In order to see the equivalence between these two definitions of $\hat{D}(\alpha)$, we first notice that

$$\exp(-\alpha^* \hat{a})|\,0\rangle = \left(1 - \alpha^* \hat{a} + \frac{1}{2}(\alpha^* \hat{a})^2 + \cdots \right)|\,0\rangle = |\,0\rangle, \tag{3.40}$$

because $\hat{a}\,|\,0\rangle = 0$. Inserting this in equation (3.37), we arrive easily at

$$\hat{D}(\alpha) = e^{-|\alpha|^2/2} e^{\alpha \hat{a}^\dagger} e^{-\alpha^* \hat{a}}. \tag{3.41}$$

We can then use the *Baker–Hausdorff formula* (see appendix A), relating the exponentials of two operators A and B, as

$$e^{A+B} = e^A e^B e^{-[A,\,B]/2}. \tag{3.42}$$

This is valid for two operators that commute with their commutator. Choosing $A = \alpha \hat{a}$, and $B = \alpha^* \hat{a}^\dagger$, we can identify the two expressions (3.39) and (3.41). Similarly, we could have defined the displacement operator as

$$\hat{D}(\alpha) = e^{|\alpha|^2/2} e^{-\alpha^* \hat{a}} e^{\alpha \hat{a}^\dagger}. \tag{3.43}$$

Furthermore, using the same formula, we could also show that

$$\hat{D}(\alpha + \beta) = \hat{D}(\alpha)\hat{D}(\beta)\exp\left\{-i\Im(\alpha\beta^*)\right\}. \tag{3.44}$$

Other properties of the displacement operator include unitarity

$$\hat{D}^{\dagger}(\alpha) = \hat{D}(-\alpha) = \hat{D}^{-1}(\alpha), \tag{3.45}$$

and the displacement properties

$$\hat{D}^{\dagger}(\alpha)\hat{a}\hat{D}(\alpha) = \hat{a} + \alpha, \quad \hat{D}^{\dagger}(\alpha)\hat{a}^{\dagger}\hat{D}(\alpha) = \hat{a}^{\dagger} + \hat{a}^*. \tag{3.46}$$

They can be demonstrated using another operator relation

$$e^{-\alpha A}Be^{\alpha A} = B - \alpha[A, B] + \frac{\alpha^2}{2!}[A, [A, B]] + \cdots \tag{3.47}$$

For $A = \hat{a}^{\dagger}$ and $B = \hat{a}$, this reduces to $e^{-\alpha\hat{a}}\hat{a}\, e^{\alpha\hat{a}^{\dagger}} = \hat{a} + \alpha$.

3.2 Field representations

Here we discuss various representations of the quantum field states in classical phase-space. In the previous chapter we introduced the Wigner function, which could provide such representation. But this is not the only choice, and other functions with similar properties can be defined. This is the case of the P-function, first used by Glauber [20] and Sudarshan [21], and the Q-function, proposed independently by Husimi [10] and Kano [22]. Furthermore, it is possible to redefine the Wigner function in more formal terms, using field operators. For symmetry, we call it the W-function. Finally, a more general representation is possible, the G-representation, which includes all these three functions as particular cases. In our review on these representations we mainly follow [2, 3].

3.2.1 P-representation

We have seen previously that the coherent states, although non-orthogonal, form an overcomplete set and satisfy a closure relation, given by equation (3.16), which provides an overcomplete set of state vectors. This suggests the use of a coherent state representation for the density operator, of the form

$$\hat{\rho} = \int P(\alpha)|\,\alpha\rangle\langle\alpha\,|\,d\alpha. \tag{3.48}$$

Here, the quantity $P(\alpha)$ represents a probability distribution. Obviously, it has to satisfy the normalisation condition

$$\int P(\alpha)\,d\alpha = 1. \tag{3.49}$$

Such a condition is linked with the normalisation condition for the density operator, as determined by $Tr.(\hat{\rho}) = 1$. Equation (3.48) provides the *P-representation* of the operator $\hat{\rho}$. It is particularly useful if we want to calculate the mean value of operators containing products of the form $\hat{a}^{\dagger m}\hat{a}^n$, as defined by

$$\langle \hat{a}^{\dagger m}\hat{a}^n \rangle = Tr.(\hat{\rho}\hat{a}^{\dagger m}\hat{a}^n). \tag{3.50}$$

From equation (3.48), we then get

$$\langle \hat{a}^{\dagger m}\hat{a}^n \rangle = \int P(\alpha)\, Tr.(|\, \alpha\rangle\langle \alpha \mid \hat{a}^{\dagger m}\hat{a}^n) d\alpha. \tag{3.51}$$

Using the eigenvalue equation, $\hat{a} \mid \alpha\rangle = \alpha \mid \alpha\rangle$, and its Hermetic conjugate: $\langle \alpha \mid \hat{a}^\dagger = \alpha^* \langle \alpha \mid$, we obtain

$$\langle \hat{a}^{\dagger m}\hat{a}^n \rangle = \int P(\alpha)\, \alpha^{*m}\alpha^n d\alpha. \tag{3.52}$$

This formula shows that $P(\alpha)$ can be seen as the probability density for the products $\alpha^{*m}\alpha^n$. However, this should be taken with caution, because the same is not valid for $\langle \hat{a}^n\hat{a}^{\dagger m}\rangle$ as shown below. We should also notice, in this respect, that any operator involving products of $\hat{a}$ and $\hat{a}^\dagger$ can be converted into a sum of terms of the type $\hat{a}^{\dagger m}\hat{a}^n$, by using the commutator $[\hat{a}, \hat{a}^\dagger] = 1$. For example, we have $\hat{a}^2\hat{a}^\dagger = \hat{a}^\dagger a^2 + 2\hat{a}$. On the other hand, $P(\alpha)$ can eventually take negative values, which is uncompatible with the concept of probability. For this reason it is sometimes called a *quasi-probability*.

Let us now derive an explicit expression for $P(\alpha)$. For this purpose, we define two coherent states $|\, \beta\rangle$ and $|\, \beta'\rangle$, corresponding to two distinct eigenvalues of $\hat{a}$. Using equation (3.48), we can write

$$\langle \beta' |\, \hat{\rho} \mid \beta\rangle = \int P(\alpha)\langle \beta' |\, \alpha\rangle\langle \alpha \mid \beta\rangle d\alpha. \tag{3.53}$$

Using equation (3.14), this can be rewritten as

$$\langle \beta' |\, \hat{\rho} \mid \beta\rangle = e^{(|\beta|^2 + |\beta'|^2)/2} \int P(\alpha)\, e^{-|\alpha|^2} e^{\beta\alpha^* + \alpha\beta'^*} d\alpha. \tag{3.54}$$

At this point, it is useful to take $\beta' = -\beta$. Writing α and β in their explicit complex form, as $\alpha = x + iy$, and $\beta = q + ip$, and

$$\beta\alpha^* + \alpha\beta'^* = \beta\alpha^* - \alpha\beta^* = 2i(xp - yq). \tag{3.55}$$

Using $d\alpha = dxdy$, we get

$$\langle -\beta \mid \rho \mid \beta\rangle e^{|\beta|^2} = \int f(x, y) e^{2i(xp - yq)} dxdy \tag{3.56}$$

with

$$f(x, y) = P(\alpha)\, e^{-|\alpha|^2} \equiv P(x, y)\, e^{-(x^2 + y^2)}. \tag{3.57}$$

We can see that the r.h.s. of equation (3.56) is the two-dimensional Fourier transform of $f(x, y)$, apart from a factor of 4. Taking the inverse transformation, and get

$$f(x, y) = 4 \int \langle -\beta \mid \rho \mid \beta\rangle e^{|\beta|^2} \frac{dqdp}{(2\pi)^2}. \tag{3.58}$$

Replacing this in equation (3.57), we obtain an explicit expression for the quasi-probability $P(\alpha)$, written in terms of the coherent states, as

$$P(\alpha) = \frac{e^{|\alpha|^2}}{\pi^2} \int \langle -\beta \mid \rho \mid \beta \rangle e^{-\beta\alpha^* + \alpha\beta^*} d\beta, \tag{3.59}$$

where $\beta\alpha^* - \beta^*\alpha = 2i(xp - yq)$. If the radiation field is described by a coherent state $\mid \alpha_0 \rangle$, we have $\hat{\rho} = \mid \alpha_0 \rangle \langle \alpha_0 \mid$, and this expression simply reduces to

$$P(\alpha) = \delta(\alpha - \alpha_0). \tag{3.60}$$

This is a two-dimensional delta function, given the complex nature of the eigenvalue α, such that $\delta(\alpha - \alpha_0) \equiv \delta(x - x_0)\delta(y - y_0)$. Another interesting example of a radiation field is that of a pure number state, $\mid n \rangle$, which has no classical limit. In this case, we have $\hat{\rho} = \mid n \rangle \langle n \mid$, leading to

$$\langle -\beta \mid \hat{\rho} \mid \beta \rangle = \langle -\beta \mid n \rangle \langle n \mid \beta \rangle = e^{-|\beta|^2} \frac{(-\beta^*\beta)^n}{n!}. \tag{3.61}$$

Replacing this in equation (3.59), we obtain

$$P(\alpha) = \frac{e^{|\alpha|^2}}{\pi^2 n!} \int (-\beta^*\beta)^n e^{-\beta\alpha^* + \alpha\beta^*} d\beta. \tag{3.62}$$

This can also be written as

$$P(\alpha) = \frac{e^{|\alpha|^2}}{\pi^2 n!} \frac{\partial^{2n}}{\partial\alpha^n \partial\alpha^{*n}} \delta(\alpha). \tag{3.63}$$

At this point, we should not forget that this expression only has a meaning after integration. For instance, using some multiplication function $F(\alpha, \alpha^*)$ we get, after integration

$$\int F(\alpha, \alpha^*) \frac{\partial^{2n}}{\partial\alpha^n \partial\alpha^{*n}} \delta(\alpha) d\alpha = \left[\frac{\partial^{2n} F(\alpha, \alpha^*)}{\partial\alpha^n \partial\alpha^{*n}} \right]_{\alpha = \alpha^* = 0}. \tag{3.64}$$

3.2.2 Q-representation

An alternative representation of the photon field can be constructed, if we define a new quasi-distribution of the form

$$Q(\alpha) = \frac{1}{\pi} \langle \alpha \mid \hat{\rho} \mid \alpha \rangle. \tag{3.65}$$

Again, using the normalisation condition $Tr.(\hat{\rho}) = 1$, and the completeness relation for coherent states, we obtain

$$Tr.(\hat{\rho}) \equiv Tr.(\hat{\rho} I) = \frac{1}{\pi} \sum_n \int \langle n \mid \alpha \rangle \langle \alpha \mid \rho \mid n \rangle d\alpha. \tag{3.66}$$

Now, using the completeness relation for the number states, $\sum_n | n\rangle\langle n | = I$, we get

$$Tr.(\rho) = \frac{1}{\pi}\int \langle\alpha \mid \rho \mid \alpha\rangle d\alpha. \tag{3.67}$$

Comparing this with equation (3.65), we conclude that

$$\int Q(\alpha)\, d\alpha = 1. \tag{3.68}$$

Let us now calculate the mean value of an operator of the type $\hat{a}^m\hat{a}^{\dagger n}$. We get

$$\langle\hat{a}^m\hat{a}^{\dagger n}\rangle = Tr.(\hat{\rho}\hat{a}^m\hat{a}^{\dagger n}) = \sum_\nu < \nu|\hat{\rho}\ \hat{a}^m\hat{a}^{\dagger n}|\nu > . \tag{3.69}$$

This can be written as

$$\begin{aligned}\langle\hat{a}^m\hat{a}^{\dagger n}\rangle &= \frac{1}{\pi}\int\sum_\nu\langle\nu \mid \hat{\rho}\ \hat{a}^m \mid \alpha\rangle\langle\alpha \mid \hat{a}^{\dagger n} \mid \nu\rangle d\alpha \\ &= \frac{1}{\pi}\int\ \alpha^m\alpha^{*n}\langle\alpha \mid \hat{\rho} \mid \alpha\rangle d\alpha.\end{aligned} \tag{3.70}$$

Using the definition of $Q(\alpha)$ in equation (3.65), we finally get

$$\langle\hat{a}^m\hat{a}^{\dagger n}\rangle = \int Q(\alpha)\alpha^m\alpha^{*n}d\alpha. \tag{3.71}$$

This shows that the new quasi-probability, $Q(\alpha)$, allows to calculate mean values of operators of the type $(a^{\dagger m}a^n)$, and plays a symmetric role when compared with $P(\alpha)$. But, in contrast with $P(\alpha)$ it cannot be negative, $Q(\alpha) \geqslant 0$, and its maximum value is $1/\pi$. Let us illustrate this with two examples. For a coherent state $\mid \alpha_0\rangle$, the corresponding Q-probability is given by

$$Q(\alpha) = \frac{1}{\pi}\mid\langle\alpha \mid \alpha_0\rangle\mid^2 = \frac{1}{\pi}\exp(-|\alpha - \alpha_0|^2). \tag{3.72}$$

On the other hand, for a number state $\mid n\rangle$, it becomes

$$Q(\alpha) = \frac{1}{\pi}\mid\langle\alpha \mid n\rangle\mid^2 = \frac{|\alpha|^{2n}}{\pi n!}e^{-|\alpha|^2}. \tag{3.73}$$

3.2.3 W-representation

Another useful quasi-probability is the *Wigner function*, $W(\alpha)$, historically the first one introduced in quantum physics. It can be defined as

$$W(\alpha) = \frac{1}{\pi^2}\int e^{-\beta\alpha^* + \beta^*\alpha}\chi(\beta)d\beta, \tag{3.74}$$

where $\chi(\beta)$ is the characteristic function defined by

$$\chi(\beta) = Tr.(\hat{\rho}e^{\beta\hat{a}^\dagger - \beta^*\hat{a}}). \tag{3.75}$$

In contrast with the previous two quasi-probabilities, $P(\alpha)$ and $Q(\alpha)$, the Wigner function can be used to evaluate mean values of the symmetrically ordered products of $\hat{a}$ and $\hat{a}^\dagger$, as shown by the relation

$$\langle \hat{a}\hat{a}^\dagger + \hat{a}^\dagger\hat{a}\rangle = 2\int W(\alpha)|\alpha|^2 d\alpha. \tag{3.76}$$

It can also be shown that $W(\alpha)$ is a Gaussian convolution of $P(\alpha)$, given by

$$W(\alpha) = \frac{2}{\pi}\int P(\beta)\, e^{-2|\alpha-\beta|^2} d\beta. \tag{3.77}$$

The Wigner function can be written in a more familiar form in terms of the position and momentum operators, $\hat{q}$ and $\hat{p}$. Using $\alpha = q + ip$ and $\beta = x + iy$, we can rewrite equation (3.74) as

$$W(\alpha) = \frac{1}{2\pi^2}\int dx \int dy e^{i(px + qy)} Tr \cdot (\hat{\rho} e^{-i(px + qy)}) d\beta. \tag{3.78}$$

Using the trace in the coordinate representation, and noting that $\exp(-ixp/2)|\, q\rangle = |\, q - x/2\rangle$, this then leads to

$$W(\alpha) = \frac{1}{2\pi\hbar}\int \langle q - x/2 \,|\, \hat{\rho} \,|\, q + x/2\rangle e^{-ixp/\hbar} dx, \tag{3.79}$$

which is formally identical to the definition used in chapter 2.

3.2.4 G-representation

Let us now consider a generalised representation $G(\alpha)$ of the density operator $\hat{\rho}$, which reduces to the previous P, Q and W-representations in particular situations. For this purpose, let us consider an operator $\hat{\Delta}(\alpha;\, \hat{a},\, \hat{a}^\dagger)$, such that

$$\hat{\rho} = \int G(\alpha)\, \hat{\Delta}(\alpha;\, \hat{a},\, \hat{a}^\dagger) d\alpha, \tag{3.80}$$

and assume that this operator $\hat{\Delta}$ is defined as

$$\hat{\Delta}(\alpha;\, \hat{a},\, \hat{a}^\dagger) = \frac{1}{\pi^2}\int e^{\Omega(\beta)}\, e^{-\beta(\alpha^* - \hat{a}^\dagger)}\, e^{\beta^*(\alpha - \hat{a})} d\beta. \tag{3.81}$$

As a first particular case, let us assume that the function $\Omega(\beta)$ appearing in this definition is equal to

$$\Omega(\beta) = -\frac{|\beta|^2}{2}. \tag{3.82}$$

Using the Baker–Hausdorff identity (3.42) with $A = \beta\hat{a}^\dagger$ and $B = -\beta^*\hat{a}$, we have $[A,\, B] = |\beta|^2$. This allows us to write equation (3.81) as

$$\hat{\Delta}(\alpha;\, \hat{a},\, \hat{a}^\dagger) = \frac{1}{\pi^2}\int e^{\beta\hat{a}^\dagger - \beta^*\hat{a}}\, e^{-\beta\alpha^* + \beta^*\alpha} d\beta. \tag{3.83}$$

Using the completeness relation for coherent states, this can be transformed into

$$\hat{\Delta}(\alpha;\hat{a},\hat{a}^{\dagger})=\frac{1}{\pi^3}\int d\alpha'\int d\beta e^{\beta^*(\alpha-\hat{a})}\,|\,\alpha'\rangle\langle\alpha'|\;e^{-\beta(\alpha^*-\hat{a}^{\dagger})}. \tag{3.84}$$

We can now use the two-dimensional Dirac delta function

$$\delta^2(\alpha-\alpha')\equiv\delta(\alpha-\alpha')\delta(\alpha^*-\alpha'^*)=\frac{1}{\pi^2}\int e^{\beta^*(\alpha-\alpha')}\,e^{-\beta(\alpha^*-\alpha')}d\beta. \tag{3.85}$$

Noting that $\int d\beta$ represents a double integral, we then get

$$\hat{\Delta}(\alpha;\hat{a},\hat{a}^{\dagger})=\frac{1}{\pi}\int\delta^2(\alpha-\alpha')|\,\alpha'\rangle\langle\alpha'|\,d\alpha'=\frac{1}{\pi}\,|\,\alpha\rangle\langle\alpha\,|. \tag{3.86}$$

Replacing this in the definition of the generalised quasi-distribution $G(\alpha)$, equation (3.80), we conclude that

$$\hat{\rho}=\int G(\alpha)\;|\alpha>\;<\alpha|d\alpha. \tag{3.87}$$

This shows that the particular choice of $\hat{\Delta}(\alpha;\hat{a},\hat{a}^{\dagger})$ defined by equation (3.82), reduces the generalised function to $G(\alpha)=P(\alpha)$, therefore recovering the P-representation. Let us now consider a different choice of $\hat{\Delta}(\alpha;\hat{a},\hat{a}^{\dagger})$, and replace equation (3.82) by

$$\Omega(\beta)=+\frac{|\beta|^2}{2}. \tag{3.88}$$

In this case, we can use again the Baker–Hausdorff identity and transform equation (3.81), into

$$\hat{\Delta}(\alpha;\hat{a},\hat{a}^{\dagger})=\frac{1}{\pi^2}\int e^{-\beta(\alpha^*-\hat{a}^{\dagger})}\,e^{\beta^*(\alpha-\hat{a})}d\beta. \tag{3.89}$$

Let us now calculate the mean value of this operator, as

$$\langle\alpha'|\,\hat{\Delta}(\alpha;\hat{a},\hat{a}^{\dagger})|\,\alpha'\rangle=\frac{1}{\pi^2}\int e^{-\beta(\alpha^*-\alpha'^*)}\,e^{\beta^*(\alpha-\alpha')}d\beta=\delta^2(\alpha-\alpha'). \tag{3.90}$$

Replacing this in equation (3.80), we conclude that

$$\langle\alpha\,|\,\hat{\rho}\,|\,\alpha\rangle=\pi\int G(\alpha')\langle\alpha'|\,\hat{\Delta}(\alpha;\hat{a},\hat{a}^{\dagger})|\,\alpha'\rangle d\alpha'=\pi G(\alpha). \tag{3.91}$$

Comparing this with equation (3.65), we conclude that this new choice of $\Omega(\beta)$ reduces to the Q-representation, and that we now have $G(\alpha)=Q(\alpha)$. Finally, let us consider the condition

$$\Omega(\beta)=0. \tag{3.92}$$

In this case, the generalised distribution $G(\alpha)$ reduces to the Wigner function, $G(\alpha)=W(\alpha)$, because equations (3.80) and (3.81) lead to

$$G(\alpha)=\frac{2}{\pi^2}e^{2|\alpha|^2}\int\langle-\beta\,|\,\rho\,|\,\beta\rangle e^{-2(\beta\alpha^*-\beta^*\alpha)}d\beta. \tag{3.93}$$

3.3 Squeezed states

Let us start with the definition of *squeezing*. It is known that, given two observables A and B, satisfying a commutation relation $[A, B] = iC$, their mean values cannot be known simultaneously, because they satisfy the uncertainty relation

$$\Delta A\, \Delta B \geqslant \frac{1}{2} \mid \langle C \rangle \mid . \tag{3.94}$$

We can say that a quantum system is in a *squeezed state* if one of these two observables (A, for instance) satisfies the condition

$$(\Delta A)^2 < \frac{1}{2} \mid \langle C \rangle \mid . \tag{3.95}$$

This means that, in a squeezed state, one of the observables has a reduced uncertainty, while the other variable (B) will necessarily have an increased uncertainly. As an example of this concept, we consider an electromagnetic field mode, with frequency ω. The electric field operator is

$$\hat{\mathbf{E}} = i\sqrt{\frac{\hbar\omega}{2\epsilon_0 V}}\, \mathbf{e}\{\hat{a}e^{-i\omega t} - \hat{a}^{\dagger}e^{i\omega t}\}. \tag{3.96}$$

Let us introduce the dimensionless position-like and momentum-like operators q and p, such that

$$\hat{q} = \frac{1}{2}(\hat{a} + \hat{a}^{\dagger}), \quad \hat{p} = \frac{1}{2i}(\hat{a} - \hat{a}^{\dagger}). \tag{3.97}$$

These operators are sometimes called the *quadratures* of the field mode. Noting that $[\hat{a}, \hat{a}^{\dagger}] = 1$, we have

$$[\hat{q}, \hat{p}] = \frac{i}{2}. \tag{3.98}$$

The electric field operator can be written in terms of these operators, as

$$\hat{\mathbf{E}} = 2\sqrt{\frac{\hbar\omega}{2\epsilon_0 V}}\, \mathbf{e}\{\hat{q}\sin(\omega t) - \hat{p}\cos(\omega t)\}. \tag{3.99}$$

According to the commutator (3.98), the uncertainty principle (3.94) becomes

$$\Delta q \cdot \Delta p \geqslant \frac{1}{4}. \tag{3.100}$$

A squeezed state of the electromagnetic field will therefore verify one of these two conditions

$$(\Delta q)^2 < \frac{1}{4}, \quad or \quad (\Delta p)^2 < \frac{1}{4}. \tag{3.101}$$

In order to illustrate this concept, let us apply it to a Fock (or photon number) state, $|\,n\rangle$. We have

$$\langle\hat{q}\rangle \equiv \langle n\,|\,\hat{q}\,|\,n\rangle = \frac{1}{2}\{\langle n\,|\,\hat{a}\,|\,n\rangle + \langle n\,|\,\hat{a}^\dagger\,|\,n\rangle\} = 0, \tag{3.102}$$

and

$$\langle\hat{q}^2\rangle \equiv \langle n\,|\,\hat{q}^2\,|\,n\rangle = \frac{1}{2}\langle n\,|\,(\hat{a}^2 + \hat{a}^{\dagger 2} + \hat{a}\hat{a}^\dagger + \hat{a}^\dagger\hat{a})\,|\,n\rangle = \frac{1}{4}(2n+1). \tag{3.103}$$

From here we calculate the variance of the position-like operator, as

$$(\Delta q)^2 \equiv \langle\hat{q}^2\rangle - \langle\hat{q}\rangle^2 = \frac{1}{4}(2n+1). \tag{3.104}$$

The same could be obtained for the other quadrature, $(\Delta p)^2$. This shows that a Fock state can never be squeezed, even for vacuum $n = 0$. Let us now consider the example of a coherent state, $|\,\alpha\rangle$. Explicit calculation of the above quantities then lead to

$$(\Delta q)^2 \equiv \langle\alpha\,|\,\hat{q}^2\,|\,\alpha\rangle - (\langle\alpha\,|\,\hat{q}\,|\,\alpha\rangle)^2 = \frac{1}{4}. \tag{3.105}$$

The same result holds for the other quadrature. This means that, again, a coherent state will never be squeezed, although it reduces the field uncertainty to its minimum, $\Delta q\ \Delta p = 1/4$. In figure 3.2 we illustrate the regions of existence (shaded regions) of a possible squeezed state. Generation of a squeezed state involves nonlinear processes, to be described later, and can formally described by a *squeezing operator*, defined as

$$\hat{S}(\xi) = \exp\left(\frac{1}{2}\xi^*\hat{a}^2 - \frac{1}{2}\xi\hat{a}^{\dagger 2}\right), \tag{3.106}$$

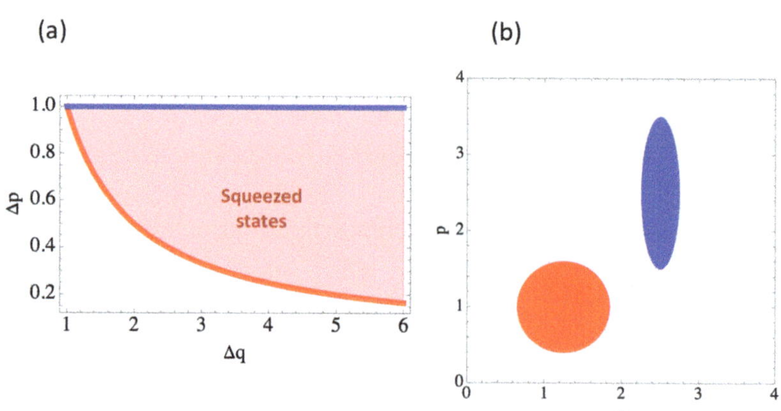

Figure 3.2. Squeezed states: (a) region of existence in the uncertainty plane $(\Delta q, \Delta p)$, and similar region when the axes are reversed; (b) comparison between a coherent state (in red) and a squeezed state (in blue), in the quadrature plane (q, p).

where ξ is a complex number. Using polar coordinates, we can write $\xi = r\exp(i\theta)$, where the amplitude r is usually called the *squeezing parameter*. It can easily be shown that the operator $\hat{S}(\xi)$ is unitarian, and satisfies

$$\hat{S}^{-1}(\xi)\hat{S}(\xi) = I, \tag{3.107}$$

where I is the identity operator. We can also find that

$$\hat{S}^{-1}(\xi) = \hat{S}^{\dagger}(\xi) = \hat{S}(-\xi). \tag{3.108}$$

Let us apply this squeezing operator to the creation and annihilation operators, $\hat{a}^{\dagger}$ and $\hat{a}$. Using the Baker–Hausdorff relation, we obtain

$$\hat{S}^{\dagger}(\xi)\hat{a}\hat{S}(\xi) = \hat{a}\cosh(r) - \hat{a}^{\dagger}e^{i\theta}\sinh(r), \tag{3.109}$$

and

$$\hat{S}^{\dagger}(\xi)\hat{a}^{\dagger}S(\xi) = \hat{a}^{\dagger}\cosh(r) - \hat{a}e^{-i\theta}\sinh(r). \tag{3.110}$$

Given a generic field state $|\,\psi\rangle$, we can generate a squeezed state $\left|\,\psi_S\right\rangle$ using this operator, as

$$|\,\psi_S\rangle = \hat{S}(\xi)|\,\psi\rangle. \tag{3.111}$$

In particular, we can generate a *squeezed vacuum*, through the operation

$$|\,\xi\rangle = \hat{S}(\xi)|\,0\rangle, \tag{3.112}$$

where the squeezing operator is applied to the vacuum state $|\,0\rangle$. Using equations (3.109)–(3.110) we can then show that, for such a squeezed vacuum state, the variance of $\hat{q}$ and $\hat{p}$ are determined by

$$(\Delta q)^2 = \frac{1}{4}\{\cosh^2(r) + \sinh^2(r) - 2\sinh(r)\cosh(r)\cos(\theta)\}, \tag{3.113}$$

and

$$(\Delta p)^2 = \frac{1}{4}\{\cosh^2(r) + \sinh^2(r) + 2\sinh(r)\cosh(r)\cos(\theta)\}. \tag{3.114}$$

For $\theta = 0$, this reduces to

$$(\Delta q)^2 = \frac{1}{4}e^{-2r}, \quad (\Delta p)^2 = \frac{1}{4}e^{2r}. \tag{3.115}$$

This shows that squeezing exists in the q-quadrature. In contrast, for $\theta = \pi$, squeezing in the p-quadrature will occur. Now, if we consider rotation of the real and imaginary axis in the complex (q, ip) plane, by an angle θ, we can define a new annihilation operator $\hat{a}'$, such that

$$\hat{a}' \equiv (\hat{q}' + i\hat{p}') = (\hat{q} + i\hat{p})e^{i\theta} = \hat{a}e^{i\theta}. \tag{3.116}$$

The resulting new field quadratures can be defined as

$$\hat{Q} = \hat{q}' e^{-r}, \quad \hat{P} = \hat{p}' e^{r}. \tag{3.117}$$

For $r > 0$, this is able to describe a reduction of the field amplitude along the new real axis, and an increase along the new imaginary axis. With this in mind, let us construct a *squeezed coherent state*, by successively applying the displacement and the squeezing operators, $\hat{D}(\alpha)$ and $\hat{S}(\xi)$, to the vacuum state $| 0\rangle$. We get

$$| \alpha, \xi\rangle = \hat{S}(\xi)\hat{D}(\alpha)| 0\rangle. \tag{3.118}$$

It can then be shown that, for such state, we have

$$\langle\hat{a}\rangle \equiv \langle\alpha, \xi | \hat{a} | \alpha, \xi\rangle = \alpha \cosh(r) - \alpha^* e^{i\theta} \sinh(r). \tag{3.119}$$

We can also calculate the uncertainty associated with the new field quadratures, defined by equation (3.117), and obtain

$$(\Delta Q)^2 = \frac{1}{4} e^{-2r}, \quad (\Delta p)^2 = \frac{1}{4} e^{2r}. \tag{3.120}$$

This shows that

$$\Delta Q \cdot \Delta P = \frac{1}{4}, \tag{3.121}$$

which means that the squeezed coherent state $| \alpha, \xi\rangle$ is an ideal squeezed state, with the squeezing parameter equal to $r = |\xi|$.

3.4 Correlations

3.4.1 Classical correlations

In order to understand the main qualitative features of the quantum theory of radiation, we consider the field and the intensity correlations. We start with the simplest case of field correlations, using as a typical example, the famous *Young's double-slit experiment*. This experiment was first performed by Thomas Young around 1800 [23], and became one of central pieces of the wave theory of light, and one of the most famous experiments in physics of all times. It is also often used as a paradigmatic *Gedankenexperiment* to discuss the principles of quantum mechanics. In particular, the delicate problems of particle (photon) path-information, and quantum decoherence [24].

For a question of clarity, it is useful to start with the classical description. In this experiment, radiation emitted from a single light source passes through two distinct slits or pinholes, and the resulting intensity pattern is observed on a screen (see figure 3.3(a)). The electric field measured at a given position $\mathbf{r}$ on the screen is the sum of two fields, coming from the two slits, located at positions $\mathbf{r}_j$, for $j = 1, 2$. Assuming that the slits are located at distances d_j from the observation point. The total field on the screen is

$$\mathbf{E}(\mathbf{r}, t) = u_1\mathbf{E}(\mathbf{r}_1, t_1) + u_2\mathbf{E}(\mathbf{r}_2, t_2), \tag{3.122}$$

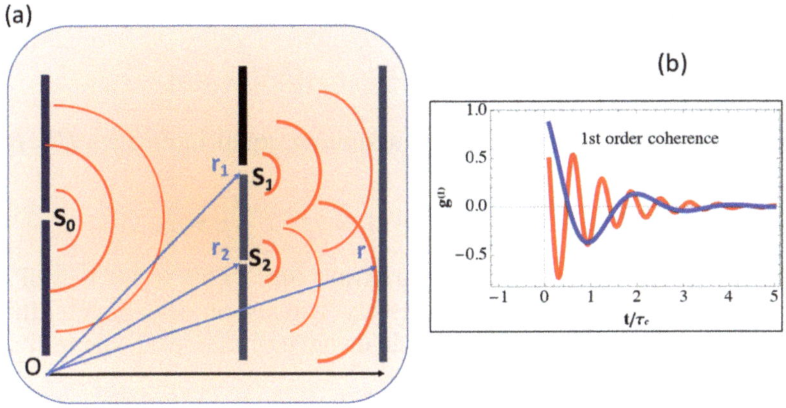

Figure 3.3. Young's double-slit experiment: (a) experimental geometry; (b) first-order coherence $g^{(1)}$, as determined by equation (3.135), for two values of the correlation time (τ_c).

where the coefficients u_j depend on the geometric configuration and are, typically, inversely proportional to the distances d_j. On the other hand, the instants t_j are retardation times, defined as

$$t_j = t - \frac{d_j}{c}, \quad d_j = |\mathbf{r} - \mathbf{r}_j|, \tag{3.123}$$

for $j = 1, 2$. The observed radiation intensity, averaged over one cycle, is then given by

$$\begin{aligned} I(\mathbf{r}, t) = \frac{\epsilon_0}{2} c \,|\, \mathbf{E}(\mathbf{r}, t)|^2 &= I_1(\mathbf{r}, t) + I_2(\mathbf{r}, t) \\ &+ \epsilon_0 c (u_1^* u_2) \Re[\mathbf{E}^*(\mathbf{r}_1, t_1) \cdot \mathbf{E}(\mathbf{r}_2, t_2)], \end{aligned} \tag{3.124}$$

where I_j is the intensity received from the slit j at the position $\mathbf{r}$ on the screen, and is

$$I_j(\mathbf{r}, t) = \frac{\epsilon_0}{2} c |u_j|^2 |\mathbf{E}(\mathbf{r}_j, t_j)|^2. \tag{3.125}$$

In a steady-state experiment, we integrate the intensity over a long period of time T, much longer than the typical wave period. The time averaged intensity, is then represented by

$$I(\mathbf{r}) = \langle I_1(\mathbf{r}, t)\rangle + \langle I_2(\mathbf{r}, t)\rangle + \epsilon_0 c (u_1^* u_2) \Re[\langle \mathbf{E}^*(\mathbf{r}_1, t_1) \cdot \mathbf{E}(\mathbf{r}_2, t_2)\rangle]. \tag{3.126}$$

Here, we can identify the field correlation function, as

$$\langle \mathbf{E}^*(\mathbf{r}_1, t_1) \cdot \mathbf{E}(\mathbf{r}_2, t_2)\rangle = \frac{1}{T}\int_0^T \mathbf{E}^*(\mathbf{r}_1, t_1) \cdot \mathbf{E}(\mathbf{r}_2, t_1 + \tau) dt_1, \tag{3.127}$$

where $\tau = t_2 - t_1$ is the retardation time difference. In the case of monochromatic radiation with frequency ω, we can use

$$\mathbf{E}(\mathbf{r}, t) = \mathbf{E}_0 \cos(\mathbf{k} \cdot \mathbf{r} - \omega t). \tag{3.128}$$

This allows us to write

$$\langle \mathbf{E}^*(\mathbf{r}_1, t_1) \cdot \mathbf{E}(\mathbf{r}_2, t_2) \rangle = \frac{| E_0 |^2}{2} \cos [k(d_1 - d_2)], \tag{3.129}$$

with $\delta \mathbf{r}_j = (\mathbf{r} - \mathbf{r}_j)$. From where we get, assuming equal intensities $I_1 = I_2 = I_0$, the expression

$$I(\mathbf{r}) = 2I_0\{1 + \cos[k(d_1 - d_2)]\} = 2I_0\{1 + \cos \omega\tau\}, \tag{3.130}$$

where $\tau = t_1 - t_2 = (d_1 - d_2)/c$ is the retardation time. This result is well known in classical physics and describes the intensity fringes on the screen due to field interference. In a more general case, the fields have phases $\varphi_j(t)$ due to fluctuations, and the correlation function take the form

$$\langle \mathbf{E}^*(\mathbf{r}_1, t_1) \cdot \mathbf{E}(\mathbf{r}_2, t_2) \rangle = \frac{|E_0|^2}{2} \langle e^{i\varphi(t_1)} e^{-i\varphi(t_1 - \tau)} \rangle \cos(\omega\tau). \tag{3.131}$$

The phase correlation function appearing in this new expression can be characterised by a correlation time τ_c, such that

$$\langle e^{i\varphi(t_1)} e^{-i\varphi(t_1 - \tau)} \rangle = \exp\left(-\frac{|\tau|}{\tau_c} \right). \tag{3.132}$$

For completely incoherent phases, we will have $\tau_c \to 0$, and the interference fringes in Young's experiment completely vanish

$$\langle \mathbf{E}^*(\mathbf{r}_1, t_1) \cdot \mathbf{E}(\mathbf{r}_2, t_2) \rangle = 0. \tag{3.133}$$

At this point, it is convenient to define a normalised *first-order coherence*, such that

$$g^{(1)}(\mathbf{r}_1, t_1; \mathbf{r}_2, t_2) = \frac{\langle \mathbf{E}^*(\mathbf{r}_1, t_1) \cdot \mathbf{E}(\mathbf{r}_2, t_2) \rangle}{[\langle |E(\mathbf{r}_1, t_1)|^2 \rangle \langle |E(\mathbf{r}_2, t_2)|^2 \rangle]^{1/2}}. \tag{3.134}$$

We can see from here that, for an arbitrary degree of coherence, we have

$$g^{(1)}(\mathbf{r}_1, t_1; \mathbf{r}_2, t_2) = \cos(\omega\tau) \exp\left(-\frac{|\tau|}{\tau_c} \right), \tag{3.135}$$

where the case of $\tau_c = 0$ corresponds to an incoherent light source. The other extreme case is that of $\tau_c \to \infty$, describing a completely coherent light source. A finite value of τ_c will characterise partial coherence. Therefore, the fringe contrast will give a measure of the temporal coherence of the light source.

3.4.2 Quantum correlations

Let us now turn to a quantum description of Young's experiment, and of the corresponding field correlations. For this purpose, we replace the classical field (3.122) by the electric field operator

$$\hat{\mathbf{E}}(\mathbf{r}, t) = \hat{\mathbf{E}}^{(+)}(\mathbf{r}, t) + \hat{\mathbf{E}}^{(-)}(\mathbf{r}, t), \tag{3.136}$$

where

$$\hat{\mathbf{E}}^{(+)}(\mathbf{r}, t) = \sum_{\mathbf{k}} \mathbf{e}_{\mathbf{k}} \hat{a}_{\mathbf{k}} e^{i\mathbf{k}\cdot\mathbf{r} - i\omega_k t}, \quad \hat{\mathbf{E}}^{(-)}(\mathbf{r}, t) = \sum_{\mathbf{k}} \mathbf{e}_{\mathbf{k}}^* \hat{a}_{\mathbf{k}}^\dagger e^{-i\mathbf{k}\cdot\mathbf{r} + i\omega_k t}. \tag{3.137}$$

We can write a quantum expression for the mean intensity, as

$$I(\mathbf{r}) = \frac{\epsilon_0}{2} c \left\langle \hat{\mathbf{E}}^{(-)}(\mathbf{r}, t) \cdot \hat{\mathbf{E}}^{(+)}(\mathbf{r}, t) \right\rangle. \tag{3.138}$$

This is formally identical to the classical expression, but now the time average of the previous definition is replaced by a quantum average. The corresponding quantum first-order degree of coherence, defined in terms of the field operators, will be

$$g^{(1)}(\mathbf{r}, t; \tau) = \frac{\left\langle \hat{\mathbf{E}}^{(-)}(\mathbf{r}, t) \cdot \hat{\mathbf{E}}^{(+)}(\mathbf{r}, t + \tau) \right\rangle}{\left[\left\langle \hat{\mathbf{E}}^{(-)}(\mathbf{r}, t) \cdot \hat{\mathbf{E}}^{(+)}(\mathbf{r}, t) \right\rangle \left\langle \hat{\mathbf{E}}^{(-)}(\mathbf{r}, t + \tau) \cdot \hat{\mathbf{E}}^{(+)}(\mathbf{r}, t + \tau) \right\rangle \right]^{1/2}}. \tag{3.139}$$

Simplifying the notation, and considering the simplest case of a single field mode, this can be reduced to

$$g^{(1)}(\tau) = \frac{\langle \hat{a}^\dagger(t) \hat{a}(t + \tau) \rangle}{\langle \hat{a}^\dagger \hat{a} \rangle}. \tag{3.140}$$

Let us apply these quantum expressions to the specific case of Young's experiment. The mean intensity measured at a given position $\mathbf{r}$ on the screen, will be given by

$$I(\mathbf{r}) = I_1 + I_1 + \epsilon_0 c\, \Re\left[\left\langle \hat{\mathbf{E}}^{(-)}(\mathbf{r}_1, t_1) \cdot \hat{\mathbf{E}}^{(+)}(\mathbf{r}_2, t_2) \right\rangle \right], \tag{3.141}$$

with

$$I_j = \frac{\epsilon_0}{2} c \left\langle \hat{\mathbf{E}}^{(-)}(\mathbf{r}_j, t_j) \cdot \hat{\mathbf{E}}^{(+)}(\mathbf{r}_j, t_j) \right\rangle, \tag{3.142}$$

for $j = 1, 2$. Let us assume that we detect a single field mode from each of the two slits. We then have

$$\hat{a} = \frac{1}{\sqrt{2}} (\hat{a}_1 + \hat{a}_2), \quad \hat{a}^\dagger = \frac{1}{\sqrt{2}} (\hat{a}_1^\dagger + \hat{a}_2^\dagger), \tag{3.143}$$

where the normalisation factor $1/\sqrt{2}$ is used to satisfy $[\hat{a}, \hat{a}^\dagger] = 1$. We should notice that $[\hat{a}_1, \hat{a}_2] = [\hat{a}_1, \hat{a}_2^\dagger] = [\hat{a}_1^\dagger, \hat{a}_2] = 0$, because these operators correspond to two different field modes. In particular, even if they have the same frequency ω, but two different wavevectors, $\mathbf{k}_1$ and $\mathbf{k}_2$. The quantum state corresponding to n photons at the detector can then be defined by the operator $\hat{a}^\dagger$ defined above, acting on the total vacuum state

$$| n \rangle = \frac{(\hat{a}^\dagger)^n}{\sqrt{n!}} | 0 \rangle_1 | 0 \rangle_2 = \frac{(\hat{a}^\dagger)^n}{\sqrt{n!}} | 0 \rangle. \tag{3.144}$$

If the slit number 2 is absent (or obstructed), the mean number of photons detected on screen associated with this field state is

$$\langle \hat{N}_1 \rangle \equiv \langle n \mid \hat{a}_1^\dagger \hat{a}_1 \mid n \rangle = \frac{1}{n!} \langle 0 \mid (\hat{a})^n \hat{a}_1^\dagger \hat{a}_1 (\hat{a}^\dagger)^n \mid 0 \rangle. \tag{3.145}$$

Using the definition (3.143), we can then show that

$$\langle \hat{N}_1 \rangle = \frac{n}{2}. \tag{3.146}$$

Similarly, if the slit number 1 is absent, we get

$$\langle \hat{N}_2 \rangle \equiv \langle n \mid \hat{a}_2^\dagger \hat{a}_2 \mid n \rangle = \frac{n}{2}. \tag{3.147}$$

Finally, in the presence of both slits, the total field operators behave like

$$\hat{E}^{(+)}(\mathbf{r}, t) \propto \hat{a}_1 e^{ikd_1} + \hat{a}_2 e^{ikd_2}, \quad \hat{E}^{(-)}(\mathbf{r}, t) \propto \hat{a}_1^\dagger e^{-ikd_1} + \hat{a}_2^\dagger e^{-ikd_2}, \tag{3.148}$$

and the resulting mean intensity is

$$I(\mathbf{r}) = \frac{\epsilon_0}{2} c \left\{ \langle n \mid \hat{N}_1 \mid n \rangle + \langle n \mid \hat{N}_2 \mid n \rangle + 2\cos[k(d_1 - d_2)] \langle n \mid \hat{a}_1^\dagger \hat{a}_2 \mid n \rangle \right\}, \tag{3.149}$$

where we obtain, for the interference term, the following value

$$\langle n \mid \hat{a}_1^\dagger \hat{a}_2 \mid n \rangle = \frac{1}{2n!} \langle 0 \mid (\hat{a}_1 + \hat{a}_2)^n \hat{a}_1^\dagger \hat{a}_2 (\hat{a}_1^\dagger + \hat{a}_2^\dagger)^n \mid 0 \rangle = \frac{n}{2}. \tag{3.150}$$

This shows that the classical result of equation (3.130) for the interference fringes is exactly recovered with the quantum description of Young's experiment. But an important difference is that, in the quantum case, the fringes still persist even for a single photon state, when $n = 1$. This means that a single photon can interfere with itself, as if this single photon had followed two different paths from the two slits at the same time.

3.4.3 Intensity correlations

In order to introduce the intensity correlations, we now turn to another famous experiment, performed by Hanbury-Brown and Twiss in 1957 [25, 26]. Here a photon beam originating from a single source is divided by a semi-transparent plate in two secondary beams, which are separately detected by two photo-detectors. This contrasts with Young's experiment, where a single detector was used. See figure 3.4 for an illustration. This experiment is important because it reveals important quantum properties of light, such as photon anti-bunching and sub-Poissonian photon distributions, as explained below. Observation of these effects was made in the late 70s of the last century [27, 28], which was an important step in quantum optics, helping to establish it as a separate discipline.

In this experiment, the intensities measured at the two detectors are $I_j(t)$, for $j = 1, 2$, and we can consider intensity correlations defined in classical terms as

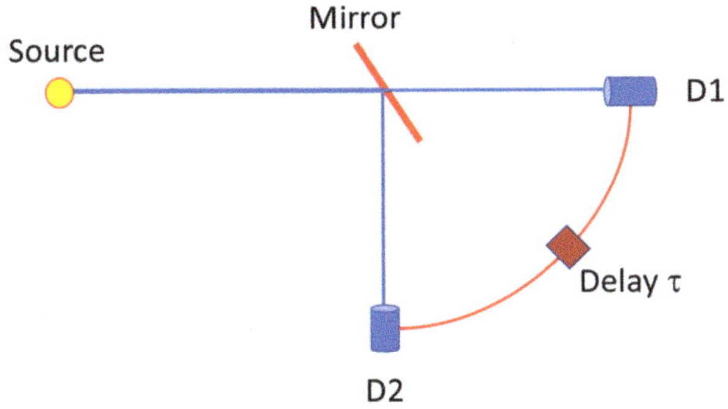

Figure 3.4. Hanbury-Brown and Twiss experiment. Experimental scheme: light emitted by a single source is divided by a semi-transparent mirror and detected by two photo-detectors D1 and D2.

$$\langle I_1(t)I_2(t)\rangle = \frac{\epsilon_0}{2}c\langle E_1^*(t)E_2^*(t+\tau)E_1(t)E_2(t+\tau)\rangle. \tag{3.151}$$

This leads us to the definition of the *second-order degree of coherence*, as

$$g^{(2)}(\mathbf{r}_1, t_1; \mathbf{r}_2, t_2) = \frac{\langle E^*(\mathbf{r}_1, t_1)E^*(\mathbf{r}_2, t_2)E(\mathbf{r}_1, t_1)E(\mathbf{r}_2, t_2)\rangle}{\langle |E(\mathbf{r}_1, t_1)|^2\rangle\langle |E(\mathbf{r}_2, t_2)|^2\rangle}. \tag{3.152}$$

Let us assume that the intensity is measured at a single position $\mathbf{r}$. For a steady state of radiation, $g^{(2)}$ will only depend on the difference between the two instants t_1 and t_2, and will not depend on their specific value. This expression can then be reduced to

$$g^{(2)}(\tau) = \frac{\langle I(t)I(t+\tau)\rangle}{\langle I\rangle^2} = \frac{\langle E^*(t)E^*(t+\tau)E(t)E(t+\tau)\rangle}{\langle E^*(t)E(t)\rangle^2}. \tag{3.153}$$

It is also well known that, from the definition of intensity, the following inequality holds

$$I^2(t_1) + I^2(t_2) \geqslant 2I(t_1)I(t_2). \tag{3.154}$$

This means that

$$\langle I^2(t)\rangle \geqslant \langle I(t)\rangle^2, \tag{3.155}$$

which implies the following property for the second-order degree of coherence

$$g^{(2)}(0) \geqslant 1. \tag{3.156}$$

Let us now turn to the quantum theory. The quantum equivalent to the definition (3.152) can be written as

$$g^{(2)}(\mathbf{r}_1, t_1; \mathbf{r}_2, t_2) = \frac{\left\langle \hat{E}^{(-)}(\mathbf{r}_1, t_1)\hat{E}^{(-)}(\mathbf{r}_2, t_2)\hat{E}^{(+)}(\mathbf{r}_1, t_1)\hat{E}^{(+)}(\mathbf{r}_2, t_2)\right\rangle}{\left\langle \hat{E}^{(-)}(\mathbf{r}_1, t_1)\hat{E}^{(+)}(\mathbf{r}_1, t_1)\right\rangle\left\langle \hat{E}^{(-)}(\mathbf{r}_2, t_2)\hat{E}^{(+)}(\mathbf{r}_2, t_2)\right\rangle}. \tag{3.157}$$

In order to illustrate the meaning of this definition, let us consider the case of intensity correlations involving a single field mode. Using an obvious simplified notation, we have

$$g^{(2)}(\tau) = \frac{\langle \hat{a}^\dagger(t)\hat{a}^\dagger(t+\tau)\hat{a}(t)\hat{a}(t+\tau)\rangle}{\langle \hat{a}^\dagger(t)\hat{a}(t)\rangle^2}. \tag{3.158}$$

Using the commutator $[\hat{a}, \hat{a}^\dagger] = 1$, we can rewrite this expression, as

$$g^{(2)}(0) = \frac{\langle \hat{a}^\dagger(\hat{a}\hat{a}^\dagger - 1)\hat{a}\rangle}{\langle \hat{a}^\dagger\hat{a}\rangle^2} = \frac{\langle \hat{a}^\dagger\hat{a}\hat{a}^\dagger\hat{a} - \hat{a}^\dagger\hat{a}\rangle}{\langle \hat{a}^\dagger\hat{a}\rangle^2}. \tag{3.159}$$

Using the photon number operator $\hat{N} = \hat{a}^\dagger\hat{a}$, this can be replaced by a simpler expression

$$g^{(2)}(0) = \frac{\left\langle \hat{N}(\hat{N} - 1)\right\rangle}{\langle \hat{N}\rangle^2} = \frac{\left\langle \hat{N}^2\right\rangle - \langle \hat{N}\rangle}{\langle \hat{N}\rangle^2}. \tag{3.160}$$

Let us apply this to a simple non-classical field state, corresponding to the Fock state $| n\rangle$. The result is

$$g^{(2)}(0) = \frac{n-1}{n}. \tag{3.161}$$

We can see that, in this case, we have $g^{(2)}(0) = 0$, for the special case of $n = 1$. On the other hand, for a large number of photons, $n \gg 1$, the intensity correlation will tend to one. This shows that, in contrast with the classical definition, the quantum second-order degree of coherence can take values that are forbidden in the classical regime, such as values between zero and one, $0 \leqslant g^{(2)}(0) \leqslant 1$. This is a clear indication that the Fock state has no classical equivalent.

Let us focus on the Hanbury-Brown and Twiss experiment, where we have two separate photon detectors. The relevant second-order coherence will be

$$g^{(2)}(\tau) = \frac{\left\langle a_1^\dagger(t)a_2^\dagger(t+\tau)a_1(t)a_2(t+\tau)\right\rangle}{\langle a_1^\dagger(t)a_1(t)\rangle\langle a_2^\dagger(t)a_2(t)\rangle} = \frac{\langle N_1(0)N_2(\tau)\rangle}{\langle N_1\rangle\langle N_2\rangle}. \tag{3.162}$$

If the radiation source is in a Fock state $| n\rangle$, as defined by equation (3.144), and if the two arms in the experimental arrangement are symmetric, we have $\langle N_1\rangle = \langle N_2\rangle = n/2$. We also have

$$\langle N_1N_2\rangle = \frac{1}{4}n(n-1). \tag{3.163}$$

This leads to the following second-order coherence

$$g^{(2)}(0) = \frac{n(n-1)}{n} = 1 - \frac{1}{n}. \tag{3.164}$$

Again, this quantity can take values in the non-classical range, between zero and one. The condition for $g^{(2)}(0) < 1$, corresponds to *photon anti-bunching*, which is a peculiar property of the quantum states of light. In contrast, the range $g^{(2)}(0) > 1$, corresponds to *photon bunching*, which is a characteristic feature of classical states.

Anti-bunching is therefore a specific property of the quantum states of light. This can be more clearly understood if we consider the existence of very small values of $g^{(2)}(0)$. In general, this implies that

$$g^{(2)}(\tau) \geqslant g^{(2)}(0). \tag{3.165}$$

It means that the probability for the arrival of two photons at the detectors separated by some finite interval $\tau > 0$, is larger than the probability for their simultaneous arrival. As a result, the successive photons tend to be detected separately, as if they repelled each other. They tend to avoid bunching. In contrast, as we have seen, in a classical field state, we tend to have

$$g^{(2)}(\tau) \leqslant g^{(2)}(0), \tag{3.166}$$

which means that we have the opposite situation of bunching at the detectors. Classical sources of radiation (such as a lamp) tend to show variations of intensity, resulting from this bunching effect. In contrast, for resonance fluorescence, the photons tend to anti-bunch, which leads to a reduction of intensity fluctuations.

It can be shown that for a thermal source of radiation, we have $g^{(2)}(0) = 2$, which is described as *super-Poissonian* light. A coherent state, $|\alpha>$, such as produced by a laser, leads to the value $g^{(2)}(0) = 1$, and is described by a Poisson distribution, as we have shown. Finally, for a Fock state, we have $g^{(2)}(0) < 1$, which corresponds to a *sub-Poissonian* photon distribution. Sub-Poissonic light was first observed around 1980, its importance for quantum optics is discussed in [29, 30].

3.5 Photon entanglement

One of the most important and intriguing features of a quantum system is *non-locality*, which is intimately related with the existence of entanglement. This has been a source of debate on the physical meaning of quantum mechanics for several decades. This debate started with the celebrated paper by Einstein, Podolski and Rosen, which states the so-called *EPR paradox* [31]. This led much later to Bell's inequalities [32] and to the experiments on these inequalities by Aspect and others [33, 34]. The debate, in a sense, is still going on, even after these conclusive experiments, due to the difficulty to reconcile causality and relativity with the observations [35]. Here, we avoid such a debate and mainly focus on the definition of entanglement, and on Bell's inequalities.

In order to discuss this problem, and its relevance to the quantum theory of radiation, we consider a two-component system, with two particles (for instance photons) in two independent spin states ± 1. If these particles were independent, the total quantum state of the system would simply be $| 1, 2\rangle = | 1\rangle | 2\rangle$. This would mean that the polarisation of one of these photons could be measured without perturbing the state of the other one. However, in Quantum Mechanics, the total state of a system is generally not so simple. For instance, if the system of two particles has total spin equal to zero, $S = 0$, its state vector can be represented by the anti-symmetric form

$$|\psi> \equiv |1,\, 2> = \frac{1}{\sqrt{2}}(|1,\, +>|2,\, -> - |1,\, ->|2,\, +>), \tag{3.167}$$

where the signs $\pm$ represent states with spin ± 1 along some given direction. In such case, the spin state of each of the individual particles is not independent from the state of the other. This means that their quantum states are *entangled.* Such an entanglement is associated with non-locality. Even if the two particles move far away from each other, their quantum states will stay interconnected, and any measurement or action made on one of them will instantaneously influence the other one, independently of their distance. This, of course, contradicts the theory of relativity, which explicitly forbids instantaneous action at a distance, and limits velocity of propagation to the speed of light c.

In order to understand this problem of entangled non-locality inside a quantum system, let us define $S_a(1)$ as the spin of particle 1 along the direction of some arbitrary unit vector $\mathbf{a}$. Formally, it can be defined as

$$S_a(1) = \mathbf{S}(1) \cdot \mathbf{a} = \hbar\bar{\bar{\sigma}}(1) \cdot \mathbf{a}, \tag{3.168}$$

where the components of $\bar{\bar{\sigma}}$ are the Pauli matrices

$$\sigma_x = \begin{bmatrix} 0 & 1 \\ 1 & 0 \end{bmatrix}, \quad \sigma_y = \begin{bmatrix} 0 & -i \\ i & 0 \end{bmatrix}, \quad \sigma_z = \begin{bmatrix} 1 & 0 \\ 0 & -1 \end{bmatrix}. \tag{3.169}$$

We can see from here that $S_a(1)$ can only take the values $\pm\hbar$. On the other hand we can say that, by definition, its mean value is

$$<S_a(1)> = <1,2|S_a(1)|1,2> = 0. \tag{3.170}$$

Such a result is obvious, if we assume that the positive and negative values of the spin are equally probable. If we now measure simultaneously the spin of particle 1 along the direction $\mathbf{a}$ and the spin of particle 2 along some other direction $\mathbf{b}$, the product of the two measured spins will be

$$S_a(1)S_b(2) = \hbar^2(\bar{\bar{\sigma}}(1) \cdot \mathbf{a})(\bar{\bar{\sigma}}(2) \cdot \mathbf{b}). \tag{3.171}$$

The mean value of this new operator is given by

$$P(\mathbf{a},\, \mathbf{b}) \equiv <S_a(1)S_b(2)> = \frac{\hbar^2}{2} < 1,2|(\bar{\bar{\sigma}}(1) \cdot \mathbf{a})(\bar{\bar{\sigma}}(2) \cdot \mathbf{b})|1,2> . \tag{3.172}$$

Using equation (3.167), this leads to

$$P(\mathbf{a},\, \mathbf{b}) = -\frac{\hbar^2}{2}(\mathbf{a} \cdot \mathbf{b}) = -\frac{\hbar^2}{2}\cos\theta, \tag{3.173}$$

where θ is the angle between the vectors **a** and **b**. In the particular case of the two spin measurements being made along the same direction, $\vec{a} = \vec{b}$, this probability reduces to $P(\mathbf{a}, \mathbf{b}) = -\hbar^2/2$, which corresponds to the existence of two opposite spin values, one equal to $+\hbar$ and the other to $-\hbar/2$, as expected.

Let us now consider a third direction of observation **c**. It is useful to compare the values of the three probabilities $P(\mathbf{a}, \mathbf{b})$, $P(\mathbf{a}, \mathbf{c})$ and $P(\mathbf{b}, \mathbf{c})$. In the particular case where the three vectors are in the same plane, with angles $\theta_{a,b}$, $\theta_{a,c}$ and $\theta_{b,c}$, such that $\theta_{a,b} = \theta_{b,c} = \theta$ and $\theta_{a,c} = 2\theta$, we can establish the following relations

$$P(\mathbf{a}, \mathbf{b}) = P(\mathbf{b}, \mathbf{c}) = -\frac{\hbar^2}{2}\cos\theta, \quad P(\mathbf{a}, \mathbf{c}) = -\frac{\hbar^2}{2}\cos 2\theta, \tag{3.174}$$

from where we obtain

$$| P(\vec{a}, \vec{b}) - P(\vec{a}, \vec{c})| = \frac{\hbar^2}{4} | \cos\theta. - \cos 2\theta | \tag{3.175}$$

This quantity can become larger than $\hbar^2/2$, and can even verify the condition

$$| P(\mathbf{a}, \mathbf{b}) - P(\mathbf{a}, \mathbf{c})| > \left[\frac{\hbar^2}{2} + P(\mathbf{b}, \mathbf{c})\right]. \tag{3.176}$$

Such a condition is equivalent to have

$$| \cos\theta - \cos 2\theta | > 1 - \cos\theta. \tag{3.177}$$

Take the particular case of $\theta = \pi/3$. We have $\cos\theta = 1/2$ and the above expression becomes $1 > 1/2$. This means that the inequality (3.176) can eventually be satisfied in the frame of quantum mechanics.

Let us now consider the possible existence of hidden variables in the system, as a consequence of some hypothetical incompleteness of the standard quantum description. In this case, the spin component $S_a(1)$ along the axis **a** is replaced by a function $(\hbar)A(\mathbf{a}, \lambda)$, where the variable λ represents the hidden parameter. Similarly, $S_b(2)$ will be replaced by $(\hbar)B(\mathbf{b}, \lambda)$. But, because the total spin of the system is assumed to be zero, we will always have

$$A(\mathbf{a}, \lambda) = -B(\mathbf{b}, \lambda). \tag{3.178}$$

On the other hand, we can assign a certain probability $p(\lambda)$ for the hidden variable to take some precise value λ. Such a probability necessarily verifies the normalisation and non-negative conditions

$$\int p(\lambda)d\lambda = 1, \quad p(\lambda) \geqslant 0. \tag{3.179}$$

Taking the average of the product of the two quantities that replaced $S_a(1)$ and $S_b(2)$ over all the possible values of the hidden variable λ, we obtain the joint probability

$$P_{\text{hidd}}(\mathbf{a}, \mathbf{b}) = \frac{\hbar^2}{4}\int p(\lambda)A(\mathbf{a}, \lambda)B(\mathbf{b}, \lambda)d\lambda. \tag{3.180}$$

In the particular case of $\mathbf{a} = \mathbf{b}$, this would reduce to

$$P_{\text{hidd}}(\mathbf{a}, \mathbf{a}) = -\frac{\hbar^2}{2}, \tag{3.181}$$

as it should. Let us now assume, as previously, the same difference between joint probabilities associated with different choices of the axis over which we measure the spin of the two particles in the system. Using this new definition, we have

$$P_{\text{hidd}}(\mathbf{a}, \mathbf{b}) - P_{\text{hidd}}(\mathbf{a}, \mathbf{c}) = \frac{\hbar^2}{2} \int p(\lambda) A(\mathbf{a}, \lambda)[B(\mathbf{b}, \lambda) - B(\mathbf{c}, \lambda)] d\lambda. \tag{3.182}$$

Using the property (3.178), we can also write

$$P_{\text{hidd}}(\mathbf{a}, \mathbf{b}) - P_{\text{hidd}}(\mathbf{a}, \mathbf{c}) = \frac{\hbar^2}{2} \int p(\lambda) A(\mathbf{a}, \lambda) A(\mathbf{b}, \lambda)[1 + A(\mathbf{b}, \lambda) B(\mathbf{c}, \lambda)] d\lambda. \tag{3.183}$$

Noting that $p(\lambda) \geqslant 0$ and that the quantities A can only take the values of ± 1, we finally arrive at

$$| P_{\text{hidd}}(\mathbf{a}, \mathbf{b}) - P_{\text{hidd}}(\mathbf{a}, \mathbf{c})| < \left[\frac{\hbar^2}{2} + P_{\text{hidd}}(\mathbf{b}, \mathbf{c})\right]. \tag{3.184}$$

This is called the *Bell's inequality*. Comparing this results with the purely quantum mechanical result of equation (3.176), we conclude that Bell's inequality is violated by quantum mechanics. Or in other words, the standard quantum theory is not compatible with the existence of hidden variables.

Violation of Bell's inequality was experimentally demonstrated in 1982, using photons where spin states are determined by the field polarisation [33, 36]. In these experiments, a two-photon radiative transition between two levels of the calcium atom, $4p^2\ {}^1S_0$ and $4s^2\ {}^1S_0$, was measured using two rotating polarisers. These atomic quantum levels have zero total angular momentum, which implies that by conservation of the total angular momentum, the angular momentum of the system of two photons is also zero. These and other similar experiments agree with the quantum mechanical expression (3.176) for joint probability measurements, and demonstrate that there are no hidden variables. As a consequence, entanglement at a distance persists between two particles of the same system, even if their quantum states are measured very far apart.

References

[1] Louisell W H 1975 *Quantum Statistical Properties of Radiation* (New York: Wiley)

[2] Walls D F and Milburn G J 1994 *Quantum Optics* (Berlin: Springer)

[3] Scully M O and Zubairy M S 1999 *Quantum Optics* (Cambridge: Cambridge University Press)

[4] Schleich W P 2001 *Quantum Optics in Phase Space* (Berlin: Wiley)

[5] Gerry C C and Knight P L 2005 *Introductory Quantum Optics* (Cambridge: Cambridge University Press)

[6] Glauber R J 1963 Coherent and incoherent states of radiation *Phys. Rev.* **131** 2766–88

[7] Zhang W-M, Feng D H and Gilmore G 1990 Coherent states: theory and some applications *Rev. Mod. Phys.* **62** 867–927
[8] Schrödinger E 1926 Der stetige Übergang von der Miko- zur Makromechanik *Naturwissenshaften* **14** 664–6
[9] Kennard E H 1927 Quantenmechanik einfacher Bewegunstypen *Z. Phys.* **44** 326–52
[10] Husimi K 1940 Some formal properties of the density matrix *Proc. Phys. Math. Soc. Jpn* 22 264–314
[11] Takahashi H 1965 *Advances in Communication Systems, Theory and Applications* ed Balakrishnan A V vol 1 (New York: Academic) pp 227–310
[12] Plebański J 1956 Wave functions of a Harmonic Oscillator *Phys. Rev.* **101** 1825–6
[13] Walls D 1983 Squeezed states of light *Nature* **306** 141–6
[14] Loudon R and Knight P L 1987 Squeezed light *J. Mod. Opt* **34** 709–59
[15] Dodonov V V 2002 Nonclassical states in quantum optics: a squeezed review of the first 75 years *J. Opt. B: Quantum Semiclass. Opt* **4** R1–33
[16] Andersen U L, Gehring T, Marquardt C and Leuchs G 2016 30 years of squeezed light generation *Phys. Scr.* **91** 053001
[17] Schnabel R 2017 Squeezed states of light and their applications in laser interferometers *Phys. Rep.* **684** 1–51
[18] Poisson S D 1837 *Probabilité des jugements en matière criminelle et en matière civile, précédées des règles générales du calcul des probabilités* (Paris: Bachelier)
[19] Handbook of Mathematical Functions with Formulas; M Abramowitz and I A Stegun 1964 *Handbook of Mathematical Functions with Formulas, Graphs, and Mathematical Tables* NBS Applied Mathematics Series 55 (Washington, DC: National Bureau of Standards)
[20] Glauber R J 1963 Photon correlations *Phys. Rev. Lett.* **10** 84–6
[21] Sudarshan E C G 1963 Equivalence of semiclassical and quantum mechanical descriptions of statistical light beams *Phys. Rev. Lett.* **10** 277–9
[22] Kano Y 1965 A new phase-space distribution function in the statistical theory of the electromagnetic field *J. Math. Phys.* **6** 1913–5
[23] Young T 1804 Bakerian lecture: experiment and calculations relative to physical optics Phil *Trans. R. Soc* **94** 1–2
[24] Tan S M and Walls D F 1993 Loss of coherence in interferometry *Phys. Rev.* A **47** 4663–76
[25] Hanbury-Brown R and Twiss R Q 1956 Correlation between photons in two coherent beams of light Nature 177 27-29; 1956 A test of a new type of stellar interferometer on Sirius *Nature* **178** 1046–8
[26] Hanbury-Brown R 1974 *The Intensity Interferometer* (London: Taylor and Francis)
[27] Kimble H J, Dagenais M and Mandel L 1977 Photon antibunching in resonant fluorescence *Phys. Rev. Lett.* **39** 691–5
[28] Short R and Mandel L 1983 Observation of sub-Poissonian photon statistics *Phys. Rev. Lett.* **51** 384–8
[29] Teich M C and Saleh B E A 1988 Photon bunching and antibunching *Progress in Optics XXVI* ed E Wolf (Amsterdam: North-Holland) 1–014 ch I
[30] Davidovich L 1996 Sub-Poissonian processes in quantum optics *Rev. Mod. Phys.* **68** 127–73
[31] Einstein A, Podolsky B and Rosen N 1935 Can quantum-mechanical description of physical reality be considered complete? *Phys. Rev.* **47** 777–80
[32] Bell J 1964 On the Einstein Podolsky Rosen paradox *Physics* **1** 195–200

[33] Aspect A, Grangier P and Roger G 1981 Experimental tests of realistic local theories via Bell's theorem *Phys. Rev. Lett.* **47** 460–4

[34] Rowe M A, Kielpinski D, Meyer V, Sackett C A, Itano W M, Monroe C and Wineland D J 2001 Experimental violation of a Bell's inequality with efficient detection *Nature* **409** 791–4

[35] Smolin L 2020 *Einstein's Unfinished Revolution* (London: Penguin Books)

[36] Aspect A, Dalibard J and Roger G 1982 Experimental test of Bell's inequalities using time-varying analyzers *Phys. Rev. Lett.* **49** 1804–7

IOP Publishing

The Quantum Nature of Light
From photon states to quantum fluids of light
J T Mendonça

Chapter 4

Photon–atom interactions

Having previously discussed the properties of the quantum field, we are now in position to discuss the interactions of this field with matter. We assume a single atom immersed in the electromagnetic field. The case of many atoms will be considered later.

In this chapter, we focus on the basic models describing the interaction of an atom with the radiation field. This includes the *quantum Rabi model*, also sometimes called the *Jaynes–Cummings model*, spontaneous emission, formation of dark states, the dressed atom concept, and a discussion of resonance fluorescence and Mollow triplet. Most of these topics are covered by the books cited in the previous chapter, to which we can add [1–3].

The simplest model of an atom in an external oscillating field is that of a two-level atom. This can be seen as the quantum version of a forced harmonic oscillator. The semi-classical description of an atom in a rotating magnetic field was first studied by Rabi, in 1936, and led to the discovery of nuclear magnetic resonance [4]. The case of a non-rotating but oscillating field was described by Bloch and Siegert in 1940 [5], and a purely quantum model for a two-level atom interacting with a single cavity mode was considered by Jaynes and Cummings in 1963 [6]. For a modern account see the review [7]. This model can be easily described in the rotating wave approximation, as described below. But, more exact solutions can also be found, as discussed in more recent years [8, 9].

4.1 Hamiltonians

We first need to establish the Hamiltonian description of the atom–field interactions. The Hamiltonian operator H contains three distinct terms

$$H = H_a + H_f + H_{\text{int}}, \tag{4.1}$$

doi:10.1088/978-0-7503-2786-2ch4

where H_a is the Hamiltonian of the atom alone, H_f is that of the radiation field, and H_{int} describes the interaction of the atom with the field. The radiation H_f was discussed previously, and can be represented in terms of the creation and annihilation operators $a_{\mathbf{k}}^{\dagger}$ and $a_{\mathbf{k}}$, for each field mode $\mathbf{k}$, as

$$H_f = \sum_{\mathbf{k}} \hbar\omega_{\mathbf{k}} \left(a_{\mathbf{k}}^{\dagger} a_{\mathbf{k}} + \frac{1}{2} \right). \tag{4.2}$$

The other two terms are discussed next. We start with the first term, H_a. This is essentially the electron Hamiltonian, because the interaction of the atom with the field is assumed to be due to nearly resonant response of the electron states of the atom to the external field. In the elementary formulation of quantum mechanics (see chapter 2), the Hamiltonian H_0 describing the electronic states of the atom satisfies the (time-independent) Schrödinger equation, as

$$H_0\psi(\mathbf{r}) = E\psi(\mathbf{r}), \quad H_0 = -\frac{\hbar^2}{2m}\nabla^2 + V(\mathbf{r}), \tag{4.3}$$

where E are the electron energy eigenvalues, $\psi(\mathbf{r})$ the corresponding eigenfunctions, and $V(\mathbf{r})$ the electrostatic potential binding the electrons to the atomic nucleus. Given the linearity of this equation, we can write the electron wavefunction as a superposition of eigenfunctions $\psi_j(\mathbf{r})$, corresponding to different eigenstates E_j, as

$$\psi(\mathbf{r}) = \sum_j b_j \psi_j(\mathbf{r}), \tag{4.4}$$

where b_j are expansion coefficients. Starting from this basic description, we can follow a second quantisation procedure, where the wavefunction $\psi(\mathbf{r})$ is replaced by an operator $\hat{\psi}(\mathbf{r})$. In this new description, the quantities b_j are replaced by operators, while the eigenfunctions $\psi_j(\mathbf{r})$ remain ordinary complex functions. We can then write, for these electron state operators and their corresponding Hermitian conjugates,

$$\hat{\psi}(\mathbf{r}) = \sum_j \hat{b}_j \psi_j(\mathbf{r}), \quad \hat{\psi}^{\dagger}(\mathbf{r}) = \sum_j \hat{b}_j^{\dagger} \psi_j^{*}(\mathbf{r}). \tag{4.5}$$

The energy operator becomes

$$H_a = \int \hat{\psi}(\mathbf{r}) H_0 \hat{\psi}^{\dagger}(\mathbf{r}) d\mathbf{r} = \sum_j E_j \hat{b}_j^{\dagger} \hat{b}_j, \tag{4.6}$$

where the orthogonality between the different eigenstates j was used. In order to understand the meaning of the operators $\hat{b}_j^{\dagger}$ and $\hat{b}_j$, let us define the electron vacuum state $|\, 0\rangle$, as

$$\hat{b}_j \,|\, 0\rangle = 0, \quad \hat{b}_j^{\dagger} \,|\, 0\rangle = |\, j\rangle, \tag{4.7}$$

with $\psi_j(\mathbf{r}) = \langle \mathbf{r} \mid j \rangle$. We notice that, in contrast with the photons (which are bosons), the electrons are fermions and, due to the *exclusion principle*, two electrons cannot occupy the same state $\mid j \rangle$ simultaneously. This means that

$$\hat{b}_j^\dagger \hat{b}_j^\dagger \mid 0 \rangle = \hat{b}_j^\dagger \mid j \rangle = \mid 0 \rangle, \quad \left(\hat{b}_j^\dagger\right)^2 = 0. \tag{4.8}$$

We then need to use *anti-commutator relations* between electron state operators, of the form

$$\left\{\hat{b}_j^\dagger, \hat{b}_k\right\} = \delta_{jk}, \quad \left\{\hat{b}_j, \hat{b}_k\right\} = \left\{\hat{b}_j^\dagger, \hat{b}_k^\dagger\right\} = 0. \tag{4.9}$$

Here, we have used the standard notation $\{A, B\} = AB + BA$. This leads to

$$\{\hat{\psi}^\dagger(\mathbf{r}), \hat{\psi}(\mathbf{r}')\} = \delta(\mathbf{r} - \mathbf{r}'), \quad \{\hat{\psi}(\mathbf{r}), \hat{\psi}(\mathbf{r}')\} = 0. \tag{4.10}$$

The quantum state of an atom with Z electrons, occupying the states $\psi_1, \psi_2, \ldots, \psi_Z$, would then be described by

$$\left| \psi_{(j)} \right\rangle = \hat{b}_1^\dagger \hat{b}_2^\dagger \ldots \hat{b}_Z^\dagger \mid 0 \rangle. \tag{4.11}$$

An alternative, and more common, description of the atom Hamiltonian (4.6) is obtained if we replace the electron operators $\hat{b}_j^\dagger$ and $\hat{b}_j$ by the *transition operator* σ, with matrix elements defined by

$$\sigma_{ij} = \mid i \rangle \langle j \mid. \tag{4.12}$$

They are associated with the atomic eigenstates $\psi_i(\mathbf{r}) = \langle \mathbf{r} \mid i \rangle$ and $\psi_j^*(\mathbf{r}) = \langle j \mid \mathbf{r} \rangle$. Using the energy eigenvalue equation $H_a \mid i \rangle = E_i \mid i \rangle$, and the identity operator $I = \sum_j \mid j \rangle \langle j \mid$, we can readily obtain

$$H_a = \sum_i E_i \mid i \rangle \langle i \mid = \sum_i E_i \sigma_{ii}. \tag{4.13}$$

This formulation is particularly useful for the two-level atom model, where $i = 1, 2$, and this expression reduces to

$$H_a = E_1 \sigma_{11} + E_2 \sigma_{22}. \tag{4.14}$$

Using the energy difference $\hbar\omega_a = (E_2 - E_1)$, assumed positive, this can be rewritten as

$$H_a = \frac{1}{2}\hbar\omega_a(\sigma_{22} - \sigma_{11}) + \frac{1}{2}(E_1 + E_2)(\sigma_{11} + \sigma_{22}). \tag{4.15}$$

Noting that, for a two-level atom, the quantity $(\sigma_{11} + \sigma_{22}) = I$ is the identity operator, we can neglect the second term in this expression as meaningless. We then get the final expression

$$H_a = \frac{1}{2}\hbar\omega_a \sigma_z, \tag{4.16}$$

with the notation

$$\sigma_z = \sigma_{22} - \sigma_{11} = |\,2\rangle\langle 2\,| - |\,1\rangle\langle 1\,|. \tag{4.17}$$

For future purposes, we also define

$$\sigma_+ \equiv \sigma_{21} = |\,2\rangle\langle 1\,|, \qquad \sigma_- \equiv \sigma_{12} = |\,1\rangle\langle 2\,|. \tag{4.18}$$

At this point, we should stress the formal analogy between a two-level atom and a spin-1/2 particle. Using this matrix representation and introducing

$$\sigma_\pm = \sigma_x \pm i\sigma_y, \tag{4.19}$$

we can easily recover the *Pauli matrices*, characterising the spin-1/2 states

$$\sigma_x = \begin{bmatrix} 0 & 1 \\ 1 & 0 \end{bmatrix}, \quad \sigma_y = \begin{bmatrix} 0 & -i \\ i & 0 \end{bmatrix}, \quad \sigma_z = \begin{bmatrix} 1 & 0 \\ 0 & -1 \end{bmatrix}. \tag{4.20}$$

From this, we get the commutation relations

$$[\sigma_+, \sigma_-] = \sigma_z, \qquad [\sigma_-, \sigma_z] = 2\sigma_-. \tag{4.21}$$

Let us now consider the interaction Hamiltonian. It is known that, neglecting spin and relativistic effects, the Hamiltonian of an electron in the electromagnetic field is given by

$$H = \frac{1}{2m}(\mathbf{p} + e\mathbf{A})^2 - eV, \tag{4.22}$$

where $(-e)$ is the electron charge, m the electron mass, and $\mathbf{p}$ its momentum. It is also known that the potentials can change from $(\mathbf{A}, V)$ to $(\mathbf{A}', V')$, without changing the fields, by using a gauge transformation, such that

$$\mathbf{A}' = \mathbf{A} - \nabla\chi, \qquad V' = V + \frac{\partial\chi}{\partial t}, \tag{4.23}$$

where χ is an arbitrary function. In the Coulomb gauge, the radiation field is represented by the vector potential $\mathbf{A}$, associated with the transverse currents, and $V = 0$. In this gauge, the interaction between the atom (or the electron inside the atom) and the radiation field is described by the Hamiltonian

$$H_{\text{int}} = \frac{e}{m}\mathbf{p} \cdot \mathbf{A} + \frac{e^2}{2m}A^2. \tag{4.24}$$

At low intensities, the quadratic term is negligible. Assuming that the wavelength of the incident radiation is much larger than the size of the atom, we develop the vector potential around the atom position $\mathbf{r}_0$, and get

$$\mathbf{A}(\mathbf{r}, t) \simeq \mathbf{A}(t)e^{i\mathbf{k}\cdot\mathbf{r}_0}[1 + i\mathbf{k} \cdot (\mathbf{r} - \mathbf{r}_0) + \cdots]. \tag{4.25}$$

For $\mathbf{r}_0 = 0$, we obtain

$$H_{\text{int}} = \frac{e}{m}\mathbf{p} \cdot \mathbf{A}(t). \tag{4.26}$$

Another equivalent expression for the interaction Hamiltonian is obtained if we choose the gauge function $\chi(\mathbf{r}, t) = \mathbf{A}(t) \cdot \mathbf{r}$. In this case, the gauge transformation (4.23) leads to the new potentials

$$\mathbf{A}' = 0, \qquad V' = \frac{\partial \mathbf{A}}{\partial t} \cdot \mathbf{r}, \tag{4.27}$$

from where we get

$$H_{\text{int}} = \frac{e}{m}\mathcal{E} \cdot \mathbf{r}, \tag{4.28}$$

where $\mathcal{E}$ is the electric field. Using the identity operator $I = \sum_j | j\rangle\langle j \,|$ twice, we can write

$$e\mathbf{r} = -\sum_{i,j} \mathcal{P}_{ij}\sigma_{ij}, \tag{4.29}$$

where $\sigma_{ij} = |\, i\rangle\langle j\,|$ is the atomic transition matrix, and $\mathcal{P}_{ij} = -\langle i \,|\, e\mathbf{r} \,|\, j\rangle$ the *electric dipole* matrix. The electric field mode operator, at the atom position $\mathbf{r}_0$ can be written as

$$\mathcal{E}_{\mathbf{k}} = iC_{\mathbf{k}}\mathbf{e}_{\mathbf{k}}(\hat{a}_{\mathbf{k}} - \hat{a}_{\mathbf{k}}^{\dagger}), \qquad C_{\mathbf{k}} = \sqrt{\frac{\hbar\omega_{\mathbf{k}}}{2\epsilon_0 V}}. \tag{4.30}$$

Inserting this in equation (4.28), we can then write the interaction Hamiltonian for a two-level atom, $(i, j) = 1, 2$ in the form

$$H_{\text{int}} = \hbar \sum_k g_{\mathbf{k}}(\sigma_+ + \sigma_-)(a_{\mathbf{k}} + a_{\mathbf{k}}^{\dagger}), \tag{4.31}$$

where the operator $\sigma_- \equiv \sigma_{12}$ brings the atom from the upper to the lower level, and $\sigma_+ \equiv \sigma_{21}$ describes the reserve transition. As for the quantity $g_{\mathbf{k}}$, it determines the coupling strength between the atom and the field, and is defined by

$$g_{\mathbf{k}} = \frac{1}{\hbar}(\mathcal{P}_{12} \cdot \mathbf{e}_{\mathbf{k}})\sqrt{\frac{\hbar\omega_{\mathbf{k}}}{2\epsilon_0 V}}. \tag{4.32}$$

Furthermore, we can use the *rotating wave approximation*, neglecting the terms that do not conserve the total energy, $\sigma_- a_{\mathbf{k}}$ and $\sigma_+ a_{\mathbf{k}}^{\dagger}$, and we reduce equation (4.31) to

$$H_{\text{int}} = \hbar \sum_{\mathbf{k}} g_{\mathbf{k}}(\sigma_+ a_{\mathbf{k}} + \sigma_- a_{\mathbf{k}}^{\dagger}). \tag{4.33}$$

We are now in a position to write the total Hamiltonian operator (4.1) in explicit form, and describe the basic processes associated with the atom–field interaction.

4.2 Quantum Rabi model

4.2.1 Basic model

We first consider the interaction of the atom with a single mode of the radiation field. The case of an infinity of modes will be described later. This is the quantum Rabi (or Jaynes–Cummings) model. For this purpose we write the total Hamiltonian as

$$\hat{H} = \hat{H}_0 + \hat{H}_{\text{int}}, \quad \hat{H}_0 = \hbar\omega\hat{a}^\dagger\hat{a} + \frac{1}{2}\hbar\omega_a\sigma_z, \tag{4.34}$$

where we have simplified the notation, dropping the subscript $\mathbf{k}$, and neglected the constant term associated with the zero-point energy. The interaction Hamiltonian becomes

$$\hat{H}_{\text{int}} = \hbar g(\sigma_+\hat{a} + \sigma_-\hat{a}^\dagger). \tag{4.35}$$

We assume that the frequency of the field mode ω is nearly equal, but not necessarily identical, to the atomic transition frequency ω_0. The detuning $\delta\omega = \omega_a - \omega \neq 0$ plays an important role in our description. See figure 4.1 for an illustration. We define the state vector of the total system (atom+field), as

$$|\psi\rangle = \sum_n \{ C_{1,n} | 1, n\rangle + C_{2,n} | 2, n\rangle \}, \tag{4.36}$$

where the state vectors $| j, n\rangle$ represent a system where the atom is in the energy state E_j, with $j = 1, 2$, and the field contains n photons with frequency ω.

It is now convenient to introduce the *interaction picture*, where the new Hamiltonian H_I is obtained by using the unitary transformation $U(t)$, such that

$$\hat{H}_I = U^\dagger(t)\hat{H}_{\text{int}}U(t), \quad U(t) = \exp\left(-\frac{i}{\hbar}\hat{H}_0 t\right). \tag{4.37}$$

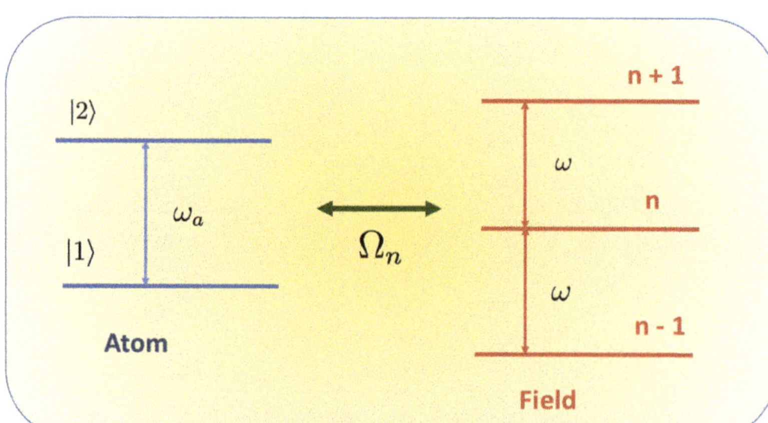

Figure 4.1. Quantum Rabi model, characteristic frequencies: (i) atomic transition frequency, $\omega_a = (E_2 - E_1)/\hbar$; (ii) field mode frequency, ω, and (iii) atom–field interaction frequency, Ω_n.

In order to proceed, we notice that the first term of $\hat{H}_0$ only acts on the operators $\hat{a}^\dagger$ and $\hat{a}$ of $\hat{H}_{\rm int}$, and the second term only acts on $\sigma_\pm$. Using the relation

$$e^{\alpha A} B e^{-\alpha A} = B + \alpha[A, B] + \frac{\alpha^2}{2}[A, [A, B]] + \cdots \tag{4.38}$$

and choosing $\alpha = i\omega t$, $A = \hat{a}^\dagger \hat{a}$ and $B = \hat{a}$, we realise that

$$e^{i\omega t \hat{a}^\dagger \hat{a}} \hat{a} e^{-i\omega t \hat{a}^\dagger \hat{a}} = \hat{a}\left(1 - \frac{\omega^2 t^2}{2} + \cdots\right) = \hat{a} e^{-i\omega t}. \tag{4.39}$$

Similarly, we have

$$e^{i\omega_a \sigma_z t/2} \sigma_\pm e^{-i\omega_a \sigma_z t/2} = \sigma_\pm e^{i\omega_a t}. \tag{4.40}$$

From this, we obtain

$$\hat{H}_I = \hbar g(\sigma_+ \hat{a} e^{i\delta\omega t} + \hat{a}^\dagger \sigma_- e^{-i\delta\omega t}), \tag{4.41}$$

where $\delta\omega = \omega_a - \omega$ is the detuning. Let us now solve the evolution equation for $|\psi\rangle$ in the interaction picture

$$i\hbar \frac{\partial}{\partial t} |\psi\rangle = H_I |\psi\rangle. \tag{4.42}$$

Using equation (4.36), we arrive at two coupled equations for the coefficients

$$\frac{\partial}{\partial t} C_{2,n} = -\frac{i}{2}\Omega_n e^{i\delta\omega t} C_{1,n+1}, \quad \frac{\partial}{\partial t} C_{1,n+1} = -\frac{i}{2}\Omega_n e^{-i\delta\omega t} C_{2,n}, \tag{4.43}$$

where

$$\Omega_n = 2g\sqrt{n+1}, \tag{4.44}$$

is the quantum expression of the *Rabi frequency*. We should keep in mind that the photon number n is proportional to the radiation intensity. Therefore, for $n \gg 1$, the Rabi frequency is proportional to the field amplitude, in agreement with the classical definition of the Rabi frequency.

In order to understand the interest and meaning of the above equations, we first consider the simple case of exact resonance $\delta\omega = 0$: if the atom is initially in the excited state $|2\rangle$, and the field in a well-defined Fock state $|n\rangle$, we should use $C_{2,n}(t=0) = \delta_{n,n_0}$ and $C_{1,n+1}(t=0) = 0$. The solution of equations (4.43) will be

$$C_{2,n_0}(t) = \cos\left(\frac{\Omega_n}{2}t\right), \quad C_{1,n_0+1}(t) = -i\sin\left(\frac{\Omega_n}{2}t\right). \tag{4.45}$$

It is now useful to introduce the *inversion of population*, defined as

$$W(t) = |C_{2,n}(t)|^2 - |C_{1,n+1}(t)|^2 = \cos(\Omega_n t). \tag{4.46}$$

This expression shows a periodic oscillation of the atomic populations (the so-called Rabi oscillations) between the two states, well known from the semi-classical theory.

What is a purely quantum effect is the persistence of such oscillations even in the case of an initial vacuum, $n = 0$. Here, the oscillations result from photon spontaneous emission from the initially excited atom, with the subsequent re-absorption.

This analysis can be easily extended to the case of a finite detuning, $\delta\omega \neq 0$, and to a more general radiation field. We can now use the initial conditions $C_{2,n}(t = 0) = C_n(0)$ and $C_{1,n+1}(t = 0) = 0$, and the solutions of equations (4.43) become

$$C_{2,n}(t) = C_n(0)\left[\cos\left(\frac{\Omega'_n}{2}t\right) - i\frac{\delta\omega}{\Omega'_n}\sin\left(\frac{\Omega'_n}{2}t\right)\right]e^{i\delta\omega t/2}, \tag{4.47}$$

$$C_{1,n+1}(t) = -iC_n(0)\frac{\Omega_n}{\Omega'_n}\sin\left(\frac{\Omega'_n}{2}t\right)e^{-i\delta\omega t/2}, \tag{4.48}$$

where we have defined a detuned Rabi frequency

$$\Omega'_n = \sqrt{\delta\omega^2 + \Omega_n^2}. \tag{4.49}$$

The inversion of population in the presence of this field is now given by

$$W(t) = \sum_n \left\{|C_{2,n}(t)|^2 - |C_{1,n+1}(t)|^2\right\}. \tag{4.50}$$

We then get

$$W(t) = \sum_n |C_n(0)|^2 \left\{\frac{\delta\omega^2}{\Omega_n'^2} + \frac{\Omega_n^2}{\Omega_n'^2}\cos(\Omega'_n t)\right\}. \tag{4.51}$$

It is useful to consider an initial coherent state $|\alpha\rangle$, such that, by definition, we have

$$C_n(t = 0) = \exp\left(-\frac{|\alpha|^2}{2}\right)\frac{\alpha^n}{\sqrt{n!}}. \tag{4.52}$$

In contrast with the above Fock state, with these new initial conditions the Rabi oscillations will collapse after some time τ_{col}, due to the frequency mixing between the different terms in equation (4.51), as illustrated in figure 4.2. Furthermore, some time after collapse, we can observe a revival of the Rabi oscillations, which looks like a quantum echo. Let us discuss the qualitative differences between the various time-scales associated with this *collapse-revival process*.

First, we have the Rabi oscillations, with period $\tau_R \simeq 1/\Omega_R$, where Ω_R is a kind of averaged Rabi frequency. Noting that, for a coherent state, the mean photon number is simply $\alpha = \langle n\rangle$, we get

$$\tau_R \simeq \frac{1}{\Omega'_{\langle n\rangle}} \equiv \frac{1}{(\delta\omega^2 + 4g^2\langle n\rangle)^{1/2}} \sim \frac{1}{2g\langle n\rangle^{1/2}}. \tag{4.53}$$

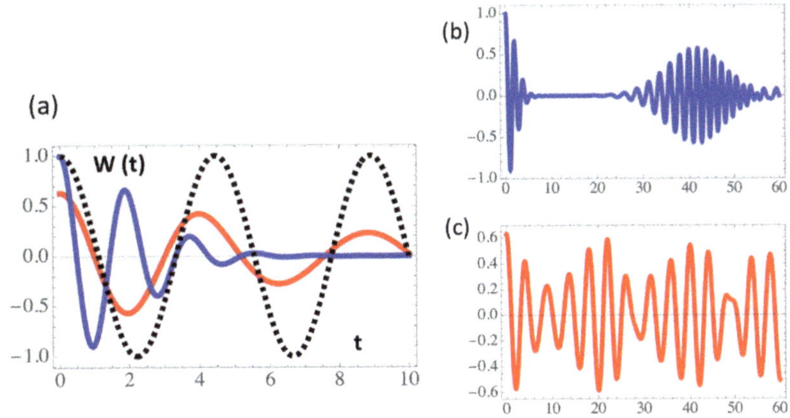

Figure 4.2. Inversion of population $W(t)$, for a field mode initially in a coherent state: (a) collapse of Rabi oscillations, for $\alpha = 1$ (in red) and $\alpha = 3$ (in blue); (b)–(c) revival of Rabi oscillations for the same two states are observed, for longer times. Undamped oscillations, for a single Fock state at $n = 1$ are also represented (dashed) for comparison.

Noting that collapse is due to phase mixing, we can estimate the collapse time as

$$\tau_{\text{col}} \simeq \frac{1}{\left[\Omega'_{\langle n\rangle+\Delta n} - \Omega'_{\langle n\rangle-\Delta n}\right]}, \tag{4.54}$$

where Δn is the photon number uncertainty. For a coherent state, this uncertainty is $\Delta n = \langle n\rangle^{1/2} = \alpha^{1/2}$. Assuming that $\Delta n \ll \alpha$, we can then write

$$\tau_{\text{col}} \simeq \frac{1}{2g}\left(1 + \frac{\delta\omega^2}{\Omega_R^2}\right)^{1/2}, \tag{4.55}$$

with $\Omega_R^2 \equiv 4g^2\langle n\rangle$. At resonance, $\delta\omega = 0$, and for a large photon number, $\alpha \gg \delta\omega/4g^2$, this timescale is nearly independent of α. We then have $\tau_{\text{col}} \simeq \tau_R\langle n\rangle^{1/2}$. Finally, let us assume revival, which occurs when the phases in equation (4.49) only differ by multiples of 2π. This means that the revival time τ_{rev} is of the order of

$$\tau_{\text{rev}} \simeq \frac{2\pi m}{\left(\Omega'_{\langle n\rangle} - \Omega'_{\langle n\rangle-1}\right)}, \tag{4.56}$$

where m is an integer representing the successive revival times. Again, taking the limit of a large mean photon number, $\langle n\rangle \gg 1$, we get

$$\tau_{\text{rev}} \simeq 2\pi m\left(1 + \frac{\delta\omega^2}{\Omega_R^2}\right)^{1/2}\frac{\Omega_R}{g^2}. \tag{4.57}$$

This shows that, for the first revival time, $m = 1$, and for exact resonance $\delta\omega = 0$, we can write

$$\tau_{\text{rev}} \sim \tau_{\text{col}}\sqrt{\langle n \rangle} \sim \tau_R\langle n \rangle. \tag{4.58}$$

Therefore, revival of the Rabi oscillations, for a two-level atom in the presence of a coherent photon state, typically occurs after a time of the order of α times the Rabi period.

4.2.2 Dressed atom

At this point it is useful to introduce the concept of dressed states, first proposed by Cohen-Tannoudji and collaborators [10, 11]. We discuss this concept in the frame of the above quantum Rabi model. We have seen that the evolution of the quantum 'atom + radiation' system is described by the Hamiltonian $\hat{H} = (\hat{H}_a + \hat{H}_f) + \hat{H}_{\text{int}}$, which can be written in terms of the pseudo-spin operators, σ_z and $\sigma_\pm$, as

$$\hat{H}_a = \frac{1}{2}\hbar\omega_0\sigma_z, \quad H_f = \hbar\omega\hat{a}^\dagger\hat{a}, \quad \hat{H}_{\text{int}} = \hbar g(\hat{a}\sigma_+ + \hat{a}^\dagger\sigma_-). \tag{4.59}$$

This last expression is valid in the rotating wave approximation. We also have seen that this simple model is able to describe radiative transitions between two different states $\left| \psi_j \right\rangle$, with $j = 1, 2$, as

$$| 2, n\rangle \rightleftarrows | 1, n + 1\rangle. \tag{4.60}$$

For convenience, these states of the system will be called here the *bare states*. Such states are orthogonal to each other, $\langle 2, n | 1, n + 1\rangle = 0$, and can be used to represent the Hamiltonian $\hat{H}$. In particular, we have

$$\hat{H}_{\text{int}}\begin{bmatrix} | 2, n\rangle \\ | 1, n + 1\rangle \end{bmatrix} = \hbar\begin{bmatrix} 0 & \Omega'_n \\ \Omega'_n & 0 \end{bmatrix}\begin{bmatrix} | 2, n\rangle \\ | 1, n + 1\rangle \end{bmatrix}. \tag{4.61}$$

This gives a 2×2 matrix representation of the Hamiltonian $\hat{H}$, such that, for a given photon number n, we get

$$\hat{H} = \hbar\begin{bmatrix} n\omega + \omega_0/2 & \Omega'_n \\ \Omega'_n & (n + 1)\omega - \omega_0/2 \end{bmatrix}. \tag{4.62}$$

From here, we can derive the energy eigenstates, which are

$$E_\pm = \left(n + \frac{1}{2}\right)\hbar\omega \pm \hbar\Omega'_n, \tag{4.63}$$

The corresponding state vectors can be written as

$$\left| \psi_+(n) \right\rangle = \cos(\Phi_n/2)| 1, n + 1\rangle + \sin(\Phi_n/2)| 2, n\rangle, \tag{4.64}$$

and

$$|\,\psi_-(n)\rangle = -\sin(\Phi_n/2)|\,1,\,n+1\rangle + \cos(\Phi_n/2)|\,2,\,n\rangle, \tag{4.65}$$

where the angle Φ_n is defined by

$$\Phi_n = \arctan\left(\frac{\Omega_n}{\delta\omega}\right). \tag{4.66}$$

More explicitly, we write

$$\sin(\Phi_n/2) = \frac{1}{\sqrt{2}}\left(\frac{\Omega'_n - \delta\omega}{\Omega'_n}\right)^{1/2}, \quad \cos(\Phi_n/2) = \frac{1}{\sqrt{2}}\left(\frac{\Omega'_n + \delta\omega}{\Omega'_n}\right)^{1/2}. \tag{4.67}$$

These new state vectors $|\,\psi_\pm\rangle$, with energies determined by equation (4.63), are called the *dressed states*. At exact resonance, $\delta\omega = 0$, they would be reduced to

$$|\,\psi_\pm(n)\rangle = \frac{1}{\sqrt{2}}(|\,1,\,n+1\rangle \pm |\,2,\,n\rangle). \tag{4.68}$$

The difference between bare and dressed states can be illustrated by an energy diagram, as shown in figure 4.3. The representation in terms of dressed states is physically appealing, and can be useful in some specific situations, such as in the calculation of the spectrum of resonance fluorescence, where the occurrence of satellite radiation lines is well understood as transitions between dressed states. Furthermore, given that $|\,2,\,n\rangle = |\,2\rangle\,|\,n\rangle$, and $|\,1,\,n+1\rangle = |\,1\rangle\,|\,n+1\rangle$, the dressed states $|\,\psi_\pm(n)\rangle$ can also be seen as *entangled states* between the atom and radiation, where the atom and field states cannot be treated separately.

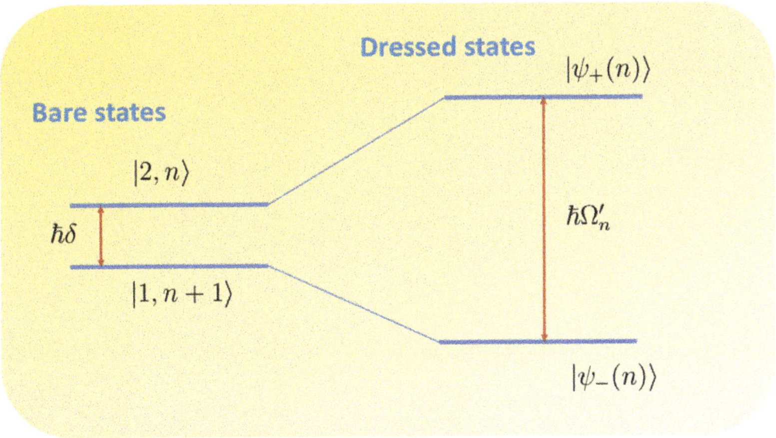

Figure 4.3. Quantum states of the atom+field system: comparison between bare states and dressed states.

4.3 Three-level atom

4.3.1 Dark states

Resonant atomic transitions can be used to manipulate the refractive index of an optical medium. Of particular relevance is the formation of *dark states*, first noted by [12], and the resulting phenomena of *electromagnetic induced transparency*, or *EIT* [13, 14].

In order to understand such manipulating processes we start by considering a three-level atom, with two possible radiative transitions, as shown in figure 4.4(a). The corresponding atom Hamiltonian is

$$\hat{H}_a = \sum_j E_j \mid j\rangle\langle j \mid, \quad (j = 1, 2, 3). \tag{4.69}$$

The frequencies of the two atomic transitions are $\omega_1 = (E_3 - E_1)/\hbar$ and $\omega_2 = (E_3 - E_2)/\hbar$, where E_3 is the upper energy level, and radiation between the two lower states E_2 and E_1 is assumed forbidden. Let us consider the resonant interaction of the atom with radiation. We only need to retain two relevant field modes, with an arbitrary direction of propagation and frequencies equal to ω_1 and ω_2. As shown previously, the interaction Hamiltonian can be written in the form

$$\hat{H}_{\text{int}} = -\frac{\hbar}{2}(\Omega_1 e^{-i\omega_1 t} \mid 3\rangle\langle 1 \mid + \Omega_2 e^{-i\omega_2 t} \mid 3\rangle\langle 2 \mid) + h.\ c., \tag{4.70}$$

where $\Omega_i = |\Omega_i|\exp(-i\varphi_i)$, for $i = 1, 2$, are the complex Rabi frequencies associated with the coupling between the atom and the two field modes, and φ_i are constant phases. We now have to solve Schrödinger's equation

$$i\hbar\frac{\partial}{\partial t} \mid \psi\rangle = (H_a + H_{\text{int}})\mid \psi\rangle. \tag{4.71}$$

For that purpose, we use the following type of solution

$$\mid \psi\rangle = \sum_{j=1,2,3} C_j(t)e^{-iE_j t/\hbar} \mid j\rangle, \tag{4.72}$$

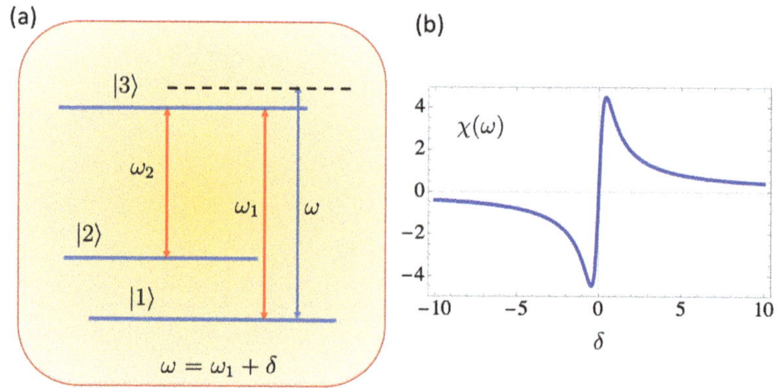

Figure 4.4. (a) Quantum states of the three-level atom, with two radiative transitions; (b) EIT susceptibility, $\chi(\omega)$ as a function of the detuning. We use a normalised quantity χ/χ_0, with $\chi_0 = N_a \mid \mathcal{P}_{13} \mid^2/\epsilon_0 \mid \Omega_2 \mid$.

where the quantities $|C_j(t)|^2$ represent the probability for the atom to occupy the energy eigenstate $|j\rangle$. Replacing this in equation (4.71) and using the above expressions for the two Hamiltonian operators $\hat{H}_a$ and $\hat{H}_{\text{int}}$, we obtain a system of coupled equations for the coefficients $C_j(t)$, of the form

$$\frac{dC_1}{dt} = \frac{i}{2}\Omega_1^* C_3, \quad \frac{dC_2}{dt} = \frac{i}{2}\Omega_2^* C_3, \quad \frac{dC_3}{dt} = \frac{i}{2}(\Omega_1 C_1 + \Omega_2 C_2). \tag{4.73}$$

Let us consider the following initial conditions, corresponding to the case where only the two lower energy levels are occupied, such that

$$|\psi(0)\rangle = C_1(0)|1\rangle + C_2(0)|2\rangle, \tag{4.74}$$

and $C_3(0) = 0$. This leads to

$$\frac{d^2C_3}{dt^2} = -\frac{1}{4}(|\Omega_1|^2 + |\Omega_2|^2)C_3, \tag{4.75}$$

with similar equations for the other two coefficients

$$\frac{d^2C_1}{dt^2} = -\frac{1}{4}(|\Omega_1|^2 C_1 + \Omega_1^*\Omega_2 C_2), \quad \frac{d^2C_2}{dt^2} = -\frac{1}{4}(\Omega_1\Omega_2^* C_1 + |\Omega_2|^2 C_2). \tag{4.76}$$

Solution of equation (4.75) is straightforward, and gives

$$C_3(t) = \frac{i}{\Omega}\sin\left(\frac{\Omega}{2}t\right)\{\Omega_1 C_1(0) + \Omega_2 C_2(0)\}, \tag{4.77}$$

where the new frequency Ω is defined as

$$\Omega = \sqrt{|\Omega_1|^2 + |\Omega_2|^2}. \tag{4.78}$$

Solving the coupled equations for the other two coefficients (4.76), we similarly get

$$C_1(t) = \frac{1}{\Omega^2}\left\{\left[|\Omega_1|^2\cos\left(\frac{\Omega}{2}t\right) + |\Omega_2|^2\right]C_1(0) - 2\Omega_1^*\Omega_2\sin^2\left(\frac{\Omega}{4}t\right)C_2(0)\right\}, \tag{4.79}$$

and

$$C_2(t) = \frac{1}{\Omega^2}\left\{-2\Omega_1\Omega_2^*\sin^2\left(\frac{\Omega}{4}t\right)C_1(0) + \left[|\Omega_2|^2\cos\left(\frac{\Omega}{2}t\right) + |\Omega_1|^2\right]C_2(0)\right\}. \tag{4.80}$$

Notice that these three coefficients oscillate a the frequency $(\Omega/2)$, which is the geometric mean of the two Rabi frequencies $|\Omega_1|$ and $|\Omega_2|$, as shown by the definition (4.78). These solutions predict the existence of situations where the upper level of the atom $|3\rangle$ is never populated, because the coefficient $C_3(0)$ can remain always equal to zero. It occurs when

$$\Omega_1 = \Omega_2, \quad C_1(0) = -C_2(0). \tag{4.81}$$

This can be achieved when the two lower levels, assumed close to each other and initially equally populated, are in phase opposition. Written in another way

$$C_{j=1,2} = |C_{j=1,2}|\, e^{-i\varphi_{j=1,2}}, \quad (\varphi_2 - \varphi_1) = \pm\pi, \tag{4.82}$$

with

$$|C_1(0)| = |C_2(0)| = \frac{1}{\sqrt{2}}. \tag{4.83}$$

In such special conditions, the solutions (4.77)–(4.80) are reduced to

$$|C_1(t)|^2 = |C_2(t)|^2 = \frac{1}{2}, \quad |C_3(t)|^2 = 0. \tag{4.84}$$

As a result, the atom acts as if it was a two-level atom and the upper energy level was not present. It nevertheless plays a relevant role in the radiation process. This means that if a laser beam propagates in a medium with atoms prepared in this initial state, the field will not perceive the existence of the atoms, even if the mode frequencies are resonant with the resonant transitions and coupling between atoms and the field modes is very strong. As a result, the atomic medium will behave as a perfectly transparent medium.

The question however remains on how to prepare the atoms to make the medium transparent. This is usually achieved by using one of the field modes as a *pump field*, while the second mode experiences transparency, and is called the *probe field*. The role of these two modes can be interchanged. This is the principle of EIT, to be considered next.

4.3.2 Electromagnetic induced transparency

EIT is described by a slightly change of the previous model, where we assume a strong pump resonant with one of the radiative transitions, let us say with frequency $\omega_2 = (E_3 - E_2)/\hbar$, and a probe field is nearly resonant with the other transition frequency $\omega = \omega_1 + \delta$, where $|\delta| \ll \omega$ represents a small detuning. As shown below, under certain conditions, the medium will become transparent to the probe. Due to the existence a finite detuning, $\delta \neq 0$, the interaction Hamiltonian of equation (4.70) will only slightly change, with ω_1 replaced by ω.

Our main purpose is to determine the polarisation of the medium, and to define the refractive index. The polarisation vector $\mathbf{P}$ is determined by the off-diagonal matrix elements of the atoms in the medium. It is then useful to consider the evolution equations for the density matrix elements ρ_{31}, ρ_{21} and ρ_{32}. They can be written as

$$\frac{d\rho_{31}}{dt} = -i\omega_1\rho_{31} - \frac{i}{2}\Omega_1(\rho_{33} - \rho_{11})e^{-i\omega t} + \frac{i}{2}\Omega_2\rho_{21}e^{-i\omega_2 t}, \tag{4.85}$$

$$\frac{d\rho_{21}}{dt} = -i\omega_2\rho_{21} - \frac{i}{2}\Omega_1\rho_{23}e^{-i\omega t} + \frac{i}{2}\Omega_2^*\rho_{31}e^{i\omega_2 t}, \tag{4.86}$$

and

$$\frac{d\rho_{32}}{dt} = -i\omega_2\rho_{32} - \frac{i}{2}\Omega_2(\rho_{33} - \rho_{22})e^{-i\omega_2 t} + \frac{i}{2}\Omega_1\rho_{12}e^{-i\omega t}. \tag{4.87}$$

These equations can be simplified, using two new quantities

$$\rho_{31} = \tilde{\rho}_{31}e^{-i\omega t}, \quad \rho_{21} = \tilde{\rho}_{21}e^{-i(\omega+\omega_2)t}. \tag{4.88}$$

These new quantities can then be described by much simpler equations, of the form

$$\frac{d\tilde{\rho}_{31}}{dt} = -i\Delta\tilde{\rho}_{31} + \frac{i}{2}\Omega_1 + \frac{i}{2}\Omega_2\tilde{\rho}_{21}, \tag{4.89}$$

and

$$\frac{d\tilde{\rho}_{21}}{dt} = -i\Delta\tilde{\rho}_{21} + \frac{i}{2}\Omega_2^*\tilde{\rho}_{31}. \tag{4.90}$$

This can also be written in matrix notation, as

$$\frac{d\bar{\rho}}{dt} = -\bar{M}\cdot\bar{\rho} + \bar{\Omega}, \tag{4.91}$$

where we have defined

$$\bar{M} = \begin{bmatrix} i\Delta & -i\Omega_2^*/2 \\ -i\Omega_2/2 & i\Delta \end{bmatrix}, \quad \bar{\rho} = \begin{bmatrix} i\tilde{\rho}_{31} \\ \tilde{\rho}_{21} \end{bmatrix}, \quad \bar{\Omega} = \begin{bmatrix} i\Omega_1/2 \\ 0 \end{bmatrix}. \tag{4.92}$$

Integration of this equation for the initial conditions $\bar{\rho}(0) = 0$, or $\tilde{\rho}_{31}(0) = \tilde{\rho}_{21}(0) = 0$, leads to the formal solution

$$\bar{\rho}(t) = \int^t \exp[-\bar{M}(t-t')]\cdot\bar{\Omega}dt' = \bar{M}^{-1}\cdot\bar{\Omega}. \tag{4.93}$$

Inverting the matrix $\bar{M}$, we get

$$\bar{M}^{-1} = \frac{2i}{|\Omega_2|^2 + 4\Delta^2}\begin{bmatrix} 2\Delta & \Omega_2^* \\ \Omega_2 & 2\Delta \end{bmatrix}. \tag{4.94}$$

Replacing this in the formal solution, we obtain

$$\rho_{31}(t) = -\frac{\Delta}{2}\frac{\Omega_1}{|\Omega_2|^2 + 4\Delta^2}e^{-i\omega t}. \tag{4.95}$$

We should notice that the complex Rabi frequency Ω_1 is proportional to the amplitude of the probe field $E(\omega)$, as by definition $\Omega_1 = \mathcal{P}_{31}E(\omega)/\hbar$, where $\mathcal{P}_{31}$ is the corresponding dipole moment. We can now use the definition of the polarisation vector, as

$$\mathbf{P}(\omega) = \epsilon_0\chi(\omega)\mathbf{E}(\omega), \tag{4.96}$$

where $\chi(\omega)$ is the susceptibility of the medium, which determines the refractive index

$$n(\omega) = \sqrt{1 + \chi(\omega)}. \tag{4.97}$$

Furthermore, the susceptibility is related to the off-diagonal density matrix elements, by the expression

$$\chi(\omega) = 2N_a \frac{\mathcal{P}_{13}\rho_{31}}{\epsilon_0 E(\omega)} e^{i\omega t}, \tag{4.98}$$

which leads to our final expression

$$\chi(\omega) = \frac{N_a}{\epsilon_0} \frac{|\mathcal{P}_{13}|^2 \delta}{(|\Omega_2|^2 + \delta^2/4)}. \tag{4.99}$$

This result shows that, when $\delta \to 0$, the susceptibility tends to zero, and the refractive index tends to 1. This means that exact optical transparency is achieved in the presence of a pump field such that $|\Omega_2|^2 \neq 0$. This can be called an induced transparency effect, because the presence of the pump field is necessary to prepare the atomic quantum states, which were assumed initially in the ground state, $\rho_{11}(0) = 1$, and $\rho_{22}(0) = \rho_{33}(0) = 0$. In this way, a material medium is made to behave as if it was empty. This is illustrated in figure 4.4(b), where the real part of $\chi(\omega)$ is represented as a function of the detuning δ, for an EIT configuration, with losses included. In our analysis we have neglected spontaneous decay, which would introduce an imaginary part in the susceptibility.

4.4 Spontaneous emission

It is well known that the existence of spontaneous emission is a purely quantum process, that explains the natural line width of the atomic spectrum. It was first conceived by Einstein in 1917, with his quantum theory of black body radiation [15], and was formulated in modern quantum terms by Weisskopf and Wigner in 1930 [16]. More complete descriptions can be found in several papers, showing in particular that the decay cannot be exactly exponential [17–19]. Here, we first describe the basic Wigner–Weisskopf theory, and present a more refined approach in the next section.

Let us now consider the *spontaneous decay* of an excited atom, from the upper to the lower energy level, with emission of radiation. In order to describe this process, we need to extend the previously model of quantum Rabi oscillations, assuming the presence of not just one, but an infinite number of field modes, with frequency $\omega_\mathbf{k}$, propagating in arbitrary directions, as defined by the unit vectors $\mathbf{k}/k$. In the interaction picture, and rotating wave approximation, the interaction Hamiltonian H_I will then take the form

$$H_I = \hbar \sum_\mathbf{k} (g_k a_\mathbf{k} e^{i\delta_k t} + h.\ c.), \tag{4.100}$$

where $\delta_{\mathbf{k}} = \omega_a - \omega_{\mathbf{k}}$ is the mode detuning. The state vector of the system 'atom + field' can be written as

$$| \psi(t)\rangle = C_2(t)| 2, 0\rangle + \sum_{\mathbf{k}} C_{1,\mathbf{k}}(t)| 1,1\rangle_{\mathbf{k}}, \tag{4.101}$$

where the excited atom is initially located in the a vacuum field, with photon numbers $n_{\mathbf{k}} = 0$. Due to spontaneous decay, the atom will go to the ground state, with only a single photon present in the field $n_{\mathbf{k}} = 1$, at some arbitrary wavevector mode $\mathbf{k}$. Replacing this in the Schrödinger equation, we then obtain

$$\frac{dC_2}{dt} = -\sum_{\mathbf{k}} g_k e^{i\delta_k t} C_{1,\mathbf{k}}, \quad \frac{dC_{1,\mathbf{k}}}{dt} = g_k e^{-i\delta_k t} C_2. \tag{4.102}$$

This should be compared with the previous quantum Rabi or Jaynes–Cummings model. Integrating the last of these two equations,

$$C_{1,\mathbf{k}}(t) = g_k \int_0^t e^{-i\delta_k t'} C_2(t')dt', \tag{4.103}$$

and replacing it in the first one, we obtain

$$\frac{dC_2}{dt} = -\sum_{\mathbf{k}} |g_k|^2 \int_0^t e^{i\delta_k(t-t')} C_2(t')dt'. \tag{4.104}$$

This equation can easily be solved after some simplifying assumptions. First, we replace the summation over the modes by an integral, such that

$$\sum_{\mathbf{k}} \to 2V \int \frac{d\mathbf{k}}{(2\pi)^3}. \tag{4.105}$$

Second, we use spherical variables, such that $d\mathbf{k} = k^2 \sin\theta dk d\theta d\varphi$. This leads to integrals of the form

$$\int d\mathbf{k} = \int_0^{2\pi} d\varphi \int_0^{\pi} \sin\theta d\theta \int_0^{\infty} k^2 dk. \tag{4.106}$$

Using the explicit expression for the coupling coefficients, we can write

$$|g_k|^2 = \frac{\omega_k}{2\hbar\epsilon_0 V} |\mathcal{P}_{12}|^2 \cos^2\theta. \tag{4.107}$$

After integration of the integral in equation (4.104) over the solid angle $d\Omega = \sin^2\theta d\theta d\varphi$, we arrive at

$$\frac{dC_2}{dt} = -\frac{|\mathcal{P}_{12}|^2}{6\pi^2\hbar\epsilon_0} \int_0^t dt' \int_0^{\infty} k^2 dk \omega_k e^{i\delta_k(t-t')}. \tag{4.108}$$

Using $k = \omega_k/c$, and integrating over the frequency spectrum, we then get

$$\frac{1}{c^3} \int_0^{\infty} \omega_k^3 e^{i\delta_k(t-t')} d\omega_k \simeq 2\pi \frac{\omega_a}{c^3} \delta(t - t'). \tag{4.109}$$

Replacing this in equation (4.108), we finally obtain

$$\frac{dC_2}{dt} = -\frac{\Gamma}{2} C_2(0), \quad \Gamma = \frac{1}{\pi \epsilon_0} \frac{\omega_a^3 |\mathcal{P}_{12}|^2}{3\pi^2 \hbar c^3}. \tag{4.110}$$

This shows that the probability of occupation of the upper atomic level will decay exponentially, according to the law

$$\rho_{22}(t) \equiv |C_2(t)|^2 = \exp(-\Gamma t), \tag{4.111}$$

where Γ defines the lifetime of the excited state, $| 2\rangle$. Replacing this in equation (4.103), and performing the integration, we get for the other coefficients

$$C_{1,\mathbf{k}}(t) = g_k \frac{1 - e^{-i\delta_k t - \Gamma t/2}}{\delta_k + i\Gamma/2}. \tag{4.112}$$

The final solution for the state vector of the system 'atom + field' can then be written as

$$| \psi(t)\rangle = e^{-\Gamma t/2} | 2, 0\rangle + | 1\rangle | \Phi(t)\rangle, \tag{4.113}$$

where the state vector for the radiation field is given by

$$| \Phi(t)\rangle = \sum_{\mathbf{k}} C_{1,\mathbf{k}}(t) | 1\rangle_{\mathbf{k}}. \tag{4.114}$$

Using equation (4.112), we obtain an asymptotic expression, valid in the limit $t \to \infty$, of the form

$$| \Phi_\infty\rangle = \sum_{\mathbf{k}} \frac{g_k}{\delta_k + i\Gamma/2} | 1\rangle_{\mathbf{k}}. \tag{4.115}$$

This means that, on a very long timescale, such that $\Gamma t \gg 1$, the system will be on a state $| \psi_\infty\rangle = | 1\rangle | \Phi_\infty\rangle$, showing that the energy initially contained in the excited atomic level $| 2\rangle$ will be randomly distributed among an infinite number of all possible field modes defined by the wavevector $\mathbf{k}$, according to a Lorentzian frequency line shape described by

$$| C_{1,\mathbf{k}}(t \to \infty) |^2 = \frac{|g_k|^2}{(\omega_a - \omega_{\mathbf{k}})^2 + \Gamma^2/4}. \tag{4.116}$$

This description is valid for an atom located in an infinite vacuum region, with no boundaries. However, if the atom is located inside an electromagnetic cavity, where only a small number of $\mathbf{k}$ modes with $\omega_{\mathbf{k}} \simeq \omega_a$ can persist, then spontaneous emission will be strongly inhibited, or even suppressed, and the lifetime $1/\Gamma$ will lose its meaning. If only a single radiation mode exists in the cavity, the situation will then reduce to that of the quantum Rabi or Jaynes–Cummings model, and the energy will oscillate between the atom and the field on a timescale of $1/\Omega_{n=1}$, as previously shown. A more sophisticated description of spontaneous emission is provided by the reduced density method, as described next.

4.5 Reduced density method

4.5.1 Master equation

We consider a simple quantum system, for instance, a single atom, surrounded by a much larger and complex system, to be called the *reservoir*. This can be described by a total Hamiltonian of the form $H = H_0 + V$, where the unperturbed Hamiltonian contains two terms, $H_0 = H_s + H_r$, the system term H_s and the reservoir term H_r, and the potential V describes the interaction between system and reservoir. The total density operator, ρ_{sr} will be described by the evolution equation

$$\frac{\partial \rho_{sr}}{\partial t} = -\frac{i}{\hbar}[H, \rho_{sr}]. \tag{4.117}$$

Any operator A of the system will be determined by

$$\langle A \rangle = Tr_{sr}\{A\rho_{sr}\} = Tr_s\{A\rho_s\}, \quad \rho_s = Tr_r\{\rho_{sr}\}, \tag{4.118}$$

where ρ_s is the *reduced density operator*. The evolution equation of this operator is usually called a *master equation*. In order to derive such an equation, we start from equation (4.117). We use the interaction picture, previously defined, introducing a unitary transformation, such that

$$\rho_{sr}(t) = e^{-iH_0t/\hbar}\tilde{\rho}_{sr}(t)e^{iH_0t/\hbar}, \tag{4.119}$$

where $\tilde{\rho}_{sr}$ is the total density operator in this new picture. Taking the time derivative of this expression, and using equation (4.117), we then get an evolution equation of the form

$$\frac{\partial \tilde{\rho}_{sr}}{\partial t} = -\frac{i}{\hbar}[\tilde{V}, \tilde{\rho}_{sr}], \tag{4.120}$$

where $\tilde{V}$ is the interaction Hamiltonian in the interaction picture, as defined by

$$\tilde{V}(t) = e^{-iH_0t/\hbar}\tilde{V}(t)e^{iH_0t/\hbar}. \tag{4.121}$$

We can now define the reduced density operator in this picture, as

$$\rho(t) = Tr_r\{\tilde{\rho}_{sr}(t)\}. \tag{4.122}$$

It can be shown that this new operator is related to the reduced density operator of equation (4.118) by the unitary transformation

$$\rho_s(t) = e^{-iH_st/\hbar}\rho(t)e^{iH_st/\hbar}. \tag{4.123}$$

Notice that, in contrast with similar transformations in equations (4.119) and (4.121), here we are using the system Hamiltonian H_s, instead of the total unperturbed Hamiltonian H_0. Let us assume that, at $t = 0$ the system and the reservoir are uncorrelated. This allows us to use the simple initial conditions

$$\tilde{\rho}_{sr}(0) = \rho(0)\rho_r. \tag{4.124}$$

Integrating equation (4.120), we then get

$$\tilde{\rho}_{sr}(t) = \tilde{\rho}_{sr}(0) - \frac{i}{\hbar}\int_0^t \left[\tilde{V}(t'), \tilde{\rho}_{sr}(t')\right]dt'. \tag{4.125}$$

This is an implicit equation for $\tilde{\rho}_{sr}(t)$. Using an iterative procedure, we can write this in a more explicit but complicated form, as

$$\tilde{\rho}_{sr}(t) = \tilde{\rho}_{sr}(0) - \frac{i}{\hbar}\int_0^t \left[\tilde{V}(t'), \tilde{\rho}_{sr}(0)\right]dt' - \frac{1}{\hbar^2}\int_0^t dt' \int_0^{t'} dt''\left[\tilde{V}(t'), \left[\tilde{V}(t''), \tilde{\rho}_{sr}(0)\right]\right] + \cdots \tag{4.126}$$

Or, in an equivalent but more formal way, using an infinite series, as

$$\tilde{\rho}_{sr}(t) = \tilde{\rho}_{sr}(0) + \sum_{n=1}^{\infty}\left(-\frac{i}{\hbar}\right)^n \int_0^t dt_1 \int_0^{t_1} dt_2 \cdots \int_0^{t_{n-1}} dt_n\left[\tilde{V}(t_1), \left[\tilde{V}(t_2), \ldots\left[\tilde{V}(t_n), \tilde{\rho}_{sr}(0)\right]\right]\right]. \tag{4.127}$$

Performing the trace over the reservoir variables, we can then obtain

$$\rho(t) = [1 + U(t)]\rho(0), \qquad U(t) = \sum_{n=1}^{\infty} U_n(t), \tag{4.128}$$

with the time propagators defined as

$$U_n(t) = \left(-\frac{i}{\hbar}\right)^n Tr_r\left\{\int_0^t dt_1 \int_0^{t_1} dt_2 \cdots \int_0^{t_{n-1}} dt_n[\tilde{V}(t_1), [\tilde{V}(t_2), \ldots[\tilde{V}(t_n), \tilde{\rho}_{sr}(0)]]]\right\}. \tag{4.129}$$

This formal solution allows us to write

$$\frac{\partial \rho}{\partial t} = \dot{U}(t)\rho(0) = \mathcal{L}(t)\rho(t), \qquad \mathcal{L}(t) = \frac{\dot{U}(t)}{1 + U(t)}. \tag{4.130}$$

This is a simple expression of the evolution equation, but difficult to solve explicitly. It is therefore useful to introduce some simplifying assumptions. First, we assume that the interaction between the system and the reservoir is weak, so that we can neglect terms of order higher than $\tilde{V}^2$ in $\mathcal{L}(t)$. This reduces equation (4.130) to

$$\frac{\partial \rho}{\partial t} = -\frac{i}{\hbar}Tr_r\left[\tilde{V}(t), \rho(0)\rho_r(0)\right] - \frac{1}{\hbar^2}\int_0^t Tr_r\left[\tilde{V}(t), \left[\tilde{V}(t'), \rho(t)\rho_r(0)\right]\right]dt'. \tag{4.131}$$

Another useful approximation corresponds to assume that the relation between the potential and the reservoir is also weak, such that we can use

$$Tr_r\left\{\tilde{V}(t), \rho_r\right\} = 0. \tag{4.132}$$

In this case, we have $U_1(t) = 0$, and equation (4.130) reduces to

$$\frac{\partial \rho}{\partial t} = -\frac{1}{\hbar^2}\int_0^t Tr_r\left[\tilde{V}(t), \left[\tilde{V}(t'), \rho(t)\rho_r(0)\right]\right]dt'. \tag{4.133}$$

This simplified form of master equation will be used to study the behaviour of an atom in a radiation reservoir, to be discussed next.

4.5.2 Atom in a reservoir

Let us assume a two-level atom, surrounded by a large reservoir, containing several harmonic oscillators with frequencies ω_k, such as for instance, the radiation field modes. We can then identify the system with the atom, using

$$H_s \equiv H_a = \frac{1}{2}\hbar\omega_a\sigma_z, \qquad \sigma_z = \sigma_{22} - \sigma_{11} = | \, 2\rangle\langle 2 \, | - | \, 1\rangle\langle 1 \, |, \tag{4.134}$$

and the interaction potential with the previously defined interaction Hamiltonian, as

$$\tilde{V}(t) = \hbar \sum_k g_k[b_k^\dagger \sigma_- e^{i\delta_k t} + \sigma_+ b_k e^{-i\delta_k t}], \tag{4.135}$$

with, $\delta_\mathbf{k} = (\omega_\mathbf{k} - \omega_a)$, $\sigma_+ = | \, 2\rangle\langle 1 \, |$ and $\sigma_- = | \, 1\rangle\langle 2 \, |$ This is written in the interaction picture, and uses the rotating wave approximation. We have also used the annihilation and creation mode operators $b_\mathbf{k}$ and $b_\mathbf{k}^\dagger$ identified with the different reservoir modes $\mathbf{k}$, and corresponding to the reservoir Hamiltonian

$$H_r = \sum_\mathbf{k} \hbar\omega_\mathbf{k} b_\mathbf{k}^\dagger b_\mathbf{k}. \tag{4.136}$$

If we insert the interaction potential (4.135) into the master equation (4.131), we obtain

$$\begin{aligned} \partial\rho/\partial t = &- i \sum_\mathbf{k} g_\mathbf{k}\langle b_\mathbf{k}^\dagger\rangle[\sigma_-, \rho(0)]e^{i\delta_\mathbf{k} t} \\ &- \int_0^t dt' \sum_{\mathbf{k}\mathbf{k}'} g_k g_{k'}\Big\{[\sigma_-\sigma_-\rho(t') + \rho(t')\sigma_-\sigma_- - 2\sigma_-\rho(t')\sigma_-]\langle b_\mathbf{k}^\dagger b_\mathbf{k}^\dagger\rangle e^{i\delta_\mathbf{k} t + i\delta_{\mathbf{k}'} t'} \\ &+ [\sigma_-\sigma_+\rho(t') - \sigma_+\rho(t')\sigma_-]\langle b_\mathbf{k}^\dagger b_{\mathbf{k}'}\rangle e^{i\delta_\mathbf{k} t - i\delta_{\mathbf{k}'} t'} \\ &+ [\sigma_+\sigma_-\rho(t') - \sigma_-\rho(t')\sigma_+]\langle b_\mathbf{k} b_{\mathbf{k}'}^\dagger\rangle e^{-i\delta_\mathbf{k} t + i\delta_{\mathbf{k}'} t'}\Big\} + h.\ c. \end{aligned} \tag{4.137}$$

Let us now assume that the reservoir is in thermal equilibrium, at a temperature T, such that the reservoir density matrix can be described as a *canonical ensemble*, and therefore use

$$\rho_r = \frac{\exp(-H_r/T)}{Tr.\{\exp(-H_r/T)\}}. \tag{4.138}$$

The temperature T is written in energy units. Replacing the reservoir Hamiltonian (4.136) in this expression, we obtain

$$\rho_r = \Pi_\mathbf{k}\left[1 - \exp\left(-\frac{\hbar\omega_\mathbf{k}}{T}\right)\right]\exp\left(-\frac{\hbar\omega_\mathbf{k}}{T}\right)b_\mathbf{k}^\dagger b_\mathbf{k}. \tag{4.139}$$

From here we can calculate the various average quantities appearing in equation (4.137), through the trace operation $\langle \cdot \rangle = Tr_r\{\rho_r(\cdot)\}$. We then get

$$\langle b_\mathbf{k}\rangle = \langle b_\mathbf{k}^\dagger\rangle = 0, \quad \langle b_\mathbf{k} b_{\mathbf{k}'}\rangle = \langle b_\mathbf{k}^\dagger b_{\mathbf{k}'}^\dagger\rangle = 0, \tag{4.140}$$

and

$$\langle b_{\mathbf{k}}^{\dagger} b_{\mathbf{k}'} \rangle = \langle n_{\mathbf{k}} \rangle \delta_{\mathbf{k}\mathbf{k}'}, \qquad \langle b_{\mathbf{k}} b_{\mathbf{k}'}^{\dagger} \rangle = (\langle n_{\mathbf{k}} \rangle + 1)\delta_{\mathbf{k}\mathbf{k}'}, \tag{4.141}$$

where $\langle n_{\mathbf{k}} \rangle$ is the mean occupation number, as given by the Bose–Einstein distribution

$$\langle n_{\mathbf{k}} \rangle = \frac{1}{\exp(\hbar\omega_{\mathbf{k}}/T) - 1}. \tag{4.142}$$

Replacing this in equation (4.137), we obtain

$$\begin{aligned} \frac{\partial \rho}{\partial t} = - \int_0^t dt' \sum_{\mathbf{k}} g_{\mathbf{k}}^2 \Big\{ &[\sigma_- \sigma_+ \rho(t') - \sigma_+ \rho(t') \sigma_-] \langle n_{\mathbf{k}} \rangle e^{i\delta_{\mathbf{k}}(t-t')} \\ &+ [\sigma_+ \sigma_- \rho(t') - \sigma_- \rho(t') \sigma_+] (\langle n_{\mathbf{k}} \rangle + 1) e^{-i\delta_{\mathbf{k}}(t-t')} \Big\} + h.\ c., \end{aligned} \tag{4.143}$$

where again $\delta_{\mathbf{k}} = \omega_{\mathbf{k}} - \omega_a$, We can now replace the sum over the modes **k** by an integral in the **k** space, as we did for the previous Weisskopf–Wigner model of spontaneous emission, and we get as a final expression, the master equation

$$\frac{\partial \rho}{\partial t} = -n_0 \frac{\Gamma}{2} [\sigma_- \sigma_+ \rho(t) - \sigma_- \sigma_+ \rho(t) \sigma_-] - (n_0 + 1) \frac{\Gamma}{2} [\sigma_+ \sigma_- \rho(t) - \sigma_- \rho(t) \sigma_+] + h.\ c. \tag{4.144}$$

Here, n_0 is the value of $\langle n_{\mathbf{k}} \rangle$ for the resonant modes satisfying $k = \omega_a/c$, and Γ is identical to the previous value for the spontaneous atomic decay, as given in equation (4.110). This equation can be written in terms of the density matrix elements as

$$\frac{\partial \rho_{22}}{\partial t} = -(n_0 + 1)\Gamma \rho_{22} + n_0 \rho_{11}, \tag{4.145}$$

where the contribution from spontaneous decay is present, as well as that of stimulated emission and absorption, and

$$\frac{\partial \rho_{21}}{\partial t} = -\left(n_0 + \frac{1}{2}\right)\Gamma \rho_{21}. \tag{4.146}$$

We should notice that $\rho_{21} = \rho_{12}^*$, and the conservation of probability $Tr.(\rho) = 1$, imply the additional relations

$$\rho_{22} + \rho_{11} = 1, \qquad \frac{\partial \rho_{22}}{\partial t} = -\frac{\partial \rho_{11}}{\partial t}. \tag{4.147}$$

When the reservoir temperature tends to zero, $T \to 0$, equation (4.142) shows that no excitations are present, $n_0 = 1$, and the above equations (4.145) and (4.146) reduce to

$$\frac{\partial \rho_{22}}{\partial t} = -\Gamma \rho_{22}, \qquad \frac{\partial \rho_{21}}{\partial t} = -\frac{\Gamma}{2} \rho_{21}. \tag{4.148}$$

This coincides with the results of our previous spontaneous decay model.

4.6 Resonant scattering

Scattering can be seen as a double process of atom–field interaction, by which an atom first absorbs one photon and then emits another photon, in a different direction and not necessarily at the same frequency. To illustrate this, we consider the process of resonant scattering, also called the resonant *Rayleigh scattering*, where the frequency of the incident radiation is equal or nearly equal to the atomic transition frequency, $\omega_{\mathbf{k}} \simeq \omega_a$.

We consider a two-level atom driven by an intense and monochromatic photon beam. This problem was first addressed by Mollow [20]. In our description we include spontaneous emission but, for simplicity, we describe the incident beam as a classical field. We describe the atom using the density operator ρ, which satisfies the master equation

$$\frac{d\hat{\rho}}{dt} = -\frac{i}{\hbar}[\hat{H}, \hat{\rho}] - \frac{1}{2}\left\{\hat{\Gamma}, \hat{\rho}\right\}, \tag{4.149}$$

where $\hat{\Gamma}$ represents the relaxation matrix associated with spontaneous decay. Here, we have used the anti-commutator $\{\hat{\Gamma}, \hat{\rho}\} = \hat{\Gamma}\hat{\rho} + \hat{\rho}\hat{\Gamma}$. In its simplest formulation, we can describe the interaction of the atom with the classical field using the Hamiltonian

$$\hat{H}_{\text{int}} = \frac{\hbar\Omega_R}{2}(\hat{\sigma}_+ + \hat{\sigma}_-). \tag{4.150}$$

Here we use the Rabi frequency in its classical limit, $\Omega_R \equiv \Omega_{<N>}$, assuming that $\langle\hat{N}\rangle \gg 1$. In this limit, and at exact resonance $\delta_{\mathbf{k}} = 0$, the dressed atom states reduce to

$$\left|\psi_\pm\right\rangle = \frac{1}{\sqrt{2}}(|\,1\rangle \pm |\,2\rangle), \tag{4.151}$$

with eigenvalues

$$\hat{H}_{\text{int}}\left|\psi_\pm\right\rangle = \pm\frac{\hbar\Omega_R}{2}\left|\psi_\pm\right\rangle. \tag{4.152}$$

Now, let us consider a finite lifetime due to spontaneous emission, $\gamma \neq 0$, such that $\langle 2\,|\,\hat{\Gamma}\,|\,2\rangle = \gamma$. In this new notation, and noting that

$$\{\hat{\Gamma}, \hat{\rho}\} = -\frac{\gamma}{2}(\hat{\sigma}_+\hat{\sigma}_-\hat{\rho} - 2\hat{\sigma}_-\hat{\rho}\hat{\sigma}_+ + \hat{\rho}\hat{\sigma}_+\hat{\sigma}_-), \tag{4.153}$$

the master equation (4.149) allows us to write for the density matrix elements

$$\frac{d}{dt}\rho_{++} = -\frac{\gamma}{2}\rho_{++} + \frac{\gamma}{4}, \tag{4.154}$$

and

$$\frac{d}{dt}\rho_{+-} = -\frac{\gamma}{4}\rho_{-+} - \frac{\gamma}{2} - \left(i\Omega_R + \frac{3\gamma}{4}\right)\rho_{+-}. \tag{4.155}$$

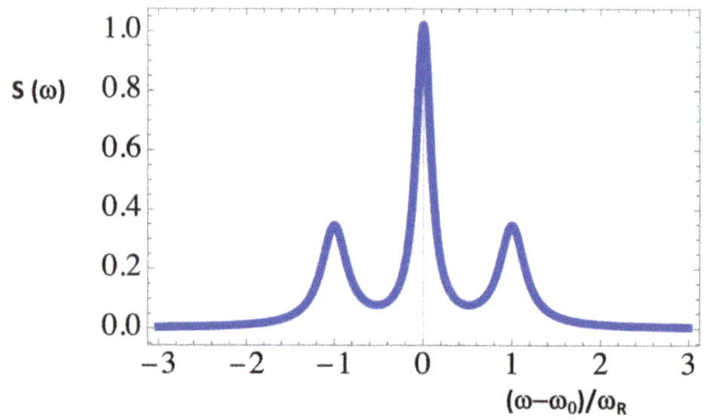

Figure 4.5. Mollow triplet: spectral representation of the spectrum $S(\omega)$ of scattered light by a two-level atom, for an intense photon beam with resonant frequency $\omega = \omega_a$.

This is complemented with $d\rho_{+-}/dt = d\rho_{-+}/dt$. In the strong field limit, we have $\Omega_R \gg \gamma$, and we can approximately write

$$\frac{d}{dt}\rho_{+-} \simeq -\left(i\Omega_R + \frac{3\gamma}{4}\right)\rho_{+-}, \tag{4.156}$$

with the solution

$$\rho_{+-}(t) = \rho_{+-}\exp\left(-i\Omega_R t - \frac{3\gamma}{4}t\right). \tag{4.157}$$

On the other hand, solving equation (4.153), we obtain

$$\rho_{++}(t) = \rho_{++}(0)\exp\left(-\frac{\gamma}{2}t\right) + \frac{1}{2}\left[1 - exp\left(-\frac{\gamma}{2}t\right)\right]. \tag{4.158}$$

This solution shows that the scattered light will contain three different frequency components. One is the *elastic scattering* component, with frequency equal to the incident frequency $\omega = \omega_{\mathbf{k}}$ and spectral width $\gamma/2$. The other two correspond to the *inelastic scattering* components, with sideband frequencies $\omega = \omega_{\mathbf{k}} \pm \Omega_R$ with spectral width $3\gamma/4$. These three components are known as the *Mollow triplet*. See figure 4.5 for an illustration of the resulting spectrum of scattered radiation, $S(\omega)$. It shows that the sidebands are similar to an amplitude modulation of the scattered signal, at a frequency given by the modified Rabi frequency Ω_R', and that the amplitude of the sideband pics, which represent the inelastic scattering, is smaller by a factor of 3 than the central pic of the triplet, representing the elastic scattering.

References

[1] Allen L and Eberly J 1975 *Optical Resonance and Two Level Atoms* (New York: Wiley)

[2] Cohen-Tannoudji C, Dupont-Roc J and Grynberg G 1988 *Processus d'Interaction entre Photons et Atomes* (Paris: Ed. du CNRS)
Cohen-Tannoudji C, Dupont-Roc J and Grynberg G 1992 *Atom-Photon Interactions* (New York: Wiley)

[3] Haken H 1986 *Light. Vol 1: Waves, photons, Atoms* (Amsterdam: North-Holland)

[4] Rabi I I 1936 On the process of space quantization *Phys. Rev.* **49** 329
Rabi I I 1937 Space quantization in a gyrating magnetic field *Phys. Rev.* **51** 652

[5] Bloch F and Siegert A J F 1940 Magnetic resonance for nonrotating fields *Phys. Rev.* **57** 522–7

[6] Jaynes E T and Cummings F W 1963 Comparison of quantum and semi-classical radiation theories with application to beam maser *Proc. IEEE* **51** 89

[7] Shore B W and Knight P L 1993 The Jaynes-Cummings model *J. Mod. Opt.* **40** 1195–238

[8] Braak D 2011 Integrability of the Rabi model *Phys. Rev. Lett.* **107** 100401

[9] Zhong H-H, Xie Q-T, Batchelor M T and Lee C H 2013 Analytical eigenstates of the quantum Rabi model *J. Phys. A: Math. Theor.* **41** 415302

[10] Cohen-Tannoudji C and Reynaud S 1977 Dressed-atom description of resonance fluorescence and absorption spectra of a multi-level atom in an intense laser beam *J. Phys.* **10** 345–63

[11] Dalibard J and Cohen-Tannoudji C 1985 Dressed-atom approach to atomic motion in laser light: the dipole force revisited *J. Opt. Soc. Am.* B **2** 1707–20

[12] Alzetta G, Gozzini A, Moi L and Orriols G 1976 An experimental method for the observation of R.F. transitions and laser beat resonances in oriented Na vapour *Il Nuovo Cimento* B **36** 5–20

[13] Harris S E, Field J E and Imamoglu A 1990 Nonlinear optical processes using electromagnetic induced transparency *Phys. Rev. Lett.* **64** 1107–10

[14] Fleischhauer M, Imamoglu A and Marangos J P 2005 Electromagnetic induced transparency: optics in coherent media *Rev. Mod. Phys.* **77** 633

[15] Einstein A 1917 Zur Quantentheorie der Strahlung *Phys.* Z **18** 121–8

[16] Weisskopf V and Wigner E 1930 Berechnung der natüauf grund der diracschen lichttheorie *Z Phys.* **92** 54–73

[17] Knight P L and Milonni P W 1976 Long-time deviations from exponential decay in atomic spontaneous emission theory *Phys. Lett.* A **56** 275–8

[18] Seke J and Herfort W 1989 Finite-time deviations from exponential decay in the Weisskopf-Wigner model of spontaneous emission *Lett. Math. Phys.* **18** 185–91

[19] Berman P R and Ford G W 2010 Unitarity, and the Weisskopf-Wigner approximation *Adv. Atom. Mol. Opt. Phys.* **59** 175–220

[20] Mollow B R 1969 Power spectrum of light scattered by two-level systems *Phys. Rev.* **188** 1969–75

Chapter 5

Boundary effects

In this chapter we describe the basic features associated with material boundaries, such as those of a bounded medium or an electromagnetic optical cavity. First of all, due the presence of boundaries, each mode in a cavity will eventually couple with the outside world, due to partial transmission, and with atoms inside the cavity or in the cavity walls. These processes will lead to photon absorption, and introduce a qualitative change in the mode itself, changing the properties of the atom–field interactions.

Here, we start analysing a single mode in a cavity and then its interaction with a single atom in the cavity. This can be described by a damped Jaynes–Cummings model, which shows that spontaneous decay of an excited state of the atom can be strongly enhanced by cavity effects. Enhancement is proportional to the quality factor Q of the cavity. This was first predicted by Purcell in 1946 [1]. In contrast, if no cavity modes are available at the atom transition frequency, spontaneous emission will be inhibited [2]. The study of the exchange of photons by one or a few atoms inside the cavity can lead photon–atom entangled states, and this is one of the main effects of the so-called *cavity QED* [3], with important applications in quantum information. Reference to QED (quantum electrodynamics) is somewhat misleading, because here we are only concerned with the non-relativistic limit of QED, where electron and positron quantum vacuum states are ignored. A more complete version of QED will be given in the third part of this book, where the possible influence of electron–positron pairs is considered.

A relevant boundary effect is associated with photon beam splitters. This simple optical devise is particularly relevant, in both classical and quantum optical circuits, as one of the basic elements of an interferometer. Here, we give a quantum description of a beam-splitter, where single photon effects are included. It is indeed possible to observe interference with single photons events, as demonstrated experimentally by Grangier *et al* in 1986 [4]. Another important property of photon

doi:10.1088/978-0-7503-2786-2ch5

beam splitters is that they are a four-ports device, capable of generating entangled photon states.

Another important effect associated with optical boundaries is the existence of quantum vacuum pressure. It was seen in chapter 3 that electromagnetic field quantisation reveals the existence of zero-point fluctuations. At first, they were considered a mere imperfection of the theory, with no physical significance, until Casimir demonstrated in 1948 that they could have important consequences. An attractive force between two parallel metallic plates in vacuum exists, due to zero-point fluctuations [5, 6]. This effect was generalised later by Lifshitz [7], who considered two parallel boundaries between different dielectric media. The Casimir effect was unequivocally demonstrated in the experiments performed by Lamoreaux in 1987 [8].

The case of temporal boundaries can also be treated in grounds similar to those of space boundaries. They give rise to the occurrence of a different kind of refraction, which can be called *time refraction*. This concept was first formulated in 2000 [9– 11], although some work on photon propagation in a time-varying medium add previously been published, both in classical [12, 13] and quantum regimes [14]. It is similar to the usual (space) refraction, in the sense that an incident optical beam can be divided into a transmitted and a reflected beam, with intensities determined by temporal Fresnel's laws [11]. Assuming two consecutive time refraction events we can produce a temporal beam-splitter, which is also a four-ports device. An obvious difference between space and time refraction is that time is a single variable, while space is described by three independent variables. In other words, time is represented by a line, while space is represented by a volume. As a result, the reflected beam due to time refraction cannot propagate backwards in time, it can only be reflected spatially, as it does in normal space refraction.

Another remarkable feature is that time refraction leads to the emission of photons from vacuum, in contrast with normal refraction. In this respect, time refraction is similar to the so-called *dynamic Casimir effect*. which occurs when one of the metallic parallel plates in the Casimir configuration are allowed to vibrate. This then leads to a kind of vacuum instability, by which pairs of photons are emitted at half the frequency of the vibrating boundary. This dynamic version of the Casimir effect was first considered by Moore in 1970 [20], for a cavity with an oscillating length, and then extended by others [21, 22] for the case of a single oscillating boundary. More recent work on this effect is reviewed in [23].

These two vacuum radiation processes, time refraction (where no spatial boundary is needed), and dynamic Casimir effect (which involves one vibrating boundary), have strong similarities, to be discussed in this chapter. Furthermore, a comparison with the *Unruh radiation*, could also be made. This hypothetical radiation process is produced by an accelerated particle moving in vacuum, and is named after the work of [24]. It was actually also predicted by others [22, 26], for that reason it is sometimes called the *Fulling–Davies–Unruh effect*. A more modern perspective is given in [27]. A comparative study of these three processes, time refraction, dynamic Casimir and Unruh effect could eventually be established [28].

The symmetric properties of space boundaries (which conserve photon energy, but not momentum) and temporal boundaries (which conserve photon momentum, but not energy) can be demonstrated if we consider the case of a moving boundary between two media. If the boundary moves with a velocity smaller than the speed of light c, it can be reduced by a Lorentz transformation to a space refraction. In this case, quantum theory will predict a static force at the boundary, which is the Casimir force. If instead, the boundary moves with a velocity faster than c, then a proper Lorentz transformation will reduce the problem to time refraction. In this case the Casimir force is replaced by the emission of radiation. These space-time symmetries will be discussed at the end of this chapter.

5.1 Cavity losses

Let us first consider the case of an imperfect optical cavity. We have previously defined the electromagnetic field modes in vacuum, or in an ideal cavity with no losses. But the existence of boundaries, such as optical mirrors, or metallic walls in the case of a microwave cavity, imply the existence of photon absorption by the walls, or through the walls, and therefore energy losses. The configuration of field modes inside the cavity can trivially be included in the previous plane wave theory, valid in unbounded space, by assuming a spatial dependence of the mode amplitudes. The existence of losses due to the absorption mechanisms is however less trivial, and will be discussed here.

For that purpose, we focus on a given cavity mode, with creation and annihilation operators simply described as $a^\dagger$ and a. We assume that the mode is coupled to a reservoir representing the loss mechanisms, which contains a large number of quantum modes $\mathbf{k}$, represented by the reservoir operators $b_\mathbf{k}^\dagger$ and $b_\mathbf{k}$. They could be the field modes outside the cavity, coupled with the inside mode by an imperfect cavity mirror, or the ensemble of oscillators representing the atoms of the metallic walls, if dissipation at the walls is considered.

This can be described by the previous model of a *system+reservoir*, where the master equation for the system was obtained, using an interaction potential of the form

$$\tilde{V}(t) = \hbar \sum_\mathbf{k} g_\mathbf{k}[b_k^\dagger a e^{i\delta_\mathbf{k} t} + a^\dagger b_\mathbf{k} e^{-i\delta_\mathbf{k} t}], \tag{5.1}$$

with the frequency detuning $\delta_\mathbf{k} = \omega_\mathbf{k} - \omega_0$, where ω_0 is the cavity mode frequency under consideration, and $\omega_\mathbf{k}$ are external mode frequencies. In the case of modes freely propagating outside the cavity, we can simply use $\omega_\mathbf{k} = kc$. We have seen that, in thermal equilibrium, the reduced density operator describing the system (in this case, the particular mode with frequency ω_0) can be written in analogy with equation (4.144), as

$$\frac{\partial \rho}{\partial t} = -\frac{\kappa}{2} n_0[aa^\dagger \rho + \rho a a^\dagger - 2a^\dagger \rho a] - (n_0 + 1)\frac{\kappa}{2}[a^\dagger a \rho + \rho a^\dagger a - 2a\rho a^\dagger], \tag{5.2}$$

where κ is the decay constant associated with the cavity losses, and n_0 is the mean photon number at frequency ω_0. The decay constant can be written in terms of the quality factor of the cavity Q, using the definition $\kappa = \omega_0/Q$. Of particular interest is the case where we can neglect the reservoir temperature. We know that, when the reservoir temperature tends to zero, we have $n_0 = 0$. The master equation for the cavity mode than reduces to

$$\frac{\partial \rho}{\partial t} = -\frac{\omega_0}{2Q}[a^\dagger a \rho + \rho a^\dagger a - 2a\rho a^\dagger]. \tag{5.3}$$

It is often more interesting to transform the master equation into a kinetic equation, which provides a different physical insight. For that purpose, we can use the P representation discussed in chapter 3, and expand the reduced density operator in a basis of coherent states α, as

$$\rho(t) = \int P(\alpha, t)| \alpha\rangle\langle\alpha | d\alpha. \tag{5.4}$$

Replacing this in equation (5.2), and using $P \equiv P(\alpha, t)$, we get

$$\begin{aligned}\int \frac{\partial P}{\partial t} | \alpha\rangle\langle\alpha | d\alpha = &-\frac{\kappa}{2}n_0 \int P[aa^\dagger | \alpha\rangle\langle\alpha | + | \alpha\rangle\langle\alpha | aa^\dagger - 2a^\dagger | \alpha\rangle\langle\alpha | a]d\alpha \\ &- (n_0 + 1)\frac{\kappa}{2} \int P[a^\dagger a | \alpha\rangle\langle\alpha | + | \alpha\rangle\langle\alpha | a^\dagger a - 2a | \alpha\rangle\langle\alpha | a^\dagger]d\alpha.\end{aligned} \tag{5.5}$$

Using the definition of a coherent state in terms of the vacuum state $| 0\rangle$, we have

$$| \alpha\rangle\langle\alpha | = e^{-|\alpha|^2} e^{\alpha a^\dagger} | 0\rangle\langle 0 | e^{\alpha^* a}. \tag{5.6}$$

This allows us to now write

$$| \alpha\rangle\langle\alpha | a = e^{-|\alpha|^2}\frac{\partial}{\partial\alpha^*}\{e^{\alpha a^\dagger} | 0\rangle\langle 0 | e^{\alpha^* a}\} = \left(\frac{\partial}{\partial\alpha^*} + \alpha\right) | \alpha\rangle\langle\alpha |. \tag{5.7}$$

Similarly, we have

$$a^\dagger | \alpha\rangle\langle\alpha | = \left(\frac{\partial}{\partial\alpha} + \alpha^*\right) | \alpha\rangle\langle\alpha |. \tag{5.8}$$

Noting also that

$$a | \alpha\rangle\langle\alpha | = \alpha | \alpha\rangle\langle\alpha |, \quad | \alpha\rangle\langle\alpha | a^\dagger = \alpha^* | \alpha\rangle\langle\alpha |, \tag{5.9}$$

we can write the sum under brackets of the first integral in equation (5.5) as

$$[aa^\dagger | \alpha\rangle\langle\alpha | + | \alpha\rangle\langle\alpha | aa^\dagger - 2a^\dagger | \alpha\rangle\langle\alpha | a] = -\left(\alpha\frac{\partial}{\partial\alpha} + \alpha^*\frac{\partial}{\partial\alpha^*} + 2\alpha\frac{\partial^2}{\partial\alpha\partial\alpha^*}\right) | \alpha\rangle\langle\alpha |. \tag{5.10}$$

Similarly, the second integral gives

$$[a^\dagger a | \alpha\rangle\langle\alpha | + | \alpha\rangle\langle\alpha | a^\dagger a - 2a | \alpha\rangle\langle\alpha | a^\dagger] = \left(\alpha\frac{\partial}{\partial\alpha} + \alpha^*\frac{\partial}{\partial\alpha^*}\right) | \alpha\rangle\langle\alpha |. \tag{5.11}$$

Replacing this in equation (5.5), we get

$$\int \frac{\partial P}{\partial t} \mid \alpha\rangle\langle\alpha \mid d\alpha = -\frac{\kappa}{2}n_0 \int P\left(\alpha\frac{\partial}{\partial\alpha} + \frac{\partial^2}{\partial\alpha\partial\alpha^*}\right) \mid \alpha\rangle\langle\alpha \mid d\alpha - (n_0 + 1)\frac{\kappa}{2} \int P\left(\alpha\frac{\partial}{\partial\alpha}\right) \mid \alpha\rangle\langle\alpha \mid d\alpha + h.\ c. \tag{5.12}$$

These new integrals can be integrated by parts, noting that $P(\alpha)$ must vanish at infinity, when we take the limit $|\alpha| \to \infty$. We then get

$$\int P\left(\alpha\frac{\partial}{\partial\alpha}\right) \mid \alpha\rangle\langle\alpha \mid d\alpha = -\int \left[\frac{\partial}{\partial\alpha}(\alpha P)\right] \mid \alpha\rangle\langle\alpha \mid d\alpha. \tag{5.13}$$

Similarly, we have

$$\int P\left(\frac{\partial^2}{\partial\alpha\partial\alpha^*} \mid \alpha\rangle\langle\alpha \mid\right) d\alpha = -\int \left[\frac{\partial^2}{\partial\alpha\partial\alpha^*}\alpha P\right] \mid \alpha\rangle\langle\alpha \mid d\alpha. \tag{5.14}$$

Replacing this in equation (5.12), and equating the coefficients of $\mid \alpha\rangle\langle\alpha \mid$ inside the integrals, we finally obtain

$$\frac{\partial P}{\partial t} = -\frac{\kappa}{2}\left[\frac{\partial}{\partial\alpha}(\alpha P) + c.\ c.\right] + \kappa n_0 \frac{\partial^2 P}{\partial\alpha\partial\alpha^*}. \tag{5.15}$$

This is a *Fokker–Planck equation*, describing the evolution of the quasi-probability $P(\alpha, t)$. A steady-state solution of this equation, $\partial/\partial t = 0$, can easily be found from

$$\frac{\partial}{\partial\alpha}\left(\alpha P + n_0\frac{\partial P}{\partial\alpha^*}\right) + c.\ c. = 0. \tag{5.16}$$

This implies that

$$\frac{1}{n_0 P}\frac{\partial n_0 P}{\partial\alpha^*} = -\frac{\alpha}{n_0}, \tag{5.17}$$

and therefore

$$P(\alpha) = \frac{1}{n_0}\exp\left(-\frac{|\alpha|^2}{n_0}\right). \tag{5.18}$$

This is a Gaussian solution, where the width is determined by the thermal number density n_0. Another interesting solution of equation (5.15) can be obtain for $\partial/\partial t \neq 0$, and takes the form

$$P(\alpha, t) = \frac{1}{\pi D(t)}\exp\left(-\frac{\mid \alpha - \alpha_0 \exp(-\kappa t/2) \mid^2}{D(t)}\right), \tag{5.19}$$

where the quantity

$$D(t) = n_0(1 - e^{-\kappa t}), \tag{5.20}$$

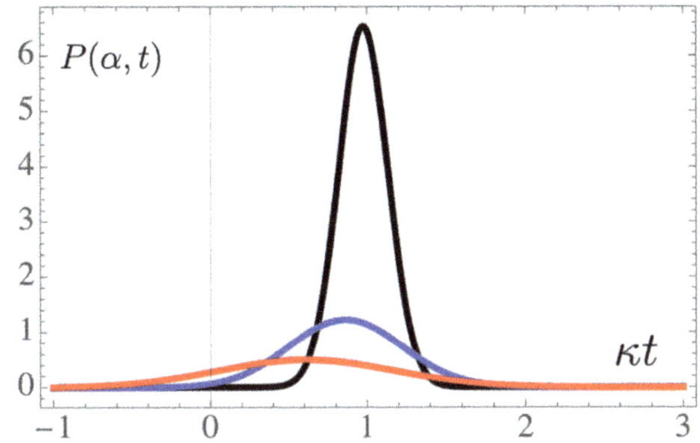

Figure 5.1. Temporal evolution of the quasi-probability $P(\alpha, t)$, as given by equation (5.19), with $\alpha_0 = 1$ and $n_0 = 1$, for increasing times $\kappa t = 0.05$ (in black), $\kappa t = 0.3$ (in blue), and $\kappa t = 1$ (in red).

defines the Gaussian width. This new solution tends to a Dirac delta function at the origin, when $t \to 0$ and $P(\alpha, 0) = \delta(\alpha - \alpha_0)$. It shows that, due to coupling with the thermal reservoir, an initial coherent state with parameter $\alpha = \alpha_0$ will disperse in time with increasing width $D(t)$. For times much longer than the mode decay time $1/\kappa$, the memory of the initial coherent parameter vanishes, and the dispersion function $D(t)$ tends to a constant. In this limit of very long times, $t \gg 1/\kappa$ the temporal solution (5.19) tends to the asymptotic state dictated by the steady-state solution (5.18). See figure 5.1 for an illustration. As a final remark, we should notice that the quantity $P(\alpha, t)$ is not a real probability, because it can take negative values, as stated previously in chapter 3.

5.2 Atom in a cavity

We now consider a single two-level atom in a lossy cavity, assuming that the atom is initially in the excited state $| e\rangle$ and the cavity is empty, with no photons. This situation is similar to that of the quantum Rabi, or Jaynes–Cummings model, with an initial photon number $n = 0$, where the cavity losses are included. In the language of photon-atom states, discussed in section 4.3, we can define here two quantum states: $| 1\rangle = | e\rangle\, | 0\rangle$, an excited atom with no photons, and $| 2\rangle = | g\rangle\, | 1\rangle$, an atom in the ground state and one photon. But to these two states we need to add a third one, $| 3\rangle = | g\rangle\, | 0\rangle$, which is the ground state of the atom-cavity system, corresponding to the possible absorption of the emitted photon due to cavity losses. This is illustrated in figure 5.2(a).

The energy difference between the first two states is minimal, of the order of the zero photon Rabi energy, $\hbar\Omega_0 = 2g\hbar$ in the absence of detuning, $\delta = \omega_0 - \omega_a = 0$. But the energy difference of these two excited estates with respect to the ground state $| 3\rangle$ is large, as dictated by $\hbar\omega_a$. The corresponding density operator is described by a 3×3 matrix. Using the reduced density approach, we can then write a master equation of the form

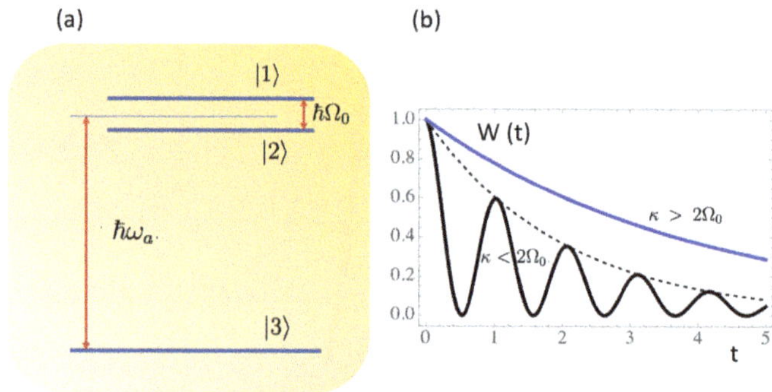

Figure 5.2. Atom in a lossy cavity: (a) energy levels; (b) damped Rabi oscillations, as represented by $W(t)$, for weak and strong damping.

$$\frac{\partial \rho}{\partial t} = -\frac{i}{\hbar}[H_{\text{int}}, \rho] - \frac{\kappa}{2}(a^{\dagger}a\rho + \rho a^{\dagger}a - 2a\rho a^{\dagger}). \tag{5.21}$$

For $\kappa \to 0$ (or equivalently, for a perfect cavity, with no losses, $Q \to \infty$) we would be reduced to the model of section 4.3. Defining the matrix elements, $\rho_{ij} = \langle i \mid \rho \mid j \rangle$, with $i, j = 1, 2, 3$, we get for the first diagonal elements

$$\frac{\partial \rho_{11}}{\partial t} = \frac{i}{2}\Omega_0(\rho_{12} - \rho_{21}), \quad \frac{\partial \rho_{22}}{\partial t} = -\frac{\partial \rho_{11}}{\partial t} - \kappa\rho_{22}. \tag{5.22}$$

This is completed with the equations for the non-diagonal elements

$$\frac{\partial \rho_{12}}{\partial t} = \frac{\partial \rho_{21}^{*}}{\partial t} = i\frac{\Omega_0}{2}W, \quad \frac{\partial \rho_{33}}{\partial t} = \kappa\rho_{22}, \tag{5.23}$$

where $W = \rho_{11} - \rho_{22}$ is the inversion of population. This reduces to the optical Bloch equations in the perfect cavity limit $\kappa = \omega_0/Q \to 0$. Introducing the difference $D = \rho_{12} - \rho_{21}$, these equations can be written in matrix form as

$$\frac{\partial}{\partial t}\begin{bmatrix} \rho_{11} \\ \rho_{22} \\ D \end{bmatrix} = \begin{bmatrix} 0 & 0 & i\Omega_0/2 \\ 0 & -\kappa & -i\Omega_0/2 \\ i\Omega_0 & -i\Omega_0 & -\kappa/2 \end{bmatrix}\begin{bmatrix} \rho_{11} \\ \rho_{22} \\ D \end{bmatrix}. \tag{5.24}$$

The eigenvalues of the 3×3 matrix in this equation are determined by

$$\left(\lambda + \frac{\kappa}{2}\right)(\lambda^2 + \kappa\lambda + \Omega_0^2) = 0, \tag{5.25}$$

with the solutions

$$\lambda_0 = -\frac{\kappa}{2}, \quad \lambda_{\pm} = -\frac{\kappa}{2}\left(1 \pm \sqrt{1 - \frac{4\Omega_0^2}{\kappa^2}}\right). \tag{5.26}$$

They show that, for weak decay rates, such that $\kappa < 2\Omega_0$, corresponding to a high quality factor $Q > \omega_a/2\Omega_0$, the matrix elements perform slightly damped Rabi oscillations, during which the spontaneously emitted photon oscillates several times between the atom and the cavity, before being finally lost. In contrast, for heavily damped cavities, $\kappa > 2\Omega_0$, or a poor quality factor $Q < \omega_a/2\Omega_0$, the Rabi oscillations will completely disappear, because the eigenvalues $\lambda_\pm$ will both be real. These two regimes are illustrated in figure 5.2(b). In the strongly damped regime, the largest decay constant is determined by λ_-, which is approximately given by

$$\lambda_- \simeq -\frac{\kappa}{2}\left[1 - \left(1 - \frac{2\Omega_0^2}{\kappa^2}\right)\right] = -\frac{\Omega_0^2}{\kappa}. \tag{5.27}$$

This can be identified with the decay rate of the atom inside the cavity, or $\lambda_- \equiv \Gamma_{cav}$. Using the explicit expressions for Ω_0 and κ, we can then write

$$\Gamma_{cav} \simeq \frac{4g^2}{\omega_a} Q. \tag{5.28}$$

Using the value for the decay rate Γ, associated with the spontaneous emission of an atom in free space, as described in section 4.5, we can then state that

$$\Gamma_{cav} = \left(\frac{6\pi^2 c^3}{\omega_a^3 V}\right) Q\Gamma. \tag{5.29}$$

For a mode with wavelength of the order of the dimensions of the cavity, say the length L, we have $\omega_a \simeq 2\pi c/L$. Noting that $V = L^3$, we can rewrite this expression as

$$\Gamma_{cav} \simeq \frac{3}{4} Q\Gamma. \tag{5.30}$$

We conclude that $\Gamma_{cav}/\Gamma \simeq Q \gg 1$, which shows that a strong enhancement of spontaneous emission takes place inside the cavity. In contrast, if no cavity mode with frequency nearly equal to atomic transition frequency ω_a is available, coupling between the atom and the photon modes will be absent and spontaneous emission is prevented.

5.3 Beam splitters

We now turn to a different boundary problem, and consider the quantum theory of photon beam splitters. Here, no atom is involved. The beam-splitter is a simple but important optical component, which has a crucial importance in quantum optics because it reveals some of the most fundamental quantum properties of light. We define *beam-splitter* as a passive apparatus, for instance a dielectric slab, which is able to divide an incoming photon beam into two parts, the reflected and the transmitted beams. For symmetry reasons, we can easily understand that the beam-splitter is a four-ports device, because it couples four different photon beams, with two incident beams arriving from opposite sides, and two transmitted beams

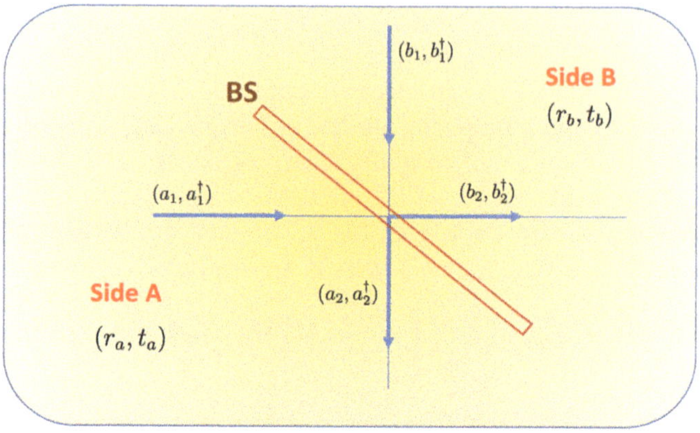

Figure 5.3. Photon beam-splitter as a four-ports device: incident beams a_1 and b_1, transmitted beams a_2 and b_2.

(see figure 5.3 for an illustration). We therefore need to describe the behaviour of four quantum modes, where two exist on side A of the device, characterised by the reflection and transmission coefficients (r_a, t_a), and the other two on the symmetric side B, with coefficients (r_b, t_b), such that $|r_a| = |r_b|$, and $|t_a| = |t_b|$. Furthermore, energy conservation implies that

$$|r_a|^2 + |t_a|^2 = |r_b|^2 + |t_b|^2 = 1. \tag{5.31}$$

The annihilation and creation operators associated with these four modes, called $(a_1, a_1^\dagger)$ and $(a_2, a_2^\dagger)$ for side A, and $(b_1, b_1^\dagger)$ and $(b_2, b_2^\dagger)$ for side B, will then satisfy the relations

$$a_2 = r_a a_1 + t_b b_1, \quad b_2 = t_a a_1 + r_b b_1, \tag{5.32}$$

and

$$a_2^\dagger = r_a^* a_1^\dagger + t_b^* b_1^\dagger, \quad b_2^\dagger = t_a^* a_1^\dagger + r_b^* b_1^\dagger. \tag{5.33}$$

These relations can be written in matrix form as

$$\begin{bmatrix} a_2 \\ b_2 \end{bmatrix} = M \begin{bmatrix} a_1 \\ b_1 \end{bmatrix}, \quad M = \begin{bmatrix} r_a & t_b \\ t_a & r_b \end{bmatrix}. \tag{5.34}$$

From here, it is useful to consider the commutation relations

$$[a_i, a_j^\dagger] = [b_i, b_j^\dagger] = \delta_{ij}, \quad [a_i, a_j] = [a_i^\dagger, a_j^\dagger] = [b_i, b_j] = [b_i^\dagger, b_j^\dagger] = 0. \tag{5.35}$$

They imply the existence of additional constraints on the beam-splitter coefficients

$$r_a^* t_b + r_b t_a^* = 0, \quad r_a^* t_a + r_b t_b^* = 0. \tag{5.36}$$

A typical assumption, corresponds to the case of a *half splitter*, such that $|r| = |t| = 1/\sqrt{2}$. Noting that reflection at the slab surface implies a phase shift of $\pi/2$, we can then transform equations (5.32) into

$$a_2 = \frac{1}{\sqrt{2}}(ia_1 + b_1), \quad b_2 = \frac{1}{\sqrt{2}}(a_1 + ib_1). \tag{5.37}$$

Let us now assume the case of a single incident photon, arriving from side A of the beam-slipper. This is described by the initial photon state

$$| \psi_{\text{in}}\rangle = | 1\rangle_a \, | 0\rangle_b = a_1^\dagger \, | 0\rangle_a \, | 0\rangle_b. \tag{5.38}$$

According to equations (5.33), the beam-splitter will then transform this state into a new state, described by

$$| \psi_{\text{out}}\rangle = (r_a a_2^\dagger + t_a b_2^\dagger) \, | 0\rangle_a \, | 0\rangle_b. \tag{5.39}$$

This expression is obtained using equations (5.33), and the additional constraints (5.36). In the particular case of a half splitter, defined by equation (5.37), this takes the form

$$| \psi_{\text{out}}\rangle = \frac{1}{\sqrt{2}}(ia_2^\dagger + b_2^\dagger) \, | 0\rangle_a \, | 0\rangle_b = \frac{1}{\sqrt{2}}\{i \, | 1\rangle_a \, | 0\rangle_b + | 0\rangle_a \, | 1\rangle_b\}. \tag{5.40}$$

This result shows that, if the initial state is a single photon arriving from side A, the half beam-splitter will create an *entangled state*, containing a reflected and a transmitted photon, emerging from both sides with equal probability. This final state cannot be reduced to a product of transmitted and reflected state vectors and, for that reason it is a photon entangled state.

Let us now consider a slightly more complicated example where, instead of a single incident photon in a well-defined Fock state, we have an incident coherent state α, described by

$$| \psi_{\text{in}}\rangle = | \alpha\rangle_a \, | 0\rangle_b = D_1(\alpha) | 0\rangle_a \, | 0\rangle_b, \tag{5.41}$$

where the displacement operator is defined by

$$D_1(\alpha) = \exp{(\alpha a_1^\dagger - \alpha^* a_1)}. \tag{5.42}$$

Noting that we can write $a_1^\dagger$ and a_1 in terms of reflected and transmitted operators, as used in equation (5.11), we rewrite this displacement operator as

$$D_1(\alpha) = \exp{\{\alpha(r_a a_2^\dagger + t_a b_2^\dagger) - \alpha^*(r_a^* a_2 + t_a^* b_2)\}}. \tag{5.43}$$

Using this expression and equation (5.41), we then get, for the output state

$$| \psi_{\text{out}}\rangle = \exp\{(r_a\alpha)a_2^\dagger - (r_a^*\alpha^*)a_2\} \exp\{(t_a\alpha)b_2^\dagger - (t_a^*\alpha^*)b_2\} \, | 0\rangle_a \, | 0\rangle_b. \tag{5.44}$$

This can easily be cast as the product of two coherent states

$$| \psi_{\text{out}}\rangle = | r_a\alpha\rangle_a \, | t_a\alpha\rangle_b. \tag{5.45}$$

This result shows that an incident coherent state is divided by the beam-splitter into two coherent states, where one is the reflected coherent state, with parameter $\alpha_r = r_a\alpha$, and the other is the transmitted coherent stated, with parameter $\alpha_t = t_a\alpha$.

This is in clear contrast with the above case of the incident photon in a Fock state, which was transformed by the beam-splitter into an entangled state. Here, there is no entanglement, and the coherent state behaves very much like a classical beam. For the half-slippers of equation (5.37), we are reduced to

$$| \psi_{out}\rangle = \left| \frac{i}{\sqrt{2}} \alpha \right\rangle_a \left| \frac{1}{\sqrt{2}} \alpha \right\rangle_b . \tag{5.46}$$

In order to explore further the quantum properties of single photon states, let us consider the case of two incident photons, coming from the two opposite sides of the beam-splitter. The initial state is now defined by

$$| \psi_{in}\rangle = | 1\rangle_a | 1\rangle_b = a_1^\dagger b_1^\dagger | 0\rangle_a | 0\rangle_b . \tag{5.47}$$

Using the transformations equations (5.33), and the constraints (5.36), we can now obtain for the output beam

$$| \psi_{out}\rangle = \{(r_a t_b) a_2^\dagger a_2^\dagger + (r_b t_a) b_2^\dagger b_2^\dagger + (r_a r_b + t_a t_b) a_2^\dagger b_2^\dagger\} | 0\rangle_a | 0\rangle_b . \tag{5.48}$$

This shows that, again, an entangled stated results from the symmetric interaction of these two photons with the beam-splitter. In the case of a half splitter, we are reduced to

$$| \psi_{out}\rangle = \frac{i}{2}(a_2^\dagger a_2^\dagger + b_2^\dagger b_2^\dagger) | 0\rangle_a | 0\rangle_b = \frac{i}{2}\{| 2\rangle_a | 0\rangle_b + | 0\rangle_a | 2\rangle_b\} . \tag{5.49}$$

We have an entangled state with equal probabilities for the reflected and transmitted photons, where the cases of simultaneous reflection or transmission of the two photons is forbidden. This anti-correlation is a clear signature of the quantum behaviour, as it contradicts the classical description. These concepts are important for the observation of quantum states of light at the single or few photons level, and can be generalised to the case of photon interferometry using two beam splitters and two mirrors. Photon anti-correlation in a beam-splitter, and single photon interference fringes using an interferometer, were first demonstrated in two experiments by Grangier *et al* [4], which strongly contributed to the validation of the quantum theory of light.

5.4 Time refraction

We now turn to temporal boundaries. Although ignored until recent years, they became relevant with the advent of ultra-fast optics. A temporal boundary can be conceived, in the same way as a spatial boundary between two optical media. Furthermore, temporal devices similar to the above spatial beam-splitter and to interferometers can be conceived. To understand this area of optics it is useful to consider first an isolated temporal transition. This corresponds to a sharp modification of the dispersion properties of an optical medium, which can be experimentally achieved in several ways, as exemplified later. This sudden change is simply described by a time-varying dielectric constant of the medium, of the form

$$\epsilon(t) = \epsilon_1 H(-t) + \epsilon_2 H(t) = \epsilon_1 + \Delta\epsilon H(t), \tag{5.50}$$

where ϵ_1 and ϵ_2 are constants, and $H(t)$ is the Heaviside function. This means that, at time $t = 0$, a sudden jump $\Delta\epsilon = \epsilon_2 - \epsilon_1$ of the dielectric constant occurs uniformly in the medium, assumed with no boundaries. This is perfectly symmetric to the case of the usual (space) boundary at $x = 0$ between two stationary media, as described by

$$\epsilon(x) = \epsilon_1 H(-x) + \epsilon_2 H(x). \tag{5.51}$$

For simplicity, we assume that the medium described by equation (5.50) is infinite and non-dispersive. For a given photon mode $\mathbf{k}$, and a given polarisation state, the electric field operator takes the form

$$\mathbf{E}_j(\mathbf{r}) = i\mathbf{e}_\mathbf{k}\sqrt{\frac{\hbar\omega_j}{2\epsilon_j}}\left\{a_j(\mathbf{k})e^{i\mathbf{k}\cdot\mathbf{r}} - a_j^\dagger(\mathbf{k})e^{-i\mathbf{k}\cdot\mathbf{r}}\right\}, \tag{5.52}$$

for $j = 1, 2$, where $\mathbf{e}_\mathbf{k}$ is the unit polarisation vector, $\omega_j = kc/n_j$ is the mode frequency and $n_j = \sqrt{\epsilon_j}$ the refractive index, before and after the transition at $t = 0$, and $a_j(\mathbf{k})$ and $a_j^\dagger(\mathbf{k})$ are the mode operators. We should also keep in mind that the corresponding displacement and magnetic field operators, are defined as

$$\mathbf{D}_j(\mathbf{r}) = \epsilon_j \mathbf{E}_j(\mathbf{r}), \quad \mathbf{B}_j(\mathbf{r}) = -\frac{i}{\omega_j}\nabla \times \mathbf{E}_j(\mathbf{r}). \tag{5.53}$$

These field operators should satisfy Maxwell's equations at all times, including the instant of transition $t = 0$. Because of the time derivatives appearing in these equations, the resulting continuity conditions for the field operators at $t = 0$, should be

$$\mathbf{D}_1(\mathbf{r}) = \mathbf{D}_2(\mathbf{r}), \quad \mathbf{B}_1(\mathbf{r}) = \mathbf{B}_2(\mathbf{r}). \tag{5.54}$$

This is valid for every position $\mathbf{r}$ in the medium and, for that reason, we should retain all the field terms evolving as $\exp(i\mathbf{k}\cdot\mathbf{r})$. From here we can derive the relations between operators valid before ($j = 1$) and after ($j = 2$) the transition time $t = 0$, as

$$a_1(\mathbf{k}) = Aa_2(\mathbf{k}) - Ba_2^\dagger(-\mathbf{k}), \quad a_1^\dagger(-\mathbf{k}) = Aa_2^\dagger(-\mathbf{k}) - Ba_2(\mathbf{k}), \tag{5.55}$$

where the coefficients of the transformation are defined by

$$A = \frac{1}{2}\frac{(1+\alpha)}{\sqrt{\alpha}}, \quad B = \frac{1}{2}\frac{(1-\alpha)}{\sqrt{\alpha}}, \tag{5.56}$$

with a new quantity α, defining the scale of the temporal transition

$$\alpha = \sqrt{\frac{\epsilon_1}{\epsilon_2}} = \frac{n_1}{n_2} = \frac{\omega_2}{\omega_1}. \tag{5.57}$$

We can immediately see that this expression implies a frequency shift associated with the temporal discontinuity. Such a frequency shift, from the initial ω_1 to the final value ω_2, results from the invariance of the photon momentum **k**, which is associated with space uniformity (see figure 5.4). This implies the existence of an invariant quantity, defined by $\omega_j n_j = kc$. Such an invariant can be seen as the manifestation of a *temporal Snell's law*, associated with the process of time refraction, and defined by the simple formula

$$\omega_1 n_1 = \omega_2 n_2. \tag{5.58}$$

It should be noted that the coefficients A and B satisfy the hyperbolic relations $A^2 - B^2 = 1$, and that the transformations (5.55) can be seen as two squeezing transformations of the form

$$a_1(\mathbf{k}) = a_2(\mathbf{k})\cosh(r) - a_2^\dagger(-\mathbf{k})\sinh(r), \quad a_1^\dagger(-\mathbf{k}) = a_2^\dagger(-\mathbf{k})\cosh(r) - a_2(\mathbf{k})\sinh(r), \tag{5.59}$$

where the squeezing parameter r is defined by

$$r = \mathrm{argcosh}(A) = \mathrm{argsinh}(B). \tag{5.60}$$

The corresponding two-mode squeezing operator can then be written as

$$S(r) = \exp\left\{r[a_1(\mathbf{k})a_1(-\mathbf{k}) - a_1^\dagger(\mathbf{k})a_1^\dagger(-\mathbf{k})]\right\}. \tag{5.61}$$

Let us describe the transformation of a Fock state $|\, n\rangle_\mathbf{k}$ under time refraction. Initially, we have the state

$$|\, \psi_{\text{in}}\rangle = |\, n\rangle_\mathbf{k}\, |\, 0\rangle_{-\mathbf{k}} = (a_1^\dagger(\mathbf{k}))^n\, |\, 0\rangle_\mathbf{k}\, |\, 0\rangle_{-\mathbf{k}}. \tag{5.62}$$

As a result of the above operator transformation, the final state which emerges from time refraction will take the form

$$|\, \psi_{\text{out}}\rangle = \sum_{s=0}^{\infty} b_s(n)|\, n+s\rangle_\mathbf{k}\, |\, s\rangle_{-\mathbf{k}}, \tag{5.63}$$

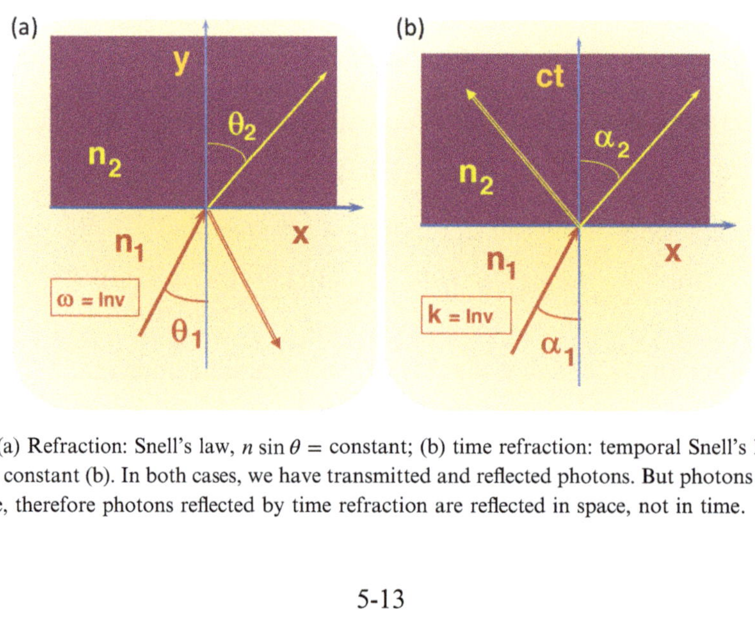

Figure 5.4. (a) Refraction: Snell's law, $n \sin\theta$ = constant; (b) time refraction: temporal Snell's law, equation (5.58), ωn = constant (b). In both cases, we have transmitted and reflected photons. But photons cannot travel back in time, therefore photons reflected by time refraction are reflected in space, not in time.

where the coefficients $b_s(n)$ are determined by

$$b_s(n) = \sqrt{1 - \left(\frac{B}{A}\right)^2} \sqrt{\frac{(n+s)!}{n!s!}} B^s A^{-(n+s)}. \tag{5.64}$$

This shows that the existence of a temporal boundary at $t = 0$, not only leads to a frequency shift determined by the conservation law (5.58), but is also associated with the creation of s photon pairs with probability $P_s(n) = |b_s(n)|^2$, propagating in opposite directions, as defined by the momentum states $\mathbf{k}$ and $-\mathbf{k}$. If initially, we have a pure photon vacuum defined by $|\psi_{in}\rangle = |0\rangle_{\mathbf{k}} |0\rangle_{-\mathbf{k}}$, after time refraction we will have two symmetric Fock states, corresponding to $|s\rangle_{\mathbf{k}} |s\rangle_{-\mathbf{k}}$. The probability for the emission of a single photon pair from vacuum will then be determined by

$$P_1(0) = |b_1(0)|^2 = \left(1 - \frac{B^2}{A^2}\right)\frac{B^2}{A^2}. \tag{5.65}$$

We can see from equations (5.56) that, for plausible values of α this probability is not negligible, as illustrated in figure 5.5. This process of pair creation therefore displays a strong similarity with the dynamic Casimir effect and Hawking radiation, and can eventually be seen in the laboratory. The first experimental evidence of time refraction was the observation made in 1989 by Yablonovich [15], of the self-blue shift of a laser pulse, due to a sudden ionisation of a gaseous medium. This can be explained by the temporal Snell's law, equation (5.58), where we use $n_1 \simeq 1$ for a diluted neutral gas, and $n_2 = \sqrt{1 - \omega_p^2/\omega^2}$, for the ionised medium, where the plasma frequency ω_p is proportional to the square root of the electron density. From then on, ionisation induced self-blue shifts of short laser pulses became a common observation in the laboratory. Some experiments, however, involve a more detailed analysis of the spatio-temporal changes of the medium [11]. In recent years, time refraction has been studied in many different media, such as epsilon-near-zero materials [16], or a Rydberg gas [17].

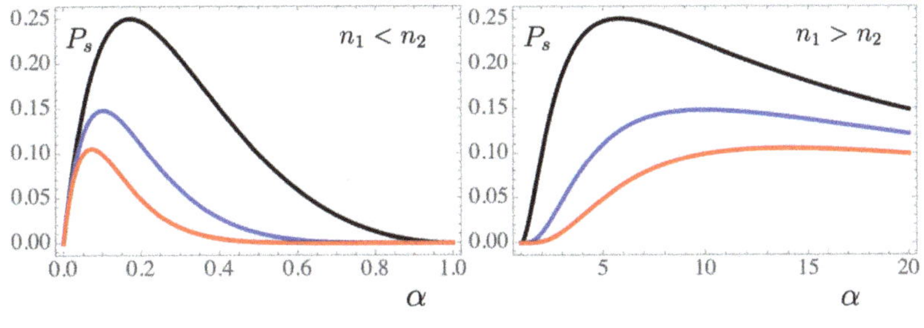

Figure 5.5. Photon pair creation from time refraction: probability P_s of the emission of s pairs of photons from vacuum, propagating in opposite directions, due to a sudden temporal change of the medium: $s = 1$ in black, $s = 2$ in blue, and $s = 3$ in red. The cases $n_1 < n_2$ and $n_1 > n_2$ are separately shown.

5.5 Temporal beam splitters

Let us now consider the temporal analogue of the photon beam splitters described in section 5.3. This can be built with two consecutive time-refraction events. In the first one, at $t = 0$, the refractive index of the medium jumps from its initial value n_0 to a new value n_1. Then, in the second event, at a time $t = \tau$, the reverse jump from n_1 to n_0 takes place. In order to discuss this process we need to consider the explicit temporal dependence of the field operator. Thus, in equation (5.52), we need to replace the operators $a_j(\mathbf{k})$ and $a_j^\dagger(\mathbf{k})$ by their temporal version, as defined by

$$a_j(\mathbf{k}, t) = a_j(\mathbf{k})\exp(-i\omega_j t), \quad a_j^\dagger(\mathbf{k}, t) = a_j^\dagger(\mathbf{k})\exp(i\omega_j t). \tag{5.66}$$

If we apply the field continuity relations, as stated in equations (5.53), but now associated with the two consecutive events, we then obtain the following relations, valid for $t = 0$,

$$a_0(\mathbf{k}, 0) = A_0 a_1(\mathbf{k}, 0) - B_0 a_1^\dagger(-\mathbf{k}, 0), \quad a_0^\dagger(-\mathbf{k}, 0) = A_0 a_1^\dagger(-\mathbf{k}, 0) - B_0 a_1(\mathbf{k}, 0), \tag{5.67}$$

and similar ones, valid for $t = \tau > 0$,

$$a_1(\mathbf{k}, \tau) = A_1 a_2(\mathbf{k}, \tau) - B_1 a_a^\dagger(-\mathbf{k}, \tau), \quad a_1^\dagger(-\mathbf{k}, \tau) = A_1 a_2^\dagger(-\mathbf{k}, \tau) - B_1 a_2(\mathbf{k}, \tau). \tag{5.68}$$

According to the previous model of time refraction, these continuity relations imply that

$$A_i = \frac{1}{2}\frac{(1 + \alpha_i)}{\sqrt{\alpha_i}}, \quad B_i = \frac{1}{2}\frac{(1 - \alpha_i)}{\sqrt{\alpha_i}}, \tag{5.69}$$

with $i = 0, 1$, where

$$\alpha_0 = \frac{n_0}{n_1} = \frac{\omega_1}{\omega_0} \equiv \alpha, \quad \alpha_1 = \frac{n_1}{n_0} = \frac{\omega_0}{\omega_1} \equiv \frac{1}{\alpha}. \tag{5.70}$$

From this we conclude that $A_1 = A_0$ and $B_1 = -B_0$, which allows us to relate the input and output operators, as

$$a_2(\mathbf{k}, \tau) = A(\tau)a_0(\mathbf{k}, 0) + B(\tau)a_0^\dagger(-\mathbf{k}, 0), \quad a_2^\dagger(-\mathbf{k}, \tau) = A^*(\tau)a_0^\dagger(-\mathbf{k}, 0) + B^*(\tau)a_0(\mathbf{k}, 0), \tag{5.71}$$

where the new coefficients of the transformation are oscillating functions of the width τ, of the temporal beam-splitter, and defined as

$$A(\tau) = \left\{\cos(\omega_1\tau) - \frac{i}{2\alpha}(1 + \alpha^2)\sin(\omega_1\tau)\right\}e^{i\omega_0\tau}, \quad B(\tau) = -\frac{i}{2\alpha}(1 - \alpha^2)\sin(\omega_1\tau)e^{i\omega_0\tau}. \tag{5.72}$$

The above relations describe the mode coupling between input and output photon states, resulting from the occurrence of a temporal beam-splitter, and reflect the influence of both the optical properties of the intermediate medium, n_1, and the temporal width of the perturbation, τ. This is very similar to what occurs with the (spatial) beam splitters considered in section 5.3, where the transmission and reflection coefficients depended, not only on the properties of the optical material, but also on the slab width.

Maximum coupling between the photon modes **k** and −**k**, propagating in opposite directions, is attained when $\cos(\omega_1\tau) = 0$, or equivalently, when $\tau = \nu\pi/\omega_1$, where ν is an integer. In this case, we can write

$$a_2(\mathbf{k}, \tau) = -\frac{i}{2\alpha}\{(1 + \alpha^2)a_0(\mathbf{k}, 0) + (1 - \alpha^2)a_0^\dagger(-\mathbf{k}, 0)\}e^{i\omega_0\tau}, \tag{5.73}$$

and

$$a_2^\dagger(-\mathbf{k}, \tau) = 2i\alpha\{(1 + \alpha^2)a_0^\dagger(-\mathbf{k}, 0) + (1 - \alpha^2)a_0(\mathbf{k}, 0)\}e^{-i\omega_0\tau}. \tag{5.74}$$

This corresponds to a *resonant mode coupling* between oppositely propagating photon modes, which maximises the influence of the temporal beam-splitter. In the anti-resonant case, defined by the condition $\sin(\omega_1\tau) = 0$, or $\tau = (\nu + 1/2)\pi/\omega_1$, we simply have $A = \exp(i\omega_0\tau)$ and $B = 0$, and mode coupling would be prevented. If we start with an initial vacuum state, defined by

$$|\,\psi_{\text{in}}\rangle = |\,0\rangle_{\mathbf{k}}\,|\,0\rangle_{-\mathbf{k}}, \tag{5.75}$$

we will observe, after the occurrence of a temporal beam-splitter, the emission of pair of photons, such that

$$|\,\psi_{\text{out}}\rangle = \sum_s C_s(\tau)|\,s\rangle_{\mathbf{k}}\,|\,s\rangle_{-\mathbf{k}}, \tag{5.76}$$

where the coefficients $C_s(\tau)$ are determined by

$$C_{s+1}(\tau) = \beta(\tau)C_s(\tau) = \beta^s(\tau)C_0(\tau), \quad C_0(\tau) = \sqrt{1 - \beta^2(\tau)}, \tag{5.77}$$

with $\beta(\tau) = B(\tau)/A(\tau)$. This means that the probability for observing the emission of a single photon pair is given by

$$P_1(\tau) = |C_1(\tau)|^2 = \beta^2(\tau)[1 - \beta^2(\tau)]. \tag{5.78}$$

This probability oscillates with τ, between a minimum of zero that corresponds to anti-resonance, and a maximum for resonant coupling (see figure 5.6). This maximum value is the same as that of a single time-refraction event, but here with a significant physical difference because the emitted photon pairs are now defined in the initial (or unperturbed) medium with refractive index n_0. This double temporal process can be extended to more complex scenarios, such as those of consecutive temporal beam splitters, separates by a distance eventually different from the beam-splitter width, $T > \tau$. A succession of identical temporal beam splitters would then be able to define a kind of *temporal cavity*. This concept of temporal cavity is actually remarkably similar to the dynamic Casimir effect, and it also has strong affinities with a *time-crystal*, to be discussed next.

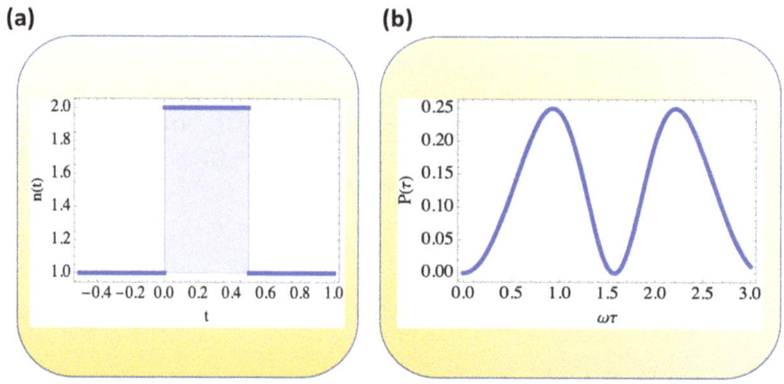

Figure 5.6. Temporal beam-splitter; (a) temporal variation of the refractive index of the medium, from $n_0 = 1$ to $n_1 = 2$ and back; (b) probability for photon-pair emission $P(\tau)$ by a temporal beam-splitter, as a function of the temporal width τ, for $\alpha = 0.5$.

5.6 Time-crystals

The concept of time-crystal was proposed by Wilczek in 2012 [18], as a periodic temporal structure that could be formed by spontaneous time symmetry breaking, in a way similar to the usual crystals, which form by spontaneous space-symmetry breaking. However, it was soon realised that a temporal structure cannot form spontaneously, because in contrast with the usual space-crystals it does not minimise the internal energy of the system, assumed initially at rest (for a review, see [19]). This means that the formation of such an oscillating structures needs the injection of energy into the system from the outside. This means that time-crystals can be formed at the expense of an outside driving source. In this sense, it is not much different from the concepts of temporal cavity or dynamic Casimir effect, because they both need a driving source.

In order to understand the formation of time-crystals, let us then extend the above time-refraction theory to the case of an oscillating medium where, as we have seen, the oscillations are imposed from the outside. In this case, the two field operators defined in equation (5.66), will be replaced by a continuum of operators such that [35]

$$a(\mathbf{k}, t) = a_{\mathbf{k}}(t)\exp(-i\phi t), \quad a^{\dagger}(-\mathbf{k}, t) = a_{-\mathbf{k}}^{\dagger}(t)\exp(i\phi(t), \tag{5.79}$$

where the field phase is now determined by a time integral

$$\phi(t) = \int^{t} \omega_{\mathbf{k}}(t')dt', \tag{5.80}$$

because the mode frequency $\omega_{\mathbf{k}}(t)$ is continuously changing in time, and not just changing in one or two sudden time steps. At every instant of time, it will satisfy the instantaneous mode dispersion relation of the medium. Therefore, we can use $\omega_{\mathbf{k}} = kc/\sqrt{\epsilon_{\mathbf{k}}}$, where $\epsilon_{\mathbf{k}}$ is the time-varying dielectric function. Similarly, the discrete continuity conditions for the operator amplitudes (5.67) are replaced by evolution equations, of the form

$$\frac{d}{dt}a_{\mathbf{k}} = \eta(t)a^{\dagger}_{-\mathbf{k}}, \quad \frac{d}{dt}a^{\dagger}_{-\mathbf{k}} = \eta^{*}(t)a_{\mathbf{k}}, \tag{5.81}$$

where the mode coupling coefficients are now determined by

$$\eta(t) = \frac{1}{2\omega_{\mathbf{k}}}\frac{d\omega_{\mathbf{k}}}{dt}\exp\left[2i\phi(t)\right]. \tag{5.82}$$

Let us assume a periodically driven dielectric function, of the form

$$\epsilon_{\mathbf{k}}(t) = \epsilon_0[1 + \delta\epsilon f(t)g(t)], \tag{5.83}$$

where ϵ_0 is the unperturbed dielectric function of the medium, $\delta\epsilon$ is the modulation amplitude, $f(t)$ a slowly varying form function, which defines the size of the time-crystal, and $g(t)$ a rapidly oscillating periodic function, with period τ_c. With generality, we can use the Fourier series

$$g(t) = \sum_{n=-\infty}^{\infty} g_n \exp(in\omega_c t), \tag{5.84}$$

where $\omega_c = 2\pi/\tau_c$ and the coefficients g_n are arbitrary. Assuming initial conditions such that $a_{\mathbf{k}}(0) = 0$ and $a_{\mathbf{k}}(0) = 0$, the solutions of equations (5.81) take the form

$$a_{\mathbf{k}}(t) = \cosh[|\bar{\eta}|], \quad a^{\dagger}_{-\mathbf{k}}(t) = -\sinh[|\bar{\eta}|t], \tag{5.85}$$

with $\bar{\eta} = \int^{t} \eta(t')dt'$. This shows that the signal associated with these modes can become very large for long times. Let us illustrate this result with a simple example. We assume a flat form function, with duration $T \gg \tau_c$, such that $f(t) = H(t) - H(t - T)$, and take an integral over a period $\Delta T \gg \tau_c$. We then get

$$\eta(t) = -\eta_0 \sin(\omega_c t) \sum_{n=-\infty}^{\infty} J_n(\kappa)\exp[-i(2\omega_0 + n\omega_c)t], \tag{5.86}$$

where ω_0 is the frequency of the photon mode in the absence of perturbations, and the parameters η_0 and κ are defined by

$$\eta_0 = \kappa\frac{\omega_c^2}{4\omega_0^2}, \quad \kappa = \delta\epsilon\frac{\omega_0}{\omega_c}. \tag{5.87}$$

This shows that, when we integrate this expression over a time interval much larger than τ_c, the only nonzero contributions will come from the modes with unperturbed frequency

$$\omega_0 = (n \pm 1)\frac{\omega_c}{2}. \tag{5.88}$$

This provides a kind of *temporal Bragg diffraction* formula, which characterises the occurrence of strong emission of radiation in the forward and backwards direction, as determined by equations (5.85). This is similar to diffraction from a crystalline structure, but now due to a temporal structure, not to a real crystal. The intense picks of radiation will occur when the frequency ratio ω_0/ω_c is equal to the values $(n \pm 1)$, for n integer or zero. The number of photons emitted by this temporal structure over a time T is then given by

$$N_{\mathbf{k}}(T) = \langle a_{\mathbf{k}}^{\dagger}(T) a_{\mathbf{k}}(T) \rangle = \sinh^2(|\bar{\eta}(T)|). \tag{5.89}$$

For a simple perturbation of the medium, as described by first harmonic in equation (5.84), and corresponds to $g_n = \delta_{1n}$, we have

$$|\bar{\eta}(T)| = \delta\epsilon \frac{T}{(n \pm 1)^2} \frac{\omega_0}{\omega_c} |J_n(\kappa)|. \tag{5.90}$$

This value decreases for higher order scattering as n^2. Equation (5.89) shows that, for short times, photon emission grows linearly, as $N_{\mathbf{k}}(T) \propto T^2$ and, for longer times, it grows exponentially with the size of the crystal, $N_{\mathbf{k}}(T) \propto \exp T$. However, this is only valid if photon losses are ignored. In the presence of photon losses, this exponential growth will eventually saturate [29].

5.7 Casimir force

We now return to static boundaries, and consider the forces due to the zero-point fluctuations of quantum vacuum. For that purpose, we assume two square parallel plates of a perfect metal, located in vacuum at a distance a in the Oz-direction (see figure 5.7). These conducting plates define an inner volume $V = aL^2$, where L is the size of the plates. In the absence of plates, the zero-point electromagnetic energy E_0 associated with this volume would be

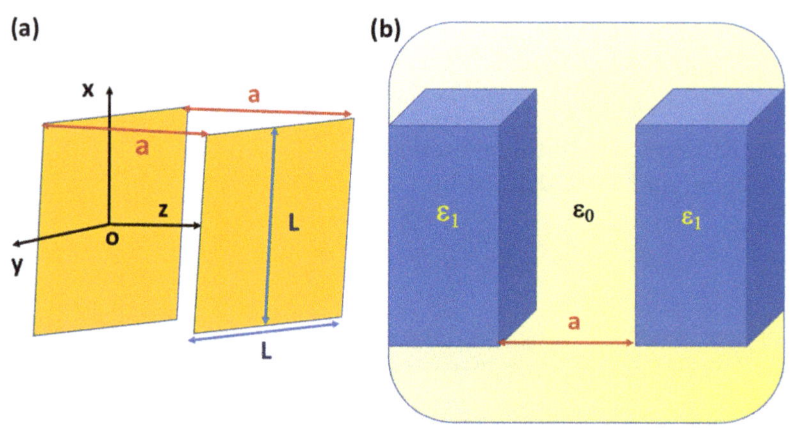

Figure 5.7. Casimir force between two boundaries: (a) two parallel plates in vacuum—original Casimir configuration; (b) two parallel boundaries between two dielectric media—Lifshitz configuration.

$$E_0 = V \int E(\mathbf{k}) \frac{d\mathbf{k}}{(2\pi)^3} = \int L^2 \frac{d\mathbf{k}_\parallel}{(2\pi)^2} \int a \frac{dk_z}{2\pi} E(\mathbf{k}). \tag{5.91}$$

The spectral energy density in the absence of photons is determined by

$$E(\mathbf{k}) = s\frac{\hbar}{2}\omega(\mathbf{k}) = \hbar\omega(\mathbf{k}), \tag{5.92}$$

where $\omega(\mathbf{k})$ is the mode frequency and $s = 2$ is the number of independent polarisation states. However, the presence of plates changes the energy content of this volume, because the number of allowed photon modes is reduced due to the new boundary conditions. The metallic boundaries imply that only discrete modes with wavenumber $k_z = \nu\pi/a$ along the z-direction will be allowed, for ν integer. This means that the integral over the spectrum is replaced by a sum, as indicated

$$\int \omega(\mathbf{k}) dk_z \to \frac{\pi}{a} \sum_{\nu=-\infty}^{\infty} \omega(\mathbf{k}_\parallel, \nu), \tag{5.93}$$

where we can now define the mode frequency as

$$\omega(\mathbf{k}_\parallel, \nu) = c|\mathbf{k}| = c\sqrt{k_\parallel^2 + \nu^2 \frac{\pi^2}{a^2}}. \tag{5.94}$$

This allows us to define the energy of the zero-point fluctuations in the inner volume defined by the plates as

$$E = s\frac{\hbar}{4} \int L^2 \frac{d\mathbf{k}_\parallel}{(2\pi)^2} \sum_{\nu=-\infty}^{\infty} \omega(\mathbf{k}_\parallel, \nu). \tag{5.95}$$

This can also be written in the form

$$E = s\frac{\hbar}{4} \int L^2 \frac{d\mathbf{k}_\parallel}{(2\pi)^2} \left\{ \omega(\mathbf{k}_\parallel, 0) + 2\sum_{\nu=1}^{\infty} \omega(\mathbf{k}_\parallel, \nu) \right\}. \tag{5.96}$$

Now, using $d\mathbf{k}_\parallel = 2\pi k dk$, where $k = |\mathbf{k}_\parallel|$ is independent of the actual direction of the parallel wavevector $\mathbf{k}_\parallel$, we can write this expression as

$$E = s\frac{\hbar}{8\pi} L^2 \int_0^\infty k dk \left\{ \omega(k, 0) + 2\sum_{\nu=1}^{\infty} \omega(k, \nu) \right\}. \tag{5.97}$$

A similar expression can be written for the unperturbed zero-point energy E_0 defined above. Using a real variable $\nu = ak_z/\pi$, we get from equation (5.91), the following

$$E_0 = s\frac{\hbar}{4\pi} L^2 \int_0^\infty k dk \int_0^\infty d\nu \omega(k, \nu). \tag{5.98}$$

It is now useful to define the auxiliary function

$$f(\nu) = \int_0^\infty \omega(\xi, \nu) d\xi, \quad \xi = \left(\frac{ak}{\pi}\right)^2. \tag{5.99}$$

With the help of this new function, we can rewrite equation (5.97) in a more compact form, as

$$E = s\frac{\hbar\pi}{8a^2}L^2\left\{\frac{1}{2}f(0) + \sum_{\nu=1}^{\infty} f(\nu)\right\}. \tag{5.100}$$

Similarly, for the unperturbed energy of equation (5.98), we get

$$E_0 = s\frac{\hbar\pi}{8a^2}L^2\int_0^{\infty} f(\nu)d\nu. \tag{5.101}$$

This allows us to calculate the energy difference per unit area

$$\epsilon = \frac{1}{L^2}(E - E_0), \tag{5.102}$$

in the following way

$$\epsilon = s\frac{\hbar\pi}{8a^2}\left\{\frac{1}{2}f(0) + \sum_{\nu=1}^{\infty} f(\nu) - \int_0^{\infty} f(\nu)d\nu\right\}. \tag{5.103}$$

We notice that the limit $\nu \to \infty$ corresponds to modes with an infinite frequency and zero wavelength, and has no physical meaning. We should therefore assume that there is an upper-limit m for the variable ν in the sum and integral of the above expression. It means that, we should use the additional assumption that

$$f(\nu \geqslant m) = 0. \tag{5.104}$$

In order to find an explicit expression for the energy difference ϵ we use the Euler–MacLaurin formula [34], which establishes a relation between the sum and the integral, as follows

$$\begin{aligned}\int_0^{\infty} f(\nu)d\nu &= \frac{1}{2}f(0) + f(1) + f(2) + \cdots \\ &\frac{1}{2}f(m) - b_2[f'(m) - f'(0)] - b_4[f'''(m) - f'''(0)],\end{aligned} \tag{5.105}$$

where the coefficients b_n are related to the *Bernoulli numbers* B_n, as defined by

$$b_n = \frac{B_n}{n!}, \quad B_n = \frac{d^n}{dx^n}\left(\frac{x}{e^x - 1}\right)\bigg|_{x=0}. \tag{5.106}$$

For the first few numbers, we have

$$B_0 = 1, \quad B_2 = \frac{1}{6}, \quad B_4 = -\frac{1}{30}. \tag{5.107}$$

Using this in equation (5.103), we obtain

$$\epsilon = s\frac{\hbar\pi}{8a^2}\{b_2[f'(m) - f'(0)] + b_4[f'''(m) - f'''(0)]\}. \tag{5.108}$$

Writing the function $\omega(k, \nu)$ in terms of the variable ξ introduced in equation (5.99), we get

$$\omega(\xi, \nu) = c|\mathbf{k}| = c\frac{\pi}{a}\sqrt{\xi + \nu^2}. \tag{5.109}$$

Using this in the definition of $f(\nu)$, we obtain

$$f(\nu) = c\frac{\pi}{a}\int_{\nu^2}^{\infty} \sqrt{\xi'}\, d\xi', \tag{5.110}$$

where $\xi' = \xi + \nu^2$. From here, we get

$$f'(0) = 0, f'''(0) = -\frac{4\pi c}{a}. \tag{5.111}$$

Assuming now a smooth upper cutoff for the function $f(\nu)$, such that $f'(m) = 0$ and $f'''(m) = 0$, equation (5.108) will reduce to

$$\epsilon = c\frac{s\hbar\pi^2}{2a^3}b_4. \tag{5.112}$$

This result shows that the energy per unit area will depend on the distance a between the two parallel plates. This means that a pressure F_C on the plates can be determined by a variation of ϵ with respect to the distance, as

$$F_C = -\left(\frac{\delta\epsilon}{\delta a}\right) = 3\frac{c}{a^4}\frac{s\hbar}{2}\pi^2 b_4. \tag{5.113}$$

Using $s = 2$ for the independent polarisation states, and the value of b_4 given by equations (5.106)–(5.107), we finally get

$$F_C = -\frac{\pi^2}{240}\frac{\hbar c}{a^4}. \tag{5.114}$$

This result was first derived by Casimir in 1948, and can be called the *Casimir pressure*, as resulting from the zero-point fluctuations of the electromagnetic vacuum. Although very small, of the order of 1.3 milli-Pascal for plates at one micron distance, it was experimentally observed in 1958. This theoretical result was generalised by Lifshitz in 1956, who considered the case of two parallel dielectric interfaces, as illustrated in figure 5.7. This new configuration is very useful when compared with the case of temporal boundaries, because a temporal beam-splitter can be seen as the exact temporal analogue of the Lifshitz configuration. The celebrated Lifshitz formula [30] is

$$\begin{aligned} F_L = -\frac{c\hbar}{32\pi^2 a^4}\int_0^{\infty} dx \int_0^{\infty} dp \Bigg\{ & \left[\frac{(s_1 + p)(s_2 + p)}{(s_1 - p)(s_2 - p)}e^x - 1\right]^{-1} \\ & + \left[\frac{(s_1 + \epsilon_1 p)(s_2 + \epsilon_2 p)}{(s_1 - \epsilon_1 p)(s_2 - \epsilon_2 p)}e^x - 1\right]^{-1}\Bigg\}\frac{x^3}{p^2}, \end{aligned} \tag{5.115}$$

with the auxiliary quantities $s_i = \sqrt{\epsilon_i - 1 + p^2}$, for $i = 1, 2$, and it was assumed that $\epsilon_0 = 1$. If we take the limit of ϵ_1 and ϵ_2 tending to infinity, which corresponds to the case of metallic plates, we get

$$F_L = -\frac{c\hbar}{16\pi^2 a^4}\int_0^\infty dx \int_0^\infty dp \frac{x^3}{p^2(e^x - 1)} = -\frac{\pi^2}{240}\frac{\hbar c}{a^3} \equiv F_C, \tag{5.116}$$

which coincides with the Casimir result.

5.8 Space-time symmetries

In this section, we give an integrated view of space-time boundaries and the corresponding symmetries, where all the above quantum vacuum processes, including time refraction, time-crystals and the Casimir force are included. For that purpose, we consider the changes in the electromagnetic energy spectrum produced by an arbitrary space-time variation of an optical medium.

In classical theory this is determined by temporal Fresnel formulae and generalised Snell's law. In quantum theory we use instantaneous Bogoliubov transformations, valid for arbitrary time dependent media, and calculate the number of photon pairs emitted from vacuum. Our approach is valid for infinite and unbounded media, where the inclusion of moving boundaries, such as those associated with optical cavities, can establish a precise link with both the Casimir and the dynamic Casimir effects. Finally, we consider vacuum radiation due to the existence of super-luminal boundaries, which can be reduced to time refraction, using appropriate Lorentz transformations [31]. For simplicity, we use ray optics, but a similar description based on wave optics can be established.

5.8.1 Ray optics

We consider an arbitrary optical medium, which can vary both in space and time. Wave propagation in this medium is described by classical fields evolving as

$$\mathbf{E}(\mathbf{r}, t) = \mathbf{A}(\mathbf{r}, t)\exp[i\phi(\mathbf{r}, t)], \tag{5.117}$$

where $\phi(\mathbf{r}, t)$ is the phase, and $\mathbf{A}(\mathbf{r}, t)$ the amplitude (not to be confused with the vector potential). Locally, at a given position $\mathbf{r}$ and time t, we can define the wavevector $\mathbf{k}$ and frequency ω, using

$$\mathbf{k} = \nabla\phi, \quad \omega = -\frac{\partial}{\partial t}\phi. \tag{5.118}$$

The field described by equation (5.117) can be considered a *wave*, if we assume that

$$\left|\frac{1}{A}\nabla A\right| \ll k, \quad \left|\frac{1}{A}\frac{\partial}{\partial t}A\right| \ll \omega. \tag{5.119}$$

This means that the field amplitude $\mathbf{A}(\mathbf{r}, t)$ is varying on a scale much larger that the wavelength $\lambda = 2\pi/|k|$, and on a time scale much longer than the period $\tau = 2\pi/\omega$.

The optical properties of the medium are determined by the refractive index, which is locally defined by

$$n(\mathbf{r}, t) \equiv n = \frac{kc}{\omega}. \tag{5.120}$$

Using equation (5.118), we can derive the *eikonal equation*, of the form

$$|\nabla\phi|^2 = \frac{n^2}{c^2}\left(\frac{\partial\phi}{\partial t}\right)^2, \tag{5.121}$$

which is the basic equation of *geometric optics* [32, 33]. From equation (5.118), we also get

$$\frac{\partial\mathbf{k}}{\partial t} = -\nabla\omega = -\left[\frac{\partial\omega}{\partial\mathbf{r}} + \frac{\partial\mathbf{k}}{\partial\mathbf{r}} \cdot \frac{\partial\omega}{\partial\mathbf{k}}\right]. \tag{5.122}$$

Here, we assume that the frequency $\omega \equiv \omega(\mathbf{r}, \mathbf{k}, t)$ is an implicit function of the position $\mathbf{r}$ and wavevector $\mathbf{k}$, as implied by the relation (5.120). This can also be stated as

$$\left(\frac{\partial}{\partial t} + \mathbf{v}_g \cdot \frac{\partial}{\partial\mathbf{r}}\right)\mathbf{k} = -\frac{\partial\omega}{\partial\mathbf{r}}, \tag{5.123}$$

where

$$\mathbf{v}_g = \frac{\partial\omega}{\partial\mathbf{k}}, \tag{5.124}$$

is the *group velocity*. Identifying $\mathbf{v}_g$ with the velocity of an optical ray, which can be seen as the classical analogue of a photon, we are then able to state the *ray equations*, as

$$\frac{d\mathbf{r}}{dt} = \frac{\partial\omega}{\partial\mathbf{k}}, \quad \frac{d\mathbf{k}}{dt} = -\frac{\partial\omega}{\partial\mathbf{r}}. \tag{5.125}$$

These are *canonical equations* of the ray, where the Hamiltonian is the frequency $\omega \equiv \omega(\mathbf{r}, \mathbf{k}, t)$, as determined by the medium dispersion. In general conditions, the ray frequency will be able to vary along time, according to

$$\frac{d\omega}{dt} = \frac{\partial\omega}{\partial t}. \tag{5.126}$$

When ω depends explicitly on time, we have *time symmetry breaking*, and the ray energy (or frequency) is not a constant of motion. The ray equations (5.125) can be written in a different and alternative form, using the *Poisson brackets*, which are the classical analogue of the quantum commutators. Given two different functions, u and v, depending on $\mathbf{r}$ and $\mathbf{k}$, we can define their Poisson bracket as

$$[u, v] = \frac{\partial u}{\partial\mathbf{r}} \cdot \frac{\partial v}{\partial\mathbf{k}} - \frac{\partial u}{\partial\mathbf{k}} \cdot \frac{\partial v}{\partial\mathbf{r}}. \tag{5.127}$$

This can easily be generalised to vector functions. In particular, we get

$$[\mathbf{r}, \mathbf{r}] = [\mathbf{k}, \mathbf{k}] = 0, \quad [\mathbf{r}, \mathbf{k}] = [\mathbf{k}, \mathbf{r}] = 1. \tag{5.128}$$

Using this, we can then rewrite the ray equations (5.125) as

$$\frac{d\mathbf{r}}{dt} = [\mathbf{r}, \omega], \quad \frac{d\mathbf{k}}{dt} = [\mathbf{k}, \omega]. \tag{5.129}$$

Notice that the total time derivative of an function $u(\mathbf{r}, \mathbf{k}, t)$ is given by

$$\frac{du}{dt} = \frac{\partial u}{\partial t} + \frac{\partial u}{\partial \mathbf{r}} \cdot \frac{d\mathbf{r}}{dt} + \frac{\partial u}{\partial \mathbf{k}} \cdot \frac{d\mathbf{k}}{dt}. \tag{5.130}$$

Using equations (5.129), we readily get

$$\frac{du}{dt} = \frac{\partial u}{\partial t} + [u, \omega], \tag{5.131}$$

equations (5.125) and (5.126) are, obviously, particular cases of this expression. Any constant of motion, or invariant $I(\mathbf{r}, \mathbf{k})$, associated with the ray trajectories, should necessarily satisfy two conditions: (i) not depend explicitly on time t, and (ii) have its Poisson bracket with the frequency ω equal to zero:

$$\frac{\partial I}{\partial t} = 0, \quad [I, \omega] = 0. \tag{5.132}$$

A particularly important case of time-symmetry breaking (when $\partial\omega/\partial t \neq 0$) occurs in the presence of a perturbation moving through the optical medium with constant velocity $\mathbf{v}$. Notice that this moving perturbation does not necessarily imply motion of the particles (atoms, molecules or charged particles), in the same way as motion of a fire front in a forest coexists with immobile trees. In this case, we can use $\omega = \omega(\mathbf{r} - \mathbf{v}t, \mathbf{k})$. Defining a canonical transformation from the initial variables $(\mathbf{r}, \mathbf{k})$ to a new pair of variables $(\mathbf{r}', \mathbf{k}')$, such that

$$\mathbf{r}' = \mathbf{r} - \mathbf{v}t, \quad \mathbf{k}' = \mathbf{k}, \tag{5.133}$$

the new ray equations (5.125) are transformed into

$$\frac{d\mathbf{r}'}{dt} = \frac{\partial \omega'}{\partial \mathbf{k}}, \quad \frac{d\mathbf{k}'}{dt} = -\frac{\partial \omega'}{\partial \mathbf{r}'}, \tag{5.134}$$

where $\omega' = \omega'(\mathbf{r}', \mathbf{k}', t)$ is the new Hamiltonian. This transformation can be performed using a *generation function*, of the form

$$F(\mathbf{r}, \mathbf{k}', t) = (\mathbf{r} - \mathbf{v}t) \cdot \mathbf{k}', \tag{5.135}$$

from where we get

$$\mathbf{r}' = \frac{\partial F}{\partial \mathbf{k}'} = \mathbf{r} - \mathbf{v}t, \quad \mathbf{k} = \frac{\partial F}{\partial \mathbf{r}} = \mathbf{k}'. \tag{5.136}$$

As for the new Hamiltonian, ω', appearing in equations (5.134), it is determined by the expression

$$\omega' = \omega + \frac{\partial F}{\partial t} = \omega(\mathbf{r}', \mathbf{k}') - \mathbf{v} \cdot \mathbf{k}'. \tag{5.137}$$

The canonical equations (5.129) are now replaced by

$$\frac{d\mathbf{r}'}{dt} = [\mathbf{r}', \omega'], \quad \frac{d\mathbf{k}'}{dt} = [\mathbf{k}', \omega'], \tag{5.138}$$

and any dynamical function associated with the ray trajectories will evolve in time according to

$$\frac{du}{dt} = \frac{\partial u}{\partial t} + [u, \omega']. \tag{5.139}$$

In the present case of a medium with perturbations moving with constant velocity $\mathbf{v}$, it is obvious that the new Hamiltonian is a constant of motion, $I_1 = \omega'$, as can be seen from equation (5.137). If we take the initial frequency and wavevector $(\omega_i, \mathbf{k}_i)$, defined at some instant in the past $t = t_i \to -\infty$, before the ray trajectory crossed the moving front, the final values $(\omega_f, \mathbf{k}_f)$ associated with a much later time $t = t_f \to +\infty$ will be such that

$$I_1 = \omega_i - \mathbf{k}_i \cdot \mathbf{v} = \omega_f - \mathbf{k}_f \cdot \mathbf{v} = \text{const}. \tag{5.140}$$

Using the relation (5.120), and defining the initial and final values of the refractive index, as n_i and n_f, we get the relation

$$\omega_i\left(1 - \frac{n_i}{c}\mathbf{s}_i \cdot \mathbf{v}\right) = \omega_f\left(1 - \frac{n_f}{c}\mathbf{s}_f \cdot \mathbf{v}\right), \tag{5.141}$$

where $\mathbf{s} = \mathbf{k}/|\mathbf{k}|$ is the unit vector along the ray trajectory and $\mathbf{s}_j$, for $j = (i, f)$, are its initial and final values. This leads to a final frequency defined as

$$\omega_f = \omega_i \frac{(1 - b_i)}{(1 - b_f)}, \quad b_j = \frac{\mathbf{k}_j \cdot \mathbf{v}}{\omega_j} = \frac{n_j}{c}(\mathbf{s}_j \cdot \mathbf{v}). \tag{5.142}$$

This is a generalisation of the famous *double-Doppler shift* associated with reflection of light at a moving mirror. But here it is generalised in two different and important ways. First, it is not associated with a pure front discontinuity, because the perturbation was assumed large with respect to the local photon wavelength. Second, the perturbation velocity $|\mathbf{v}|$, being defined as a kinematic quantity independent on the actual motion of the particles of the medium, is not bounded by the speed of light c and can eventually be super-luminal. This will have an impact on the physics of quantum vacuum, as seen below. Notice that the final frequency is independent on the shape and width of the moving front, and only depends on its velocity.

A second invariant along the ray trajectories can be obtained by noting that ω' only depends on the coordinates parallel to the front velocity $\mathbf{v}$. Using equation (5.133) and equation (5.137), we arrive at

$$\frac{d\mathbf{k}_\perp}{dt} = -\frac{\partial \omega'}{\partial \mathbf{r}_\perp} = 0, \quad \mathbf{I}_2 = \mathbf{k}_\perp = \text{const.} \tag{5.143}$$

Defining the angle of incidence θ, such that

$$\mathbf{k} \cdot \mathbf{v} = kv \cos\theta, \tag{5.144}$$

we can write the two invariants in equations (5.140) and (5.143), as

$$I_1 = \omega\left[1 - \frac{v}{c} n \cos\theta\right], \quad I_2 = k \sin\theta. \tag{5.145}$$

This allows us to write the *generalised Snell's laws* [11], as

$$n_i \sin\theta_i = \left(\frac{\omega_f}{\omega_i}\right) n_f \sin\theta_f, \tag{5.146}$$

where

$$\frac{\omega_f}{\omega_i} = \frac{1 - \beta_i \cos\theta_i}{1 - \beta_f \cos\theta_f}, \tag{5.147}$$

where $\beta_j = n_j(v/c)$, for $j = i, f$. We conclude that, in general, both the frequency ω and the angle θ change along the ray trajectories. This process is sometimes called *photon acceleration.* The two above expressions can be combined to give

$$\frac{n_i \sin\theta_i}{(1 - \beta_i \cos\theta_i)} = \frac{n_f \sin\theta_f}{(1 - \beta_f \cos\theta_f)}. \tag{5.148}$$

Notice that this expression reduces to the usual Snell's law, when the initial and final values of the refractive index coincide, $n_i = n_f$, independently of the value of $\mathbf{v}$. We then have an overall photon energy conservation, $\omega_i = \omega_f$. The same is true for $\mathrm{v} = 0$ and $n_i \neq n_f$, and we have energy conservation at all points of the trajectory, $\omega = \text{const}$. In contrast, for an infinite velocity, $|v| \to \infty$, we get

$$\tan\theta_f = \tan\theta_i, \quad \theta_f = \theta_i \tag{5.149}$$

and, at the same time, the frequency changes according to

$$\omega_f = \omega_i \frac{n_i}{n_f}, \tag{5.150}$$

as shown from equations (5.145). This is nothing but time refraction, which we have already described in quantum terms in this chapter. A continuous transition can then be established with the generalised Snell's law of equations (5.145) or (5.148) between the usual space refraction corresponding to $|v| = 0$, and time refraction,

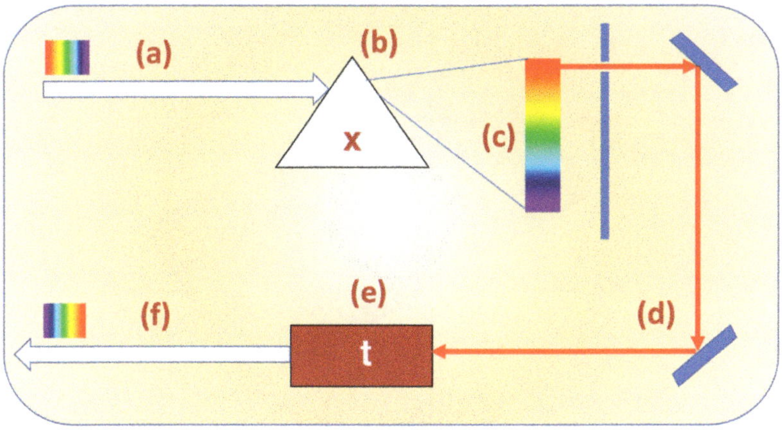

Figure 5.8. Symmetry between space and time: decomposition of white light (a) in a static optical device (b) (Newton's prism) creates a rainbow (c); another static device (pinhole) selects a nearly monochromatic component (d); finally, a temporal optical device (e) recreates the entire white light spectrum (f). This imaginary experiment illustrates the symmetry between refraction (prism) and time refraction (time-varying medium).

which corresponds to the opposite limit of $|v| \to \infty$. As intermediate cases, we have the moving fronts with a finite velocity. Such a symmetry between space and time is illustrated in figure 5.8. The first part of the experimental chain is just the (now) trivial spectral decomposition of white light using a prism, as first demonstrated by Newton. This is a purely static process, which can extract monochromatic light from the white continuum. The inverse process is a temporal arrangement which is capable of regenerating the entire spectrum of white light from a single optical component. This was demonstrated by Alfano and Shapiro in 1970 [34], who were able to generate a supercontinuum laser source starting from nearly monochromatic laser pulses, propagating in nonlinear optical media.

5.8.2 Super-luminal

We have seen that, in what concerns the angles of propagation and the frequencies, the actual shape of the moving front is secondary, as long as the ray (or the photons) have time to cross that front completely. It is therefore useful to consider just a generic sharp front, with a small width. It could be argued, based on the geometric optics approximations stated in equations (5.119) that the above generalised Snell's law is not valid in the infinitely sharp boundary. However, it can be shown, using a full wave description, that these laws stay valid in general conditions. The reason is that they are a consequence of energy and momentum conservation laws.

Let us then consider a sharp optical boundary, moving with velocity v along the Ox-direction (see figure 5.9), and abruptly changing the refractive index from n_0 to n_1, as described by

$$n(x, t) = n_1 + \delta n\ h[(t - x/v)/\tau], \tag{5.151}$$

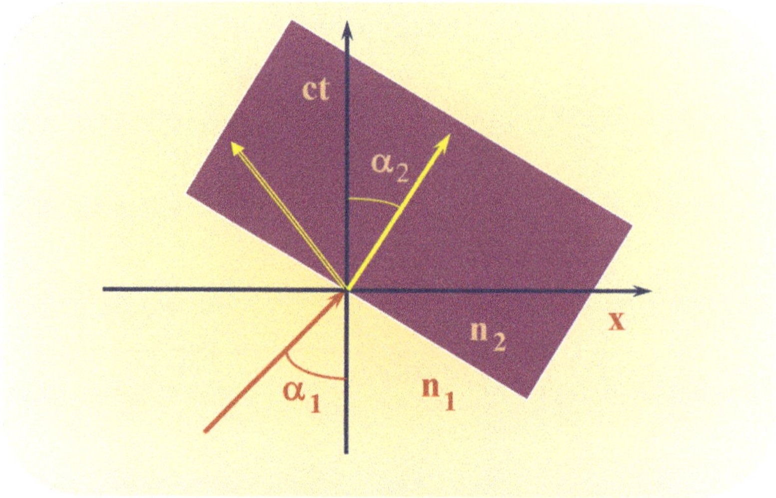

Figure 5.9. Space-time refraction at a moving front. For larger values of the front velocity $|v|$, the reflected signal, for $t > 0$, can be on medium n_1.

where $\delta n = n_2 - n_1$. Here, h is a continuous function in the range [0, 1], tending eventually to a Heaviside function. It should be noticed that the velocity v describes a delay in the change of refractive index between different points of space and does not necessarily refer to the actual velocity of the particles in the medium, as noted before. This is valid for ionisation fronts as well as for other dielectric perturbations of the medium. The particles can be at rest in the laboratory frame for arbitrarily large values of $|v|$. If this quantity is smaller than c, then we can make a Lorentz transformation to the front frame, and reduce the problem to purely static refraction. In the moving frame, there will be no frequency shift but, in the laboratory frame the frequency will be shifted due to the double Lorentz transformation to and from the moving frame. It is therefore not surprising that equation (5.142) resembles a double-Doppler shift.

But, if the front velocity $|v|$ becomes larger than c, we can use a similar but qualitatively different Lorentz transformation from the laboratory frame S into a new frame S', with relative velocity, $u_\infty = -c^2/v < c$. This means that we can move into a frame with infinite velocity v', still using a proper Lorentz transformation. From the velocity transformation law, we get

$$v' = Lim_{(u \to u_\infty)} \frac{u + v}{1 + (uv)/c^2} \to \infty. \tag{5.152}$$

In this new frame, the optical properties of the medium will be uniform in space, but there is a sharp temporal change. The moving boundary became purely temporal, and our previous time refraction model can be applied. This is important because, in the new frame S', the super-luminal boundary is perceived as a simple temporal change with spatial uniformity.

The temporal boundary can then be perceived as a four port device [9, 10], coupling two initial electromagnetic field modes, as $\exp(ik_i'x')$ and $\exp(ik_a'x')$, for $t' < 0$, with two final modes $\exp(ik_t'x')$ and $\exp(ik_r'x')$, existing for $t' > 0$. In the moving frame, the wavenumbers of the two pairs of modes are symmetrical, $k_a' = -k_i'$, and $k_r' = -k_t'$, as a result of momentum conservation. But the same is not true in the laboratory frame S, where an asymmetry is introduced by different Doppler shifts, leading to

$$k_t(x, t) = k_i \frac{1 - s_i\beta/n_0}{1 - s_t\beta/n(x, t)}, \qquad k_r(x, t) = -k_i \frac{1 - s_i\beta/n_0}{1 - s_r\beta/n(x, t)}, \tag{5.153}$$

where $\beta = v_\infty/c = -c/u$, and the signs are $s_i = k_i/|k_i|$, $s_t = k_t/|k_t|$ and $s_r = k_r/|k_r|$. This means that, if in the frame S' the change of the refractive index occurs simultaneously at all points of space, in the frame S there is a delay between different points of space, which then introduces a delay in the change of the wavenumber, as expressed by the space and time dependence of the transmitted and reflected wavenumbers, k_t and k_r.

Moreover, in the S' frame the refractive index has a different value, as a result of relativistic phase invariance [11], and it will be given by

$$n' = \frac{n - \beta}{1 - s\beta n}. \tag{5.154}$$

Finally, we can write the squeezing parameter associated with this time refraction process as $r'(t') = \alpha r(t - x/u)$. The number of photon pairs emitted from vacuum will then be determined by $< N_k(t - x/u) > = \sinh^2[\alpha r(t - x/u)]$. Notice that, for $s\beta n \to 1$, the change in the refractive index n' can, in principle, be arbitrarily large and, as a result, the number of emitted photons. These resonant conditions, which define a kind of *Cherenkov cone*, are specially interesting for experimental purposes. Since the number of photons is a relativistic invariant, the same occurs in any reference frame, including the laboratory frame S.

It should also be added that time refraction is a powerful method to generate entangled photon states [36]. The canonical entanglement entropy for photon pairs generated from an initial vacuum is given by

$$\mathcal{S} = [1 + <N>]\ln[1 + <N>] - <N>\ln <N>. \tag{5.155}$$

This expression shows that we have maximal entangled states near the resonances associated with these Cherenkov cones. The quantity $\mathcal{S}$ is the Shannon entropy associated with the increase of N photon pairs. For a thermal state, which is relevant to possible experiments, this entanglement measure cannot be applied, but a robust entanglement near the resonance cones seems to exist.

5.8.3 Vacuum processes

The different processes occurring in quantum vacuum can now be compared for a moving optical boundary, in the following way. Going back to figure 5.8, we can see that, when the velocity of the boundary is sub-luminal $v < c$, a Lorentz transformation will reduce it to a boundary at rest, where the usual (space) refraction takes place. In this case there is never any possibility of photon emission from vacuum to take place, and the only remaining quantum process is Casimir pressure, which results from an unbalanced zero-point fluctuation energy on the two sides of the boundary. This is particularly clear from the Lifshitz approach.

The situation drastically changes when the boundary moves with super-luminal velocity $v > c$. The existence of such boundaries is possible, even when the particles of the background medium are at rest, as for instance in the case of ionisation fronts. In this case it is still possible to find a proper Lorentz transformation that now reduces the problem to pure time refraction. Now, photon pair production will occur, as discussed above. Parametric instability of vacuum occurs for a modulated time refraction process, when the frequency of modulation is twice as large as the mode frequency. Given the optical similarity between a fixed cavity with a time dependent medium, and a vacuum cavity with a vibrating mirror, this is equivalent to having a time-crystal, or a dynamical Casimir effect, taken in the broad sense.

Finally, we should also mention Unruh radiation, which is associated with accelerated particles, or the language of this section, with a Lorentz transformation that is time dependent. This is also the case of accelerated but sub-luminal boundaries, where the boundary scatters virtual photons from a thermal vacuum and transforms it into radiation, making them become real photons. Comparison between the Unruh radiation with time-crystals and with the dynamical Casimir effect, which undeniably similar processes, seems to make sense [28]. The various processes of quantum vacuum are summarised in figure 5.10.

Gravitation implies curvature of the space-time and, in optical terms, it is equivalent of material optical medium, with a refractive index reflecting such curvature. Optical analogues of gravitational phenomena can then be conceived, but also quantum optical processes in curved spacetime present striking similarities

Vacuum effects at a moving front

Velocity	Lorentz transform. (u < c)	Quantum vacuum
v < c	v' = 0	Refraction Casimir force
v > c	v' = ∞	Time-refraction Cherenkov Radiation
v (t)	u (t), v' = ∞	Dynamical Casimir Time-crystals
v (t)	u (t), v' = ∞	Unruh radiation (Hawking radiation)

Figure 5.10. Quantum vacuum processes associated with an optical boundary, moving with velocity v in the laboratory frame.

with those of quantum optics. This is the case of the Hawking radiation by a black hole horizon, and the closely related Unruh radiation by an accelerated particle. These two processes are related by the equivalence principle, which states that gravitation is equivalent to acceleration. See an illustration of gravitational effects next.

5.9 Curved space-time

We now briefly discuss photon propagation in curved space-time. Our aim here is to show that time refraction can also take place in the gravitation context. This is particularly relevant to possible emission of radiation by an expanding universe, with particular emphasis on inflation, where expansion is assumed to take place at an exceptionally high time rate. Speculative as these processes may be, they can eventually be simulated by nonlinear and quantum optical models in the laboratory [37, 38].

Electromagnetic field quantisation in a gravitational field can be found in books [39, 40], and in a number of review papers, such as [41–44]. The evolution of the wave amplitude radiation in a curved space-time, A, can be described by a Klein–Gordon equation of the form (see appendix D for details)

$$\left(\Box + k_p^2 + \xi R\right)A = 0, \tag{5.156}$$

where $R = R_\nu^\nu$ is the Ricci scalar, and ξ is the minimal coupling constant. We consider a special class of motion, corresponding to a spatially homogeneous and uniformly expanding universe. This is described by the *FWK metric* (from Friedman–Robertson–Walker), as defined by the interval

$$ds^2 = g_{\mu\nu}dx^\mu dx^\nu = dt^2 - a^2(t)dr^2, \tag{5.157}$$

where $dr^2 = dx^2 + dy^2 + dz^2$. The universe *scale function*, $a(t)$, scales the flat metric space dr^2 along the expansion process. It is useful to introduce the *conformal time*, τ, as

$$\tau(t) = \int_{t_0}^{t} \frac{dt}{a(t)}, \tag{5.158}$$

where t_0 is an arbitrary time. We obviously have $dt = a(t)d\tau$. Expressing the scale function in terms of this new temporal variable, we get the interval

$$ds^2 = a^2(\tau)[d\tau^2 - dr^2]. \tag{5.159}$$

This is a flat space-time metric multiplied by a time-dependent scale, such that

$$g_{\alpha\beta} = a^2(t)\eta_{\alpha\beta} = \frac{1}{a^2(t)}\eta^{\alpha\beta}, \tag{5.160}$$

where $\eta_{\alpha\beta} = diag.\ [1, -1, -1, -1]$ is the Minkowski metric. Equation (5.159) defines the *conformally flat metric*. It is useful to note that $g_{00} = a^2$, $g_{ii} = -a^2$ and $\sqrt{-g} = a^4$. Let us return to the Klein–Gordon equation (5.156), and rewrite it as

$$(\Box + m^2)A = 0, \quad m^2 = k_p^2 + \xi R. \tag{5.161}$$

It is useful to introduce the scaled field function $\chi = a(\tau)A$. Noting that $\sqrt{-g}\,m^2 A = m^2 a^3 \chi$, and writing the d'Alembert operator for this metric as

$$\Box = \partial_t^2 + \left(3\frac{\dot{a}}{a}\right)\partial_t - \frac{1}{a^2}\partial_{\mathbf{r}}^2, \tag{5.162}$$

where $\dot{a} = da/dt$, and $\partial_{\mathbf{r}}^2 = \sum_i \partial_{x^i}^2$, we get

$$(\partial_\tau^2 - \partial_{\mathbf{r}}^2)\chi + \left(m^2 a^2 - \frac{\ddot{a}}{a}\right)\chi = 0, \tag{5.163}$$

with $\ddot{a} = d^2a/dt^2$. At this point, it is useful to expand χ in a Fourier integral, as

$$\chi(\mathbf{r}, \tau) = \int \chi_{\mathbf{k}}(\tau) \exp(i\mathbf{k}\cdot\mathbf{r}) \frac{d\mathbf{k}}{(2\pi)^3}. \tag{5.164}$$

For each Fourier component, we get from equation (5.163), the oscillator equation

$$\frac{d^2}{d\tau^2}\chi_{\mathbf{k}} + \omega^2(\tau)\chi_{\mathbf{k}} = 0, \tag{5.165}$$

with the time dependent frequency

$$\omega^2(\tau) = k^2 + m^2 a^2 - \frac{\ddot{a}}{a}. \tag{5.166}$$

Using equation (5.161) and using the curvature $R = 6(\ddot{a}/a^3)$, we get

$$\omega^2(\tau) = k^2 + k_p^2 a^2 - (1 - 6\xi)\frac{\ddot{a}}{a}. \tag{5.167}$$

Notice that, for conformal coupling, corresponding to $\xi = 1/6$, and neglecting mass effects, $k_p^2 = 0$, this is reduced to the harmonic oscillator with constant frequency, $\omega = |k|$. Let us first assume the case where the scale function is constant, $a = a_0$. The harmonic oscillator associated with each bosonic mode $\mathbf{k}$ can be quantised, in the usual way. This leads to the field operator

$$\hat{\chi}(\mathbf{r}, \tau) = \int [u_{\mathbf{k}}(\mathbf{r}, \tau)\hat{a}_{\mathbf{k}} + u_{\mathbf{k}}(\mathbf{r}, \tau)^* \hat{a}_{\mathbf{k}}^\dagger] \frac{d\mathbf{k}}{(2\pi)^3}, \tag{5.168}$$

where $u_{\mathbf{k}}(\mathbf{r}) = u_{\mathbf{k}} \exp(i\mathbf{k}\cdot\mathbf{r})$, and the mode operators $\hat{a}_{\mathbf{k}}$ and $\hat{a}_{\mathbf{k}}^\dagger$ satisfy the bosonic commutation relations

$$[\hat{a}_{\mathbf{k}}, \hat{a}_{\mathbf{k}'}^\dagger] = \delta(\mathbf{k} - \mathbf{k}'), \quad [\hat{a}_{\mathbf{k}}, \hat{a}_{\mathbf{k}'}] = 0, \quad [\hat{a}_{\mathbf{k}}^\dagger, \hat{a}_{\mathbf{k}'}^\dagger] = 0. \tag{5.169}$$

This is compatible with the requirement that

$$[\chi(\mathbf{r}_1, \tau), \chi'(\mathbf{r}_2, \tau)] = i\delta(\mathbf{r}_1 - \mathbf{r}_2), \tag{5.170}$$

which is satisfied under the normalisation condition for the Wronskian

$$W(u_{\mathbf{k}}, u_{\mathbf{k}}^*) \equiv u_{\mathbf{k}}' u_{\mathbf{k}}^* - u_{\mathbf{k}} u_{\mathbf{k}}'^* = 2i. \tag{5.171}$$

Under such conditions, we can define the vacuum state $| \, 0\rangle$, using $\hat{a}_{\mathbf{k}} \, | \, 0\rangle = 0$ for any value of $\mathbf{k}$, and a Fock state with n quanta (photons, or generally, bosons) using

$$| \, n\rangle = \frac{1}{\sqrt{n!}} (\hat{a}_{\mathbf{k}}^{\dagger})^n \, | \, 0\rangle. \tag{5.172}$$

Let us now assume that, at some instant $t = 0$, the gravitational field suffers a sudden change (a time-refraction event) determined by the scale function

$$a(\tau) = a_0 + (a_1 - a_0)H(\tau), \tag{5.173}$$

where $H(\tau)$ is the Heaviside function. In this case, for $\tau > 0$, equation (5.168) should be replaced by a new expression

$$\hat{\chi}(\mathbf{r}, \tau) = \int \left[v_{\mathbf{k}}(\mathbf{r}, \tau)\hat{b}_{\mathbf{k}} + v_{\mathbf{k}}(\mathbf{r}, \tau)^* \hat{b}_{\mathbf{k}}^{\dagger} \right] \frac{d\mathbf{k}}{(2\pi)^3}. \tag{5.174}$$

Equating the terms containing the same Fourier exponent $\exp(i\mathbf{k} \cdot \mathbf{r})$, in equations (5.168) and (5.174), we get

$$\left[v_{\mathbf{k}}\hat{b}_{\mathbf{k}} + v_{\mathbf{k}}^* \hat{b}_{-\mathbf{k}}^{\dagger} \right] = [u_{\mathbf{k}}\hat{a}_{\mathbf{k}} + u_{\mathbf{k}}^* \hat{a}_{-\mathbf{k}}^{\dagger}]. \tag{5.175}$$

Using a linear combination

$$u_{\mathbf{k}} = \alpha_{\mathbf{k}} v_{\mathbf{k}} + \beta_{\mathbf{k}} v_{\mathbf{k}}^* \tag{5.176}$$

such that $|\alpha_{\mathbf{k}}|^2 - |\beta_{\mathbf{k}}|^2 = 1$, we get the *Bogoliubov transformations* between operators, as

$$\hat{b}_{\mathbf{k}} = \alpha_{\mathbf{k}}\hat{a}_{\mathbf{k}} + \beta_{\mathbf{k}}^* \hat{a}_{-\mathbf{k}}^{\dagger}, \quad \hat{b}_{\mathbf{k}}^{\dagger} = \alpha_{\mathbf{k}}^* \hat{a}_{\mathbf{k}}^{\dagger} + \beta_{\mathbf{k}} \hat{a}_{-\mathbf{k}}^{\dagger}. \tag{5.177}$$

The mean number of particles created by this temporal discontinuity is then equal to

$$\langle \hat{N}_{\mathbf{k}} \rangle \equiv \langle 0 \, | \, \hat{N}_{\mathbf{k}}(\tau = 0^+) | \, 0\rangle = \langle 0 \, | \, \hat{b}_{\mathbf{k}}^{\dagger} \hat{b}_{\mathbf{k}} \, | \, 0\rangle = |\beta_{\mathbf{k}}|^2. \tag{5.178}$$

Using the continuity of $\hat{\chi}(\mathbf{r}, \tau)$ and of its first derivative, we then get

$$\alpha_{\mathbf{k}} = \frac{1}{2} \left(\frac{\omega_0 + \omega_1}{\sqrt{\omega_0 \omega_1}} \right)_{\mathbf{k}}, \quad \beta_{\mathbf{k}} = \frac{1}{2} \left(\frac{\omega_0 - \omega_1}{\sqrt{\omega_0 \omega_1}} \right)_{\mathbf{k}}, \tag{5.179}$$

where $\omega_0 \equiv \omega_{\mathbf{k}}(t = 0^-)$ and $\omega_1 \equiv \omega_{\mathbf{k}}(t = 0^+)$ are calculated using equation (5.166). This leads to particle creation of bosons of mode $\mathbf{k}$, as determined by

$$\langle \hat{N}_{\mathbf{k}} \rangle = \frac{1}{4} \left[\frac{(\omega_0 - \omega_1)^2}{\omega_0 \omega_1} \right]_{\mathbf{k}}. \tag{5.180}$$

This discontinuous time-refraction process can be extended to the continuous change of space-time curvature, along the lines followed in optical processes. This approach is particularly important in the case if the *de Sitter universe*, where

$$a(t) = a_0 \exp(Ht), \tag{5.181}$$

where the Hubble parameter is assumed constant. The tangent-hyperbolic model, of the form

$$a(t) = a_0 \frac{1}{2}[1 + \tanh(Ht)], \tag{5.182}$$

which is a straightforward generalisation of the *de Sitter model*, could be useful for a more accurate description of *inflation*. Let us consider the de Sitter vacuum in more detail. We can see that the conformal time τ, defined by equation (5.158), becomes

$$\tau(t) = \int^t \exp(-Ht')dt' = -\frac{1}{H}\exp(-Ht), \tag{5.183}$$

and $a(\tau) = -a_0/(H\tau)$. From here we can see that, when time t varies from minus to plus infinity, the conformal time varies from $-\infty$ to 0. On the other hand, when $t \to -\infty$ (and $\tau \to -\infty$), we have

$$\frac{\dot{a}}{a} = -\frac{1}{\tau}, \quad \frac{\ddot{a}}{a} = +\frac{2}{\tau^2} \to 0, \tag{5.184}$$

which means that, initially, this universe is asymptotically flat. Let us consider the evolution of low frequency modes during inflation. The mode frequency, for $k_p^2 = 0$ and $\xi = 0$, becomes

$$\omega^2(\tau) = k^2 - \frac{\ddot{a}}{a} = k^2 - 2H^2a^2. \tag{5.185}$$

In terms of the proper frequency $\omega_{\text{prop}} = \omega/a$, and proper wavenumber $k_{\text{prop}} = k/a$, we have a constant frequency, given by

$$\omega_{\text{prop}}^2(\tau) = k_{\text{prop}}^2 - 2H^2. \tag{5.186}$$

This shows a lower cutoff for the critical wavenumber defined by $k_{\text{crit}}^2 = 2H^2$. Below this cutoff, the modes are purely growing. These growing modes are such that $\chi_{\mathbf{k}} \propto a$, so that the corresponding field operators $\hat{A}_{\mathbf{k}}$ are static. This leads to zero-point fluctuations, given by

$$\langle 0 \mid \hat{A}_{\mathbf{k}}^\dagger \hat{A}_{\mathbf{k}} \mid 0 \rangle = |u_{-\mathbf{k}}|^2 = \frac{\hbar H^2}{k^2 V}, \tag{5.187}$$

where $V_{\text{prop}} = Va^3$ is the proper volume. We conclude that, for a given proper volume, the fluctuations will increase exponentially in time. If the boson modes are gravitons instead of photons, this could explain the formation of large macroscopic structures (galaxies?), originating from microscopic quantum fluctuations.

References

[1] Purcell E M 1946 Spontaneous emission probabilities at radio frequencies *Phys. Rev.* **69** 681
[2] Kleppner D 1981 Inhibiter spontaneous emission *Phys. Rev. Lett.* **47** 233–6

[3] Walther H, Varcor B T H, Englert B-G and Becker T 2006 Cavity quantum electrodynamics *Rep. Prog. Phys.* **69** 1325

[4] Grangier P, Roger G and Aspect A 1986 Experimental evidence for a photon anticorrelation effect on a beam splitter: a new light on single-photon interferences *Europhys. Lett.* **1** 173–9

[5] Casimir H B G 1948 On the attraction between two perfectly conducting plates *Proc. Ron. Ned. Akad. Wetensch.* **51** 793–5

[6] Milton K A 2004 The Casimir effect: recent controversies and progress *J. Phys. A: Math. Gen.* **37** R209–77

[7] Lifshitz E M 1956 The theory of molecular attractive forces between solids *Sov. Phys. JETP* **2** 73–83

[8] Lamoreaux S K 1997 Demonstration of the Casimir force in the 0.6 to 6 μm Range *Phys. Rev. Lett.* **78** 5–8

[9] Mendonça J T, Guerreiro A and Martins A M 2000 Quantum theory of time refraction *Phys. Rev.* A **62** 033805

[10] Mendonça J T, Martins A M and Guerreiro A 2003 Temporal beam splitter and temporal interference *Phys. Rev.* A **68** 043801

[11] Mendonça J T 2001 *Theory of Photon Acceleration* (Bristol: Institute of Physics Publishing)

[12] Morgenthaler F R 1958 Velocity modulation of electromagnetic waves *IRE Trans. Microw. Theory Tech.* **6** 167–72

[13] Holberg D and Kunz K 1966 Parametric properties of fields in a slab of time-varying permittivity *IEEE Trans. Antennas Propag.* **14** 183–94

[14] Dodonov V V, Klimov A B and Nikonov D E 1993 Quantum phenomena in nonstationary media *Phys. Rev.* A **47** 4422–9

[15] Yablonovitch E 1989 Accelerating reference frame for electromagnetic waves in a rapidly growing plasma: Unruh-Davies-Fulling-DeWitt radiation and the nonadiabatic Casimir effect *Phys. Rev. Lett.* **62** 1742–5

[16] Zhou Y, Alam M Z, Karimi M, Upharm J, Reshef O, Liu C, Willner A E and Boyd R W 2020 Broadband frequency translation through time refraction in a epsilon-near-zero material *Nat. Commun.* **11** 2180

[17] Mendonça J T 2020 Temporal optical processes in a Rydberg gas *J. Phys. B: At. Mol. Opt. Phys* **53** 164004

[18] Wilczek F 2012 Quantum time crystals *Phys. Rev. Lett.* **109** 160401

[19] Sacha K and Zakrzewski J 2018 Time crystals: a review *Rep. Prog. Phys.* **81** 016401

[20] Moore G T 1970 Quantum theory of the electromagnetic field in a variable-length one-dimension cavity *J. Math. Phys.* **11** 2679–91

[21] DeWitt B S 1975 Quantum field theory in curved spacetime *Phys. Rep.* **19** 295–357

[22] Fulling S A and Davies P C W 1976 Radiation from a moving mirror in two-dimensional space-time *Proc. R. Soc.* A **348** 393–414

[23] Dodonov V V 2020 Fifty years of the dynamical Casimir effect *Physics* **2** 67–104

[24] Unruh W G 1976 Notes on black hole evaporation *Phys. Rev.* D **14** 870–92

[25] Fulling S A 1973 Nonuniqueness of canonical field quantization in Riemannian spacetime *Phys. Rev.* D **7** 2850–62

[26] Davies P C W 1975 Scalar production in Schwarzschild and Rindler metrics *J. Phys. A: Math. Gen.* **8** 609–16

[27] Svidzinsky A, Azizi A, Ben-Benjamin J S, Scully M O and Unruh W 2021 Unruh and Cherenkov radiation from a negative frequency perspective *Phys. Rev. Lett.* **126** 063603

[28] Mendonça J T, Brodin G and Marklund M 2008 Vacuum effects in a vibrating cavity: time refraction, dynamic Casimir effects and effective Unruh acceleration *Phys. Lett.* A **372** 5621–4
[29] Mendonça J T, Brodin G and Marklund M 2011 The influence of temporal coherence on the dynamic Casimir effect *Phys. Lett.* A **375** 2665–9
[30] Landau L D and Lifshitz E M 1960 *Electrodynamics of Continuous Media* (Oxford: Pergamon)
[31] Guerreiro A, Mendonça J T and Martins A M 2005 New mechanism of vacuum radiation from non-accelerated moving boundaries *J. Opt. B: Quantum Semiclass. Opt* **7** S69–76
[32] Weinberg S 1962 Eikonal method in magnetohydrodynamics *Phys. Rev.* **126** 1899
[33] Kline M and Kay I W 1965 *Electromagnetic Theory and Geometrical Optics* (New York: Wiley)
[34] Alfano R R and Shapiro S L 1970 Observation of self-phase modulation and small-scale filaments in crystals and grasses *Phys. Rev. Lett.* **24** 592–4
[35] Mendonça J T and Guerreiro A 2005 Time refraction and the quantum properties of vacuum *Phys. Rev.* A **72** 063805
[36] Guerreiro A, Ferreira A and Mendonça J T 2011 Production of bright entangled photons from moving optical boundaries *Phys. Rev.* A **83** 052302
[37] Barceló C, Liberati S and Visser M 2005 Analogue gravity *Living Rev. Relativ.* **8** 12
[38] Philbin T G, Kuklewicz C, Robertson S, Hill S, König F and Leonhardt U 2008 Fiber-optical analog of the event horizon *Science* **319** 1367–70
[39] Birrell N D and Davies P C W 1982 *Quantum Fields in Curved Space* (Cambridge: Cambridge University Press)
[40] Mukhanov V and Winitzki S 2007 *Introduction of Quantum Effects in Gravity* (Cambridge: Cambridge University Press)
[41] Brout R, Massar S, Parentani R and Spindel P 1995 A primer for black hole quantum physics *Phys. Rep.* **260** 329
[42] Ford L H 1997 Quantum field theory in curved spacetime (arXiv:gr-qc/9707062)
[43] Jacobson T 2005 Introduction to quantum fields in curved spacetime and the Hawking effect *Lectures on Quantum Gravity* ed A Gomberoff and D Marolf (Boston, MA: Springer)
[44] Hollands S and Wald R M 2015 Quantum fields in curved spacetime *Phys. Rep.* **574** 1–35

Part II

Quantum fluids of light

This second part covers old and new areas of physics. First, it describes the quantum theory of lasers, which is already well covered in the existing literature. Second, it considers less common topics, such as Bose–Einstein condensation of photons, and discusses the subtle relations between laser and condensation. Another important aspect of the quantum fluids of light is the formation of light vortices. This is related to photon orbital angular momentum (POAM), which has been studied in quantum and nonlinear optics in recent years. We consider POAM in vacuum, optical media and plasmas. We also survey the concept of superradiance, which is a collective process associated with the spontaneous emission of radiation. Finally, we discuss the new topic of superfluid light.

IOP Publishing

The Quantum Nature of Light
From photon states to quantum fluids of light
J T Mendonça

Chapter 6

Laser

The laser, acronym of *light amplification of stimulated radiation*, is one of the greatest scientific inventions of the 20th century. It is one of the main examples, although not usually recognised as such, of a collective state of light. Because of its relevance, we discuss in some detail the principles of the laser, in particular, its semi-classical and quantum models.

The laser is an oscillator, and also an amplifier, in the optical frequency range, where the amplifying medium (a gas or a solid, with an inversion of population between two quantum level) is connected to a feedback system. This allows the electromagnetic signal to propagate repeatedly in the amplifying medium.

Shallow and Townes [1], and independently Prokhorov [2], in 1958, were able to demonstrate that a Fabry–Perot cavity [3] could be used to provide the feedback process. The first laser (the Rubi laser) was built and operated by Maiman in 1960 [4]. It was based on a Rubi crystal, optically pumped by a flash lamp. Many other types of laser were developed since, using a variety of active materials, covering a large spectral domain from the infrared, to the visible and XUV spectrum. Among the most iconic examples, we should mention the helium–neon laser, the first laser using a gas, built by Javan and collaborators in 1961 [5], the semiconductor laser, first created by Robert Hall and collaborators in 1962 [6] and the CO_2 laser, another gas laser and the most powerful continuous laser, developed by Patel in 1964 [7]. A detailed discussion of the laser principles and applications can be found in a number of excellent books, for instance [8–11].

Here, we discuss three different laser models, of increasing complexity, and establish a relation between them. They are: (i) the phenomenological model, (ii) the semi-classical model, and (iii) the quantum model. The phenomenological model is based on balance equations for the atomic populations, and steams directly from the ideas of Planck and Einstein, the so-called old quantum theory. The semi-classical model describes light with Maxwell's classical theory, and atoms with quantum

doi:10.1088/978-0-7503-2786-2ch6

mechanics. Finally the quantum model describes both photons and atoms with quantum fields. The quantum theory of lasers was developed by several researchers in the 1960s [12–14].

6.1 Balance equations

6.1.1 Thermal radiation

Let us first consider the thermal energy spectrum. We have seen that, in a cubic box with side L, the field components with wavevector $\mathbf{k}$ should satisfy the conditions $k_i = 2\pi n_i/L$, where $(i = x, y, z)$ represent the Cartesian components and and n_i are integers. Each field mode inside the this volume is then characterised by a choice of the values of (n_x, n_y, n_z). As a result, the number of modes $\Delta n(\vec{k})$, defined in the infinitesimal interval $(\vec{k}, \vec{k} + d\vec{k})$ satisfy the relation

$$d\mathbf{k} = \left(\frac{2\pi}{L}\right)^3 dn_x dn_y dn_z. \tag{6.1}$$

Noting that we have two independent polarisation states, we obviously get $\Delta n(\mathbf{k}) = 2dn_x dn_y dn_z$. We can write $\Delta n(\mathbf{k}) = n_\mathbf{k} d\mathbf{k}$, where $n_\mathbf{k}$ is the *density of modes.* We obtain

$$n_\mathbf{k} d\mathbf{k} = 2\left(\frac{L}{2\pi}\right)^3 d\vec{k} = 2V\frac{d\mathbf{k}}{(2\pi)^3}, \tag{6.2}$$

where $V = L^3$ is the volume. Performing an integration over the solid angle $\int d\Omega = 4\pi$, we can then establish the number of modes in the interval between $(k, k + dk)$, as

$$n_k dk = V\frac{k^2}{\pi^2}dk, \tag{6.3}$$

where n_k is the number of modes per unit wavenumber. The corresponding number of modes per unit frequency, is

$$n_\omega d\omega = V\frac{\omega^2}{\pi^2 c^3}d\omega, \tag{6.4}$$

where $\omega = kc$ was assumed. We can extend the volume of the box V to infinity, and convert the sum over discrete modes into an integration over k, or ω, using the conversion

$$\sum_\mathbf{k} \to \frac{V}{\pi^2}\int k^2 dk = \frac{V}{\pi^2 c^2}\int \omega^2 d\omega. \tag{6.5}$$

As we have seen before, each field mode has energy $E_n = \hbar\omega(n + 1/2)$. If these modes are in thermal equilibrium with a surrounding medium at a temperature T, the probability for a given mode to have a given number of photons n is determined by the Boltzmann distribution

$$P_n(\omega) = \frac{1}{Z} \exp(-E_n/T), \tag{6.6}$$

where we have written the temperature in energy units. Here, Z is the partition function, defined by the normalisation condition $\sum_n P_n = 1$, as

$$Z = \sum_{n=0}^{\infty} \exp(-E_n/T). \tag{6.7}$$

When we replace this in equation (6.6), the zero point energy $\hbar\omega/2$ cancels out, and we get

$$P_n(\omega) = \frac{e^{-nx}}{\sum_{n=0}^{\infty} e^{-nx}} = (1 - e^{-x})e^{-nx}, \tag{6.8}$$

where we have used $x = \hbar\omega/T$. The mean number of photons per field mode with frequency ω, is then given by

$$\langle n \rangle = \sum_{n=0}^{\infty} nP_n = (1 - e^{-x}) \sum_{n=0}^{\infty} ne^{-nx}. \tag{6.9}$$

This can also be written as

$$\langle n \rangle = -(1 - e^{-x}) \frac{d}{dx} \sum_{n=0}^{\infty} e^{-nx}. \tag{6.10}$$

Using the geometric series summation, we get

$$\frac{d}{dx} \sum_{n=0}^{\infty} e^{-nx} = \frac{d}{dx} \frac{1}{1 - e^{-x}} = \frac{e^{-x}}{(1 - e^{-x})^2}. \tag{6.11}$$

This then leads us to the so-called *Planck function*

$$\langle n \rangle = \frac{e^{-x}}{1 - e^{-x}} = \frac{1}{e^x - 1}. \tag{6.12}$$

Let us now define the mean energy density in the frequency range $(\omega, \omega + d\omega)$, as

$$W_T(\omega)d\omega = \hbar\omega \langle n \rangle n_\omega d\omega. \tag{6.13}$$

Using equations (6.12) and (6.4), we finally obtain

$$W_T(\omega) = \frac{\hbar\omega^3}{\pi^2 c^3} \frac{1}{\exp(\hbar\omega/T) - 1}. \tag{6.14}$$

This is the famous *Planck formula* for the energy density of electromagnetic radiation, emitted by a body at temperature T.

6.1.2 Einstein coefficients

We now assume that thermal equilibrium between radiation and matter is attained by some elementary mechanisms of emission and absorption of radiation by atoms, satisfying the resonant condition $\hbar\omega = E_2 - E_1$, where ω is the photon frequency and E_1 and E_2 the internal energy states of the atoms, such that $E_2 > E_1$. Let us consider the number of atoms per unit volume in these two energy states N_1 and N_2. Following Einstein [15], we assume that three distinct processes of emission and absorption take place. First, the absorption of photons by atoms, associated with a transition from the lower to the upper energy level, as determined by

$$\left(\frac{dN_2}{dt}\right)_{\text{abs}} = B_{12}N_1W(\omega). \tag{6.15}$$

This is obviously proportional to the number of atoms in the lower level N_1, and to the intensity of radiation $W(\omega)$, with a constant of proportionality B_{12} to be defined. In addition to the absorption process, there is emission of photons from atoms in the upper level, with a subsequent decrease of population of this level, as described by

$$\left(\frac{dN_2}{dt}\right)_{\text{em}} = -A_{21}N_2 - B_{21}N_2W(\omega). \tag{6.16}$$

The two terms here describe two different processes. The first one describes *spontaneous emission*, and the second is related to *stimulated emission*. This second term is the exact reverse of the absorption mechanism, and can be described by classical physics. In contrast, the first term has no classical analogue and was introduced to match the Planck formula. Two other coefficients, A_{21} and B_{21} for spontaneous and stimulated emission, are introduced (see figure 6.1 for an illustration). The total rate of change of the atomic population in the upper energy level, is then given by a balance equation, of the form

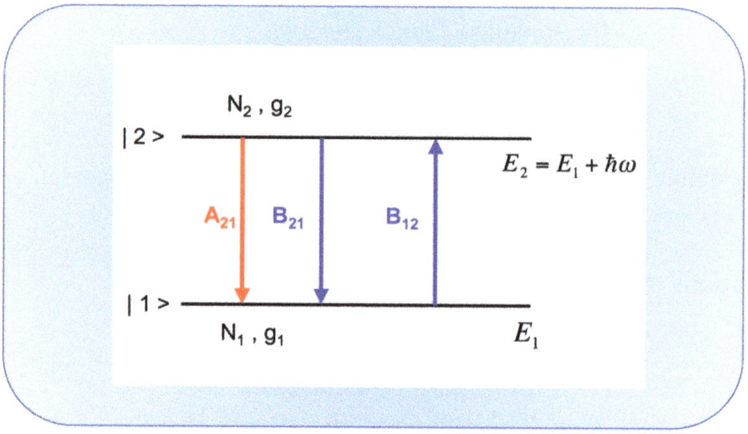

Figure 6.1. Einstein coefficients: absorption, induced emission and spontaneous emission of radiation by an atom with two quantum levels.

$$\frac{dN_2}{dt} = -[A_{21} + B_{21}W(\omega)]N_2 + B_{12}N_1W(\omega). \tag{6.17}$$

Noting that the total number of atoms $N = N_1 + N_2$ remains constant, we can also write

$$\frac{dN_2}{dt} = -\frac{dN_1}{dt}. \tag{6.18}$$

Thermal equilibrium will correspond to the stationary condition, $d/dt = 0$, such that

$$[A_{21} + B_{21}W(\omega)]N_2 = B_{12}N_1W(\omega). \tag{6.19}$$

This leads to the thermal energy distribution

$$W_T(\omega) = \frac{A_{21}}{(N_1/N_2)B_{12} - B_{21}}. \tag{6.20}$$

Using the Boltzmann formula for the atomic populations, and neglecting degeneracy of the energy levels, we can assume that

$$\frac{N_1}{N_2} = \exp\left(\frac{\hbar\omega}{T}\right), \tag{6.21}$$

which then leads to

$$W_T(\omega) = \frac{A_{21}}{B_{12}} \frac{1}{\exp(\hbar\omega/T) - (B_{21}/B_{12})}. \tag{6.22}$$

This expression reduces to Planck's law (6.14) if the emission and absorption coefficients are taken equal to

$$\frac{A_{21}}{B_{12}} = \frac{\hbar\omega^3}{\pi^2c^3}, \quad \frac{B_{21}}{B_{12}} = 1. \tag{6.23}$$

Such a result is only completely explained by the quantum theory radiation. We can now establish a balance equation for the total number of photons n exchanged with the atoms, through these three Einstein processes. The result is

$$\frac{d}{dt}n = (N_2 - N_1)BW(\omega) + AN_2, \tag{6.24}$$

where we have used $A = A_{21}$ and $B = B_{12} = B_{21}$. We can see that

$$\frac{d}{dt}n = \frac{dN_1}{dt} = -\frac{dN_2}{dt}. \tag{6.25}$$

Assuming a small frequency domain $\Delta\omega$ around the atomic transition, it is possible to derive the number of photons per field mode, $n(\omega)$, such that $\Delta\omega W(\omega) = \hbar\omega n(\omega)$. Replacing this in equation (6.24), we obtain the photon rate equation, as

$$\frac{d}{dt}n(\omega) = (N_2 - N_1)w\ n(\omega) + AN_2, \tag{6.26}$$

where we have defined a new quantity, w, as

$$w = \frac{\pi^2 c^3 A}{V\omega^2 \Delta\omega}, \tag{6.27}$$

and where $1/A$, is the spontaneous lifetime. The photon rate equation (6.26) can be used to build a phenomenological laser model, as shown later in this chapter. We should notice, in particular, that the number of photons at a given field mode can increase with time if $N_2 > N_1$, or in other words, if there is an *inversion of population*. Optical methods leading to an inversion of population are discussed next.

6.1.3 Optical pumping

It was Kastler who first suggested in 1950 [16] (see the review [17]), the idea of modifying the population of the atomic energy levels using optical beams. This is the so-called *optical pumping*. This idea was explored 1955 by Basov and Prokhorov for microwaves, and became later the main method to perform laser pumping. To understand this process, we consider the interaction of intense electromagnetic radiation at the frequency ω, with a gas of atoms with two energy levels, such that $E_2 - E_1 = \hbar\omega_{21} \simeq \hbar\omega$, and calculate the changes in populations N_1 and N_2 due to the presence of radiation. Our starting point is the balance equation (6.17), written as

$$\frac{dN_2}{dt} = -AN_2 + B(N_1 - N_2)W(\omega), \tag{6.28}$$

where we have used $B_{12} = B_{21} = B$, corresponding to equal degeneracy, $g_1 = g_2$. Noting that the total number of atoms is conserved, $N = N_1 + N_2$, this is equal to

$$\frac{dN_2}{dt} = BNW(\omega) - [A + 2BW(\omega)]N_2. \tag{6.29}$$

This can be easily integrated, if we assume that the spectral energy density remains constant. This assumption is only valid if the fraction of energy absorbed by the atoms is negligible with respect to the total electromagnetic energy. As a plausible initial condition, we assume that the upper level is initially empty, $N_2(0) = 0$. This is valid for a gas in equilibrium at a temperature T, such that $\hbar\omega_{21} \gg T$. Integration then gives

$$N_2(t) = \frac{BNW(\omega)}{A + 2BW(\omega)}\{1 - \exp[-(A + 2BW(\omega))t]\}. \tag{6.30}$$

This shows that, for short times, the population of the upper level grows linearly with time, as $N_2(t) \simeq BNW(\omega)t$, whereas for very long times, such that $t \gg [A + 2BW(\omega)]^{-1}$, it tends to a constant

$$N_2(\infty) = \frac{BNW(\omega)}{A + 2BW(\omega)}. \tag{6.31}$$

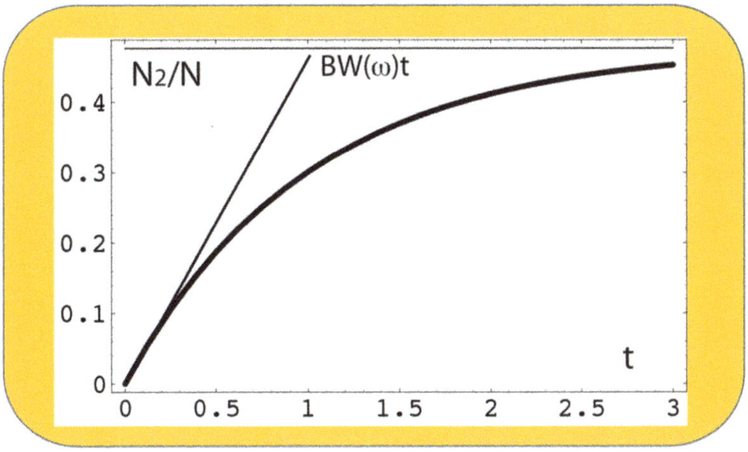

Figure 6.2. Temporal evolution of the upper level population, $N_2(t)$, for a two-level atom pumped by resonant radiation with energy density $W(\omega)$. The value of $A/2BW(\omega) = 0.1$ was used.

This asymptotic value depends on the radiation intensity. For large intensities such that $W(\omega) \gg A/2B$, we get $N_2(\infty) \simeq N/2$. It means that, for intense radiation, the maximum value for the upper level population is equal to half of the total number of atoms $N_2 = N_1 = N/2$. Therefore, it is not possible by this process to attain an inversion of population such that $N_2 > N_1$. The evolution of N_2 is represented in figure 6.2. It should also be noticed that saturation of the radiative transition, corresponding to $N_2 = N_1$, is more difficult to achieve in the optical domain than in microwaves, because the Einstein coefficient of spontaneous emission, A increases with the frequency. This results in a faster decay of the upper level, due to spontaneous emission (fluorescence). This is obvious from equation (6.31), which for $A \gg 2BW(\omega)$, leads to $N_2(\infty) \simeq N[BW(\omega)/A] \ll N$.

Let us now consider the optical pumping with a three-level atom system, as proposed by Basov and Prokhorov. This will lead to an inversion of population, as shown next. We call |1> the lowest energy level, and |2> and |3> the two excited states. The electromagnetic pump with energy density $W_p(\omega)$ is assumed nearly resonant with the radiative transition between the lowest and the highest states, such that $E_3 - E_1 = \hbar\omega_{31} \simeq \hbar\omega$. Given that saturation is very difficult to achieve in the optical domain, we assume that $N_3 \ll N$, where $N = N_1 + N_2 + N_3$ is the total number of atoms (per unit volume). On the other end, the atoms on level |3> are allowed to decay radiatively into the intermediate level |2>, assuming dipolar electric transitions, and emit photons with frequency $\omega_{32} = (E_3 - E_2)/\hbar$. This is illustrated in figure 6.3.

We can show that, in such conditions, inversion of population between the two excited states, |2> and |3>, is eventually attained with appropriate Einstein coefficients. We start with the balance equations for the three atomic populations, as

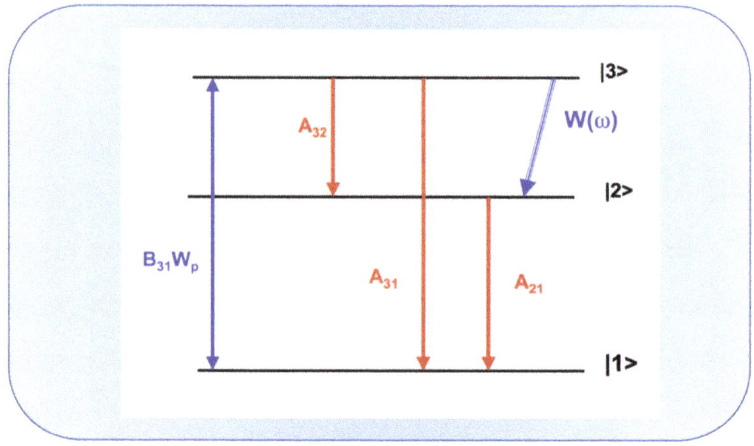

Figure 6.3. Schematic representation of optical pumping of a three-level atom.

$$\frac{dN_1}{dt} = A_{21}N_2 + A_{31}N_3 - B_{31}W_p(N_1 - N_3),$$
$$\frac{dN_2}{dt} = -A_{21}N_2 + A_{32}N_3 + B_{32}W(N_3 - N_2), \quad (6.32)$$
$$\frac{dN_3}{dt} = -(A_{31} + A_{32})N_3 + B_{31}W_p(N_1 - N_3) - B_{22}W(N_3 - N_2).$$

Given the conservation of the total number of atoms, N, these equations verify the condition

$$\frac{dN_1}{dt} + \frac{dN_2}{dt} + \frac{dN_3}{dt} = 0. \quad (6.33)$$

We now look for stationary solutions of the system (6.32). The first two equations give

$$A_{21}N_2 + A_{31}N_3 = B_{31}W_p(N_1 - N_2),$$
$$A_{21}N_2 - A_{32}N_3 = B_{32}W(N_3 - N_2). \quad (6.34)$$

For convenience, we introduce the parameter r, such that

$$r = \frac{1}{N}B_{31}W_p(N_1 - N_3), \quad (6.35)$$

and represents the fraction of atoms transferred by optical pumping onto the upper level |3>. Equations (6.34) can then be rewritten as

$$A_{21}N_2 + A_{31}N_3 = Nr,$$
$$(A_{21} + B_{32}W)N_2 = (A_{32} + B_{32}W)N_3. \quad (6.36)$$

These equations can easily be solved for N_2 and N_3, leading to

$$N_2 = \frac{Nr(A_{32} + B_{32}W)}{A_{21}(A_{31} + A_{32}) + B_{32}W(A_{21} + A_{31})}, \tag{6.37}$$

and

$$N_3 = \frac{Nr(A_{21} + B_{32}W)}{A_{21}(A_{31} + A_{32}) + B_{32}W(A_{21} + A_{31})}. \tag{6.38}$$

These expressions show that inversion of population $N_3 > N_2$ can be attained if $A_{21} > A_{32}$. Or equivalently, if the atoms in the intermediate state |2> decay more rapidly into the lower level |1> than the atoms in the upper level |3> decay into this level intermediate level |2>. When this inequality is satisfied it is possible, at least in principle, to amplify radiation at the frequency $\omega_{32} = (E_3 - E_2)/\hbar$, with an increase of energy density $W(\omega)$, for $\omega = \omega_{32}$. But equations (6.37)–(6.38) also show that, increasing the value of $W(\omega)$ the difference $(N_3 - N_2)$ will decrease and eventually saturate.

Based on this three-level optical pumping, or other alternative mechanisms of inversion of population, it is then possible to build an optical amplifier, a *laser* (light amplification of stimulated emission of radiation). But, in order to obtain the laser effect, we need to add a feedback process, which is provided by an optical cavity, as described next.

6.2 Laser cavity

6.2.1 Cavity modes

The main components of a laser are represented in figure 6.4. They are, the active medium containing the excited atoms with an inversion of population, the optical cavity defining the electromagnetic mode and providing the radiation feedback, the

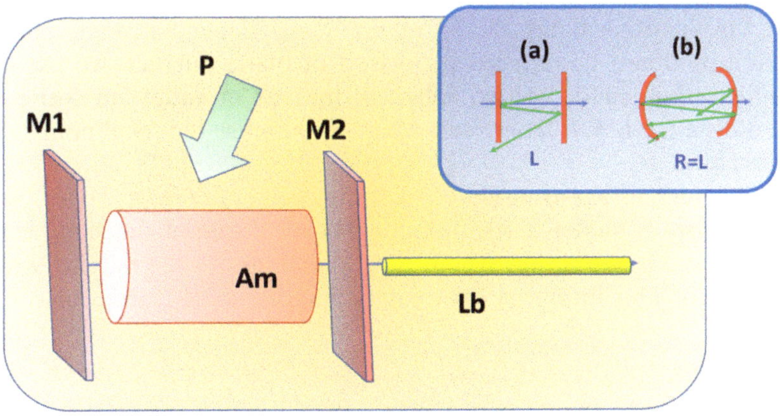

Figure 6.4. Main components of a laser: *M*1 and *M*2 the cavity mirrors, *Am* the active medium, *P* the pump beam, and *Lb* the output laser beam. Also represented, an optical cavity with (a) parallel plane mirrors, and (b) spherical mirrors.

pump system which injects energy from outside and provides the inversion mechanism, and the output optical beam. Here we consider the cavity modes.

Let us then consider an optical cavity defined by two plane parallel mirrors, and separated by a distance L, as shown also in the figure. Assuming that the reflection coefficient of the mirrors is nearly equal to 1, standing-wave modes can be excited. These modes are such that an integer number q of half-wavelengths can fit the cavity length L, or $\lambda_q = 2L/q$, where λ_q is the wavelength of the standing-wave mode. The frequency interval between two consecutive modes is then given by

$$\Delta\omega = \frac{2\pi c}{\lambda_q} - \frac{c}{\lambda_{q-1}} = \pi \frac{c}{L}. \tag{6.39}$$

This frequency interval is therefore equal to the inverse of the time spent by light to travel twice from one mirror to the other. For a cavity of 1 m this corresponds to a frequency interval of 150 MHz. The interval $\Delta\omega$ is, in general, much shorter than the natural bandwidth of the radiative transition where the inversion of population takes place. This means that it is possible to excite simultaneously several cavity modes, as discussed later.

Another important question is related with the cavity resonance width, $\Delta\omega_q$, which is inherent to each of the cavity modes $\lambda_q = 2\pi c/\omega_q$. This can be defined in terms of the *quality factor* of the cavity

$$Q_q = \frac{\omega_q}{\Delta\omega_q}. \tag{6.40}$$

This quantity is determined by the ratio between the energy contained in the cavity and the power losses of electromagnetic radiation. These losses result mainly from: (i) *reflection losses* due to the imperfection of the mirrors; (ii) *diffraction losses* at the mirrors borders, leading to energy leaking to the outside. It should also be noted that, in a laser, the reflection losses can never be reduced to zero, because one of the cavity mirrors has to be partially transparent, in order to allow for the laser beam formation. On the other hand, the diffraction losses depend strongly on the mirror configuration, and can be reduced if, instead of plane mirrors, we use a *confocal cavity* configuration, made of two spherical mirrors of equal curvature radius R, located at a distance $L = R$ from each other. In this case the focal points of the two mirrors coincide.

Finally, in order to avoid mode competition inside the laser cavity, we need to reduce the number of modes to a minimum. The ideal case of single mode operation is attained when $\Delta\omega_q \sim \Delta\omega_{\text{trans}}$, where $\Delta\omega_{\text{trans}}$ determines the natural width of the atomic transition. This implies a laser cavity length L of the order of $\pi c/\Delta\omega_{\text{trans}}$.

6.2.2 Mode losses

Apart from the already noted losses due to imperfect reflectivity and to diffraction at the mirrors, we also have dissipative losses due to absorption or scattering in the optical medium. The existence of all these lossy mechanisms can be associated with a finite quality factor Q and, ultimately, the finite lifetime of trapped photons. This will lead to a finite coherence time for the quantum field states, as shown here. Losses can be described using the concept of *cavity quasi-mode*, which was first developed in the frame of laser cavities by Fox and Li in 1961 [18], and then extended and applied by others [19–21].

We assume a static dielectric medium in a one-dimensional optical cavity, which is defined by two parallel mirrors, placed at positions $z = 0$ and $z = L$. The transverse mode structure is marginal to our discussion, and is ignored. But it can be easily accommodated in the formalism, if necessary. The electric field of a cavity mode ν can be described as

$$\mathcal{E}(z, t) = \mathcal{E}_\nu \sin(k_\nu z)\exp(-i\omega_\nu t) + c.\ c., \tag{6.41}$$

with mode wavenumbers and frequencies

$$k_\nu = 2\pi\frac{\nu}{L}, \qquad \omega_\nu = \frac{k_\nu c}{\sqrt{\epsilon(\omega_\nu)}}, \tag{6.42}$$

where ν is an integer, and $\epsilon(\omega)$ is the dielectric function of the medium. In a quantum description, this is replaced by the field operator

$$\hat{\mathcal{E}}(z, t) = [\hat{\mathcal{E}}_\nu(t)\exp(ik_\nu z) + h.\ c.] \tag{6.43}$$

with

$$\hat{\mathcal{E}}_\nu(t) = iC_\nu[\hat{a}_\nu(t) - \hat{a}_{-\nu}^\dagger(t)] \tag{6.44}$$

with the normalisation factor $C_\nu = \sqrt{\hbar\omega_\nu/2\epsilon(\omega_\nu)}$. We should notice that this cavity mode results from the superposition of two counter-propagating photon states, as described in equation (6.43). The corresponding destruction and creation operators can be defined as

$$\hat{a}_\nu(t) = \hat{A}_\nu \exp(-i\omega_\nu t), \qquad \hat{a}_{-\nu}^\dagger = \hat{A}_{-\nu}^\dagger \exp(i\omega_\nu t) \tag{6.45}$$

These modes are, by definition, monochromatic. But, in a lossy cavity, with quality factor Q_ν, the modes acquire a finite spectral width, $\gamma_\nu = \omega_\nu/Q_\nu$. We can still use a similar field description, but the modes are no longer pure field modes, and have to be replaced by the concept of cavity quasi-modes. The operators $\hat{a}_\nu(t)$ and $\hat{a}_{-\nu}^\dagger(t)$ become *effective mode operators*, which are now called *quasi-mode operators*. They are defined by a superposition of internal field modes, with nearly equal frequencies $\omega \sim \omega_\nu$, as

$$\hat{a}_\nu(t) = \int g_\nu(\xi)\tilde{a}_\nu(\xi, t)\frac{d\xi}{2\pi}, \qquad \hat{a}_{-\nu}^\dagger(t) = \int g_\nu^*(\xi)\tilde{a}_{-\nu}^\dagger(\xi, t)\frac{d\xi}{2\pi}, \tag{6.46}$$

where $\xi = (\omega - \omega_\nu)$ can be seen as an internal variable, and $g_\nu(\xi)$ defines the spectral shape function to be defined. It is obvious from here that the effective mode operators $\hat{a}_\nu(t)$ and $\hat{a}_\nu^\dagger(t)$ can be considered as external or macroscopic operators, whereas the field modes $\tilde{a}_\nu(\xi, t)$ and $\tilde{a}_{-\nu}^\dagger(\xi, t)$ are the internal or microscopic ones. These internal operators can similarly be decomposed in amplitude and phase as

$$\tilde{a}_\nu = \tilde{A}_\nu \exp[-i\phi(t)], \quad \tilde{a}_{-\nu}^\dagger = \tilde{A}_{-\nu}^\dagger \exp[i\phi(t)], \tag{6.47}$$

where the phase varies in time according to $\phi(t) = \omega t \equiv (\omega_\nu + \xi)t$. Comparing this with equation (6.45), we conclude that

$$\hat{A}_\nu = \int g_\nu(\xi)\tilde{A}_\nu(\xi)\exp(-i\xi t)\frac{d\xi}{2\pi}, \quad \hat{A}_{-\nu}^\dagger = \int g_\nu^*(\xi)\tilde{A}_{-\nu}^\dagger(\xi)\exp(i\xi t)\frac{d\xi}{2\pi}. \tag{6.48}$$

For each internal mode ξ we define a photon number operator, in the usual way, as

$$\tilde{N}(\xi) = \tilde{a}_\nu^\dagger(\xi, t)\tilde{a}_\nu(\xi, t) = \tilde{A}_\nu^\dagger(\xi)\tilde{A}_\nu(\xi). \tag{6.49}$$

We can then define a photon number operator for the effective cavity mode ν, as

$$N_\nu(t) = \int \frac{d\xi}{2\pi} \int \frac{d\xi'}{2\pi} g_\nu^*(\xi) g_m(\xi') \tilde{A}_\nu^\dagger(\xi)\tilde{A}_\nu(\xi'). \tag{6.50}$$

At this point it should be noted that the spectral shape function $g_\nu(\xi)$ can be chosen in such a way as to satisfy the normalisation condition

$$\int |g_\nu(\xi)|^2 \frac{d\xi}{2\pi} = 1. \tag{6.51}$$

This particular choice will enable the effective mode operators to obey the commutation relation $[a_\nu, a_{\nu'}^\dagger] = \delta_{\nu,\nu'}$, given the usual commutation relations valid for the internal field mode operators

$$[\tilde{a}_\nu(\xi), \tilde{a}_{\nu'}^\dagger(\xi')] = \delta_{\nu,\nu'}\delta(\xi - \xi'). \tag{6.52}$$

Until now, we have assumed the shape function $g_\nu(\xi)$ as an arbitrary complex function, obeying the normalisation condition (6.51). A natural choice, which describes a Lorentzian mode profile, with half-width γ_ν, is given by

$$g_\nu(\xi) = \frac{\sqrt{2\gamma_\nu}}{(\xi^2 + \gamma_\nu^2)^{1/2}}. \tag{6.53}$$

It is obvious that the particular case $|g_\nu(\xi)|^2 = 2\pi\delta(\xi)$ would lead to $\hat{A}_\nu = \tilde{A}_\nu$ and $\hat{A}_{-\nu}^\dagger = \tilde{A}_{-\nu}^\dagger$, and corresponds to the ideal case of a perfect cavity, valid in the limit $\gamma_\nu \to 0$.

Before concluding this short discussion on cavity quasi-modes, let us consider the field Hamiltonian H. Following the usual definition, we can obviously write it as $H = \sum_\nu H_\nu$, with

$$H_\nu = \hbar \int (\omega_\nu + \xi)\left[\tilde{N}_\nu(\xi) + \frac{1}{2}\right] d\xi. \tag{6.54}$$

This allows us to derive the Heisenberg equations for the field operators, as

$$\frac{d}{dt}\tilde{a}_\nu(\xi, t) = \frac{1}{i\hbar}[\tilde{a}_\nu(\xi, t), H_\nu] = -i(\omega_\nu + \xi)\tilde{a}_\nu(\xi, t) \tag{6.55}$$

with a similar expression for $\tilde{a}^\dagger_{-\nu}(\xi, t)$. This could also be recovered directly from equations (6.47).

6.3 Phenomenological laser model

Let us assume that inversion of population between two atomic levels is obtained by some unspecified process, and let us study the photon rate equation for a given cavity mode, coupled with the balance equations for the populations N_1 and N_2. We return to the photon rate equation (6.26), but neglect the spontaneous emission term AN_2, which is assumed irrelevant. On the other hand, we add a damping term describing the photon losses in the cavity. We then write

$$\frac{d}{dt}n = (N_2 - N_1)w\, n - \gamma_c\, n, \tag{6.56}$$

where γ_c is the cavity damping rate. The evolution equation for the upper level population is

$$\frac{dN_2}{dt} = w_{21}N_1 - w_{12}N_2 - (N_2 - N_1)wn. \tag{6.57}$$

Comparing this with equation (6.28), we see an additional term $w_{21}N_1$, to account for the inversion mechanism, which pumps the atoms from the lower to the upper level. Furthermore, in the decay term, $-w_{12}N_2$, we can include other possible de-excitation mechanisms, apart from the neglected spontaneous emission, for instance, de-excitation due to atomic collisions. A similar equation can be established for the lower level population, of the form

$$\frac{dN_1}{dt} = w_{12}N_2 - w_{21}N_1 + (N_2 - N_1)wn. \tag{6.58}$$

We can see from here that

$$\frac{d}{dt}(N_1 + N_2) = 0, \tag{6.59}$$

which is compatible with a two-level atom model, where the total number of atoms $N = N_1 + N_2$ remains constant. Subtracting equations (6.57) and (6.58) we establish the evolution equation for the population difference $D = N_2 - N_1$, as

$$\frac{dD}{dt} = N(w_{21} - w_{12}) - D(w_{21} + w_{12}) - 2wDn. \tag{6.60}$$

Noting that w_{21} and w_{12} are transition rates and, consequently, are proportional to the inverse of a transition time, we introduce a time constant τ, such that

$$\tau = \frac{1}{(w_{21} + w_{12})}. \tag{6.61}$$

We also notice that, in the absence of laser radiation ($n = 0$), equation (6.60) predicts an equilibrium for the population difference, $D = D_0$, such that

$$D_0 = \frac{(w_{21} - w_{12})N}{(w_{21} + w_{12})} = (w_{21} - w_{12})N\tau. \tag{6.62}$$

This new quantity is called the *unsaturated inversion*, for reasons to become apparent below. Using the definitions of τ and D, we can rewrite equation (6.60) in the form

$$\frac{dD}{dt} = \frac{1}{\tau}(D_0 - D) - 2wDn. \tag{6.63}$$

This equation is coupled to equation (6.56), rewritten in the form

$$\frac{d}{dt}n = Dw\, n - \gamma_c\, n. \tag{6.64}$$

These are the two basic equations of our phenomenological laser model. In order to understand this model, let us first consider the steady-state solutions. From equation (6.63), we get steady-state inversion of population, $D = \bar{D}$, such that

$$\bar{D} = \frac{D_0}{1 + 2\tau wn}. \tag{6.65}$$

This quantity differs from the unsaturated inversion D_0, because of the saturation term in the denominator, which is proportional to the photon mode number, and therefore to the laser intensity. It is obvious that saturation leads to a decrease in the steady-state value of D. On the other hand, we have two possible steady solutions of equation (6.64). One is trivial, $n = 0$, and the other determined by

$$\bar{D}w = \frac{D_0 w}{1 + 2\tau wn} = \gamma_c, \tag{6.66}$$

which gives $n = n_0$, with

$$n_0 = \frac{(D_0 w - \gamma_c)}{2\tau\gamma_c w}. \tag{6.67}$$

We can see from this that a sufficiently high pumping rate is needed, $w_{21} > w_{12}$, in order to satisfy the physical condition $n_0 > 0$. This will imply that

$$D_0 \geqslant D_{\text{crit}} \equiv \frac{\gamma_c}{w}. \tag{6.68}$$

Below this critical value D_{crit}, no laser effect is possible. This threshold is however misleading, because it assumes that $n_0 \geqslant 0$, a condition that can only be fulfilled in the absence of spontaneous emission, as shown later. Only using the quantum theory can we establish a rigorous account of the laser threshold.

Let us now study the approximate time-dependent solutions for the above laser equations. As a simplifying assumption, we assume that the inversion of population

D evolves on a time scale much longer than the transition time τ. This allows us to neglect the term $dD/dt \simeq 0$ in equation (6.63), which leads to the approximate solution

$$D(t) \simeq \frac{D_0}{1 + 2\tau w n(t)}. \tag{6.69}$$

Replacing this in equation (6.64), we obtain a closed equation for the photon number n, given by

$$\frac{dn}{dt} = \frac{D_0 w n}{1 + 2\tau w n} - \gamma_c n. \tag{6.70}$$

Expanding the denominator, we then arrive at a simple nonlinear equation of the form

$$\frac{dn}{dt} = an - bn^2 \tag{6.71}$$

with

$$a = D_0 w - \gamma_c, \quad b = 2D_0 \tau w^2. \tag{6.72}$$

This shows that, for $a > 0$, the photon number $n(t)$ initially grows exponentially, as $n(t) \simeq n(0)\exp(at)$. But, at later times, the nonlinear term slows down the exponential growth, and eventually saturates the photon number at $n_0 = a/b$. A more exact solution of the nonlinear equation (6.70), describing such a saturation process, is given by

$$n(t) = \frac{a\, n(0)\exp(at)}{a + b\, n(0)[\exp(at) - 1]}. \tag{6.73}$$

This solution is represented in figure 6.5, illustrating the saturation process.

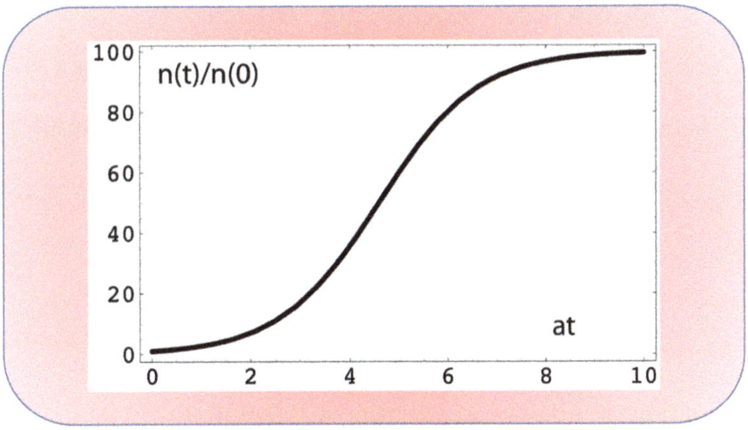

Figure 6.5. Photon number evolution, as given by equation (6.73). The value of $bn(0)/a = 0.01$ was used.

6.4 Relaxation oscillations

Let us now assume small perturbations around the steady state. This can be described using

$$D = \bar{D} + \eta, \qquad n = n_0 + \epsilon, \tag{6.74}$$

where $\bar{D}$ and n_0 were determined above, and the perturbations are described by $\eta \ll \bar{D}$ and $\epsilon \ll n_0$. Replacing this in equations (6.64), we get

$$\frac{d\epsilon}{dt} = w(\bar{D} + \eta)(n_0 + \epsilon) - \gamma_c(n_0 + \epsilon). \tag{6.75}$$

The constants $\bar{D}$ and n_0 define the steady state, $dn/dt = 0$, and therefore $\bar{D}w - \gamma_c = 0$. This equation can then be reduced to

$$\frac{d\epsilon}{dt} = w\eta(n_0 + \epsilon). \tag{6.76}$$

We can now linearise this equation with respect to the perturbations, and reduce it further to,

$$\frac{d\epsilon}{dt} = wn_0\eta. \tag{6.77}$$

We can apply a similar perturbation procedure to equation (6.63), which gives

$$\frac{d\eta}{dt} = \frac{1}{\tau}(D_0 - \bar{D} - \eta) - 2w(\bar{D} + \eta)(n_0 + \epsilon). \tag{6.78}$$

Noting that the equilibrium quantities satisfy the relation

$$\frac{1}{\tau}(D_0 - \bar{D}) = 2w\bar{D}n_0, \tag{6.79}$$

and neglecting the nonlinear term, $\eta\,\epsilon$, we can then write equation (6.78) as

$$\frac{d\eta}{dt} = \frac{\eta}{\tau} - 2w\bar{D}\epsilon. \tag{6.80}$$

The linearised coupled equations (6.77) and (6.80) can easily be solved. Elimination of ϵ leads to

$$\frac{d^2\eta}{dt^2} + \omega_0^2\eta + \gamma\frac{d\eta}{dt} = 0, \tag{6.81}$$

where we have defined

$$\omega_0 = w\sqrt{2n_0\bar{D}}, \qquad \gamma = 1/\tau. \tag{6.82}$$

This is the equation of a damped oscillator, with solutions

$$\eta(t) = Aexp(-\gamma t/2)\cos(\omega t + \varphi). \tag{6.83}$$

The constants of integration A and φ are the initial amplitude and phase, and the oscillator frequency is defined by

$$\omega = \sqrt{\omega_0^2 - \gamma^2/4}. \tag{6.84}$$

We conclude that, near steady state, $\bar{D}$ and n_0, the perturbed population difference $D(t)$ oscillates around equilibrium $D = \bar{D}$ with frequency ω, and returns to equilibrium on a time scale of order $\tau = 1/\gamma$. Similarly, the laser intensity, represented by the photon number $n(t)$, oscillates around n_0 at the same frequency, as

$$n(t) = n_0\left[1 + \frac{w}{\omega}\sin(\omega t + \varphi)\right]. \tag{6.85}$$

These solutions show that, any small perturbation around equilibrium will oscillate and vanish on a time scale of τ. Such oscillations can be observed in steady-state laser operation. For instance, for a Ruby laser, we typically have $2\pi/\omega \sim 10\ \mu s$, and $\tau \sim 1\ ms$. Other modes of laser operation are considered next.

6.5 Short laser pulses

6.5.1 Q-switching

One way to achieve short laser pulse operation is to rotate one of the mirrors of the optical cavity. This is called a *Q-switching process*, because only for a very small fraction of the rotation period will the cavity possess a large Q-factor. In other words, the cavity losses will only be negligible for a very small fraction of time, when the two cavity mirrors become parallel to each other. This means that the mirror rotation is a practical way to switch off the time the quality factor $Q = \omega/\gamma_c$ of the laser cavity. Here, Q and γ_c become time-dependent quantities. There are many different processes providing Q-switching.

Because stimulated emission is inhibited when the cavity mirrors are not aligned, we expect a sudden burst of laser radiation only at the aligned position. Much higher laser intensities emitted during a short period are obtained, as compared with steady-state operation. In the present situation, the term in $1/\tau$ in the above inversion equation (6.63) can be neglected, because we are dealing with much faster time-scales, and this equation reduces to

$$\frac{dD}{dt} \simeq -2wDn. \tag{6.86}$$

On the other hand, in the photon number equation, we replace the instantaneous value of $D(t)$ by its initial value D_i, leading to

$$\frac{dn}{dt} \simeq (D_i w - \gamma_c)\, n, \tag{6.87}$$

which has the approximate solution

$$n(t) \simeq n_i \exp(\alpha t), \quad \alpha \equiv (D_i w - \gamma_c). \tag{6.88}$$

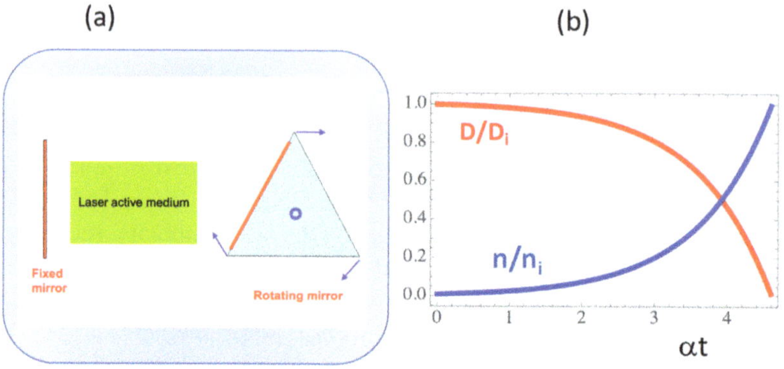

Figure 6.6. Q-switching laser: (a) geometry of the cavity. A rotating mirror provides the change in its quality factor. (b) Normalised inversion of population D/D_i and normalised photon number n/n_i, as functions of normalised time αt, as described by equations (6.88) and (6.90).

Replacing this in equation (6.86), and integrating, we obtain

$$D(t) \simeq D_i\left[1 + \frac{2wn_i}{\alpha}(1 - e^{\alpha t})\right]. \tag{6.89}$$

This approximate solution is only valid in the initial stage of the laser pulse, during which the number of photons grows exponentially at a constant rate α, and the inversion of population only suffers a small temporal decay. At later times, however, this picture has to be changed. When $\exp(\alpha t)$ becomes significantly larger than 1, equation (6.89) becomes

$$D(t) \simeq D_i\left(1 - \frac{2wn_i}{\alpha}e^{\alpha t}\right). \tag{6.90}$$

The inversion of population described by this new expression will decrease to zero, $D \simeq 0$, on a time scale of $t = t_1$, when the laser intensity attains its pick intensity, is given by

$$n_{\max} = n_i e^{\alpha t_1} \simeq \frac{\alpha}{2w}. \tag{6.91}$$

In order to describe the evolution of the photon number $n(t)$ for later times $t > t_1$, we go back to equation (6.81), with $D \simeq 0$. The resulting solution is

$$n(t) \simeq n_{\max} \exp[-\gamma_c(t - t_1)]. \tag{6.92}$$

According to this new expression, the photon number decays on a time scale determined by the cavity losses $1/\gamma_c$. These approximate solutions describe a very strong peak of high laser intensity, and are illustrated in figure 6.6(b).

6.5.2 Mode locking

Another method of short laser pulse operation is the so-called *mode locking*, where the laser oscillates in a multi-mode regime. Record values of short pulse duration, of the order of a few femtosecond have been attained using this method. The output electric field can written as

$$\mathcal{E}(t) = \sum_{\nu=1}^{N} \mathcal{E}_\nu \exp[-i(\omega_\nu t + \phi_\nu)], \tag{6.93}$$

where we have N laser modes with amplitudes $\mathcal{E}_\nu$, oscillating at the frequencies ω_ν, with relative phase shifts determined by ϕ_ν. In the incoherent case, where the laser modes are independent from each other and the phases ϕ_ν as randomly distributed in the interval $[0, 2\pi]$, the total laser intensity is given by

$$I_{\text{tot}} = \sum_{\nu=1}^{N} |\mathcal{E}_\nu|^2 \simeq N|\mathcal{E}_0|^2, \tag{6.94}$$

where we assume that all the mode amplitudes are all similar, $\mathcal{E}_\nu \simeq \mathcal{E}_0$. However, the active laser has nonlinear properties and, for large amplitudes, the modes can couple to each other. In this case, the field phases tend to lock into the same value $\phi_\nu \simeq \phi_0$, and the total field becomes

$$\mathcal{E}(t) = \mathcal{E}_\nu e^{i\phi_0} \sum_{\nu=1}^{N} \exp(-i\omega_\nu t). \tag{6.95}$$

As we have seen, the mode separation frequency is determined by $\Delta\omega = \pi c/L$, where L is the cavity length. We can then write the mode frequencies as $\omega_\nu = \omega_0 - \nu\Delta\omega$. This leads to

$$\mathcal{E}(t) = \mathcal{E}_0 e^{-i(\omega_0 t + \phi_0)} \sum_{\nu=0}^{N-1} \exp(i\pi\nu ct/L). \tag{6.96}$$

Performing the geometric series summation, we get

$$\mathcal{E}(t) = \mathcal{E}_0 e^{-i(\omega_0 t + \phi_0)} \frac{\sin(N\alpha/2)}{\sin(\alpha/2)}, \tag{6.97}$$

where $\alpha = \pi ct/L$. The corresponding laser intensity is then

$$I(t) = |\mathcal{E}_0|^2 \frac{\sin^2(N\alpha/2)}{\sin^2(\alpha/2)}. \tag{6.98}$$

This corresponds to a periodic laser pulse output with maximum intensity

$$I_{\max} = |\mathcal{E}_0|^2 N^2 = NI_{\text{tot}}, \tag{6.99}$$

and single pulse duration decrease with $1/N$. This means that, for $N \gg 1$, short laser pulses are periodically emitted with very large intensities and very short durations. Typical values are individual pulse durations of the order of a few femtoseconds, emitted with a periodicity of a few picoseconds [22].

6.6 Amplified spontaneous emission

In the simple laser model used so far, we have neglected the influence of spontaneous emission, assuming that it plays a negligible role in the lasing process. But it is important to notice that, in an amplifying medium where $D > 0$, spontaneous emission is always present and can also be amplified, disturbing sometimes the output of the laser amplifier. In order to describe this process, we go back to the photon balance equation, and include a spontaneous emission term. Neglecting the photon cavity losses, this equation becomes

$$\frac{dn}{dt} = Dwn + AN_2, \tag{6.100}$$

where N_2 is the number population of the upper level in the amplifying medium. This equation has to be modified to account for losses, noting that only a small fraction of the spontaneously emitted photons will propagate inside the medium and will be able to contribute to the amplification process. The largest part will escape laterally from the laser cavity. It is therefore useful to rewrite the above equation as

$$\frac{dn}{dt} = Dwn + A\left(\frac{\Omega}{4\pi}\right)N_2, \tag{6.101}$$

where Ω is the solid angle, out of the total solid angle 4π, corresponding to propagation in the forward direction, and staying inside the cavity. This quantity can be estimated as $\Omega = S/L^2$, where S is the transverse section of the amplifier, and L its length. Assuming that $g = Dw$, and that N_2 is independent of n, which is valid for sufficiently low intensities, we can integrate equation (6.101), and get

$$n(t) = \left(\frac{A\Omega}{4\pi}\right)\frac{N_2}{g}(e^{gt} - 1), \tag{6.102}$$

where we have assumed that $n(0) = 0$. For sufficiently long times, such that $gt \gg 1$, the amplified spontaneous emission grows exponentially, as $\exp(gt)$. If the lower level population stays low during the amplification process, such that $N_2 \gg N_1$, we can approximately write

$$\frac{N_2}{g} \equiv \frac{N_2}{w(N_2 - N_1)} \simeq \frac{1}{w} = \frac{\omega^2 \Delta\omega}{\pi^2 c^3 A}. \tag{6.103}$$

We then get the approximate formula

$$n(t) \simeq \frac{\omega^2 \Delta\omega}{4\pi^3 c^3}\Omega e^{gt}. \tag{6.104}$$

Noting that $\Delta\lambda = 2\pi c \Delta\omega/\omega^2$, we can rewrite this expression in terms of the wavelength, as

$$n(t) = \frac{\omega^4 \Delta\lambda}{8\pi^4 c^4}\Omega e^{gt} = 2\frac{\Delta\lambda}{\lambda^4}\Omega e^{gt}. \tag{6.105}$$

The corresponding the intensity $I(t) = \hbar\omega cn(t)$, will be

$$I(t) = 2\frac{hc^2}{\lambda^5}\Delta\lambda\ \Omega e^{gt}. \tag{6.106}$$

This gives a rough estimate of the intensity associated with amplified spontaneous emission, expected at the output. We should notice that, although this amplified light contributes also to the depopulation of the upper quantum level, it is random in phase and in polarisation, and cannot add in a coherent way to the final output beam. The temporal coherence of this amplified spontaneous emission is essentially different from that of laser radiation.

6.7 Susceptibility

Let us now consider the behaviour of an electromagnetic wave propagating in a material medium, with a large number of atoms. We have seen that the interaction of photons with a single atom is determined by the transition probability from one quantum state to another, with emission or absorption of a photon. From a macroscopic point of view, light propagation is characterised by the refractive index $n(\omega)$, related with the wavenumber k, as $\omega = kc/n(\omega)$. In general, this quantity is complex, $n(\omega) = n' + in''$, where the imaginary part describes damping ($n'' < 0$), or amplification ($n'' > 0$). On the other hand, the refractive index can be defined in terms of the susceptibility $\chi(\omega)$, as $n(\omega)^2 = 1 + \chi(\omega)$. The susceptibility is defined by the ratio between the electric field $\mathcal{E}$ and the resulting polarisation field $\mathbf{P}$ inside the medium. Let us then consider an electric field of the form $\mathcal{E}(t) = \mathcal{E}_0 \cos(\omega t)$. In isotropic media, the polarisation is determined by

$$\mathbf{P}(t) = \frac{1}{2}\epsilon_0 \mathbf{e}\mathcal{E}_0[\chi(\omega)e^{-i\omega t} + \chi(-\omega)e^{i\omega t}], \tag{6.107}$$

where $\mathbf{e}$ is the unit polarisation vector. In anisotropic media we would have to replace the function $\chi(\omega)$ by a susceptibility tensor. Noting that the susceptibility is also complex in general, $\chi(\omega) = \chi' + i\chi''$, we get

$$\chi' = n'^2 - n''^2 - 1, \quad \chi'' = 2n'n''. \tag{6.108}$$

The semi-classical theory can be used to calculate the real and imaginary parts of both the susceptibility and the refractive index. We know that the polarisation of a single atom is given by

$$\mathcal{P}(t) = -e < \psi(t)|z|\psi(t) >, \tag{6.109}$$

where the direction of the incident electric field is assumed in the Oz direction, $\mathbf{e} = \mathbf{e}_z$. For a two-level atom, such that

$$|\psi(t) > = C_1 e^{-iE_1 t/\hbar}|\psi_1 > + C_2 e^{-iE_2 t/\hbar}|\psi_2 >, \tag{6.110}$$

with a transition frequency $\omega_0 = (E_2 - E_1)/\hbar$, very close to the field frequency ω, we have

$$p(t) = -e[C_1^*(t)C_2(t)z_{12} + c.\,c.], \qquad z_{12} = \langle\psi_1|z|\psi_2\rangle\,. \tag{6.111}$$

Solving for the coefficients $C_1(t)$ and $C_2(t)$, as we did before, we arrive at

$$\mathcal{P}(t) = \frac{2\omega_0}{(\omega_0^2 - \omega^2)}\frac{e^2}{\hbar}|z_{12}|^2 E_0 \sin(\omega t). \tag{6.112}$$

For a medium with N atoms per unit volume, the polarisation field will be given by

$$\mathbf{P}(t) = \sum_{j=1}^{N} p_j(t) = N\mathcal{P}(t). \tag{6.113}$$

The atoms are usually oriented in a random way with respect to the direction of the electric field. This means that the axis Oz chosen to describe the atom can be at an arbitrary angle θ with respect to the field $\mathcal{E}_0\mathbf{e}_z$. We then need to replace, in the above results, $|z_{12}|^2$ by $|z_{12}|^2\cos^2\theta$. Averaging over all the possible angles, we get $\langle\cos^2\theta\rangle = 1/3$. The electric susceptibility of the medium can then be written as

$$\chi(\omega) = \frac{2}{3}\frac{e^2 N}{\epsilon_0\hbar}\frac{\omega_0}{(\omega_0^2 - \omega^2)}|z_{12}|^2. \tag{6.114}$$

Or, in a more compact form

$$\chi(\omega) = \omega_p^2\frac{f_{12}}{(\omega_0^2 - \omega^2)}, \tag{6.115}$$

where we have used

$$\omega_p^2 = \frac{e^2 N}{3\epsilon_0 m}, \qquad f_{12} = 2\frac{m\omega_0}{\hbar}|z_{12}|^2. \tag{6.116}$$

The quantity ω_p is called the plasma frequency of the medium, and f_{12} is the *oscillator strength* of the atomic transition. If, instead of a two-level atomic model, we consider the possible existence of several radiative transitions in the atom, with transition frequencies given by ω_{ij}, sufficiently close to the field frequency ω, the susceptibility of the medium becomes

$$\chi(\omega) = \omega_p^2\sum_{i,j}\frac{f_{ij}}{(\omega_{ij}^2 - \omega^2)}, \tag{6.117}$$

where f_{ij} are now the oscillator strength of the relevant transitions.

6.8 Semi-classical laser theory

We are now in conditions to discuss the semi-classical model of the laser. We start from Maxwell's equations, and write the electric field $\mathcal{E}$ in terms of the current density $\mathbf{J}$ and the displacement vector $\mathbf{D}$, as

$$\nabla \times (\nabla \times \mathcal{E}) = -\mu_0 \frac{\partial}{\partial t}\left(\vec{J} + \frac{\partial \vec{D}}{\partial t}\right), \tag{6.118}$$

where $\mathbf{D} = \epsilon_0 \mathcal{E} + \mathbf{P}$. We use Ohm's law, $\mathbf{J} = \sigma \mathcal{E}$, were σ is the conductivity of the medium. And we consider the field inside an optical cavity, with the axis along Oz. Assuming a transverse field, such that $\nabla \cdot \mathcal{E} = 0$, and neglecting the perpendicular field structure, such that $\mathcal{E}(\mathbf{r}, t) = \mathcal{E}(z, t)$, this equation is reduced to

$$\left(\frac{\partial^2}{\partial z^2} - \frac{1}{c^2}\frac{\partial^2}{\partial t^2}\right)\mathcal{E} - \mu_0 \sigma \frac{\partial \mathcal{E}}{\partial t} = \mu_0 \frac{\partial^2 \mathbf{P}}{\partial t^2}. \tag{6.119}$$

We then assume that the fields $\mathcal{E}$ and $\mathbf{P}$ are linearly polarised in the transverse plane, and forget their vectorial character. We consider monochromatic wave solutions, with frequency ω and wavenumber k, of the form

$$\mathcal{E}(z, t) = \frac{1}{2}\mathcal{E}_0(z, t)e^{i(kz-\omega t)+i\varphi(z, t)} + c.\ c., \tag{6.120}$$

where the amplitude $\mathcal{E}_0$ and phase $\varphi(z, t)$ are both slowly varying functions of space and time. When the conductivity is negligible, $\sigma \simeq 0$, they are real. Similarly, the polarisation P can be described by

$$P(z, t) = \frac{1}{2}P_0(z, t)e^{i(kz-\omega t)+i\varphi(z, t)} + c.\ c. \tag{6.121}$$

The first order derivative of the electric field with respect to the space variable is

$$\frac{\partial \mathcal{E}}{\partial z} = \frac{1}{2}\left(ik + i\frac{\partial \varphi}{\partial z} + \frac{1}{\mathcal{E}_0}\frac{\partial \mathcal{E}_0}{\partial z}\right)\mathcal{E}_0 e^{i(kz-\omega t)+i\varphi(z, t)} + c.\ c., \tag{6.122}$$

with a similar expression for the time derivative. On the other hand, for a slow variation of amplitude and phase (the so-called *envelope approximation*), we can write

$$\left(\frac{\partial}{\partial z} - \frac{1}{c}\frac{\partial}{\partial t}\right)\mathcal{E} \simeq \frac{i}{2}\left(k + \frac{\omega}{c}\right)\mathcal{E}_0 e^{i(kz-\omega t)+i\varphi(z, t)} + c.\ c = 2ik\mathcal{E}, \tag{6.123}$$

where $\omega = kc$ was used. Replacing this in the wave equation (6.119), we get

$$2ik\left(\frac{\partial}{\partial z} + \frac{1}{c}\frac{\partial}{\partial t}\right)\mathcal{E} \simeq \mu_0 \sigma \frac{\partial \mathcal{E}}{\partial t} - \mu_0 \omega^2 P. \tag{6.124}$$

Using equation (6.122), a similar time derivative expression, and splitting the real and imaginary parts, we get the evolution equations for the amplitude $\mathcal{E}_0$, and phase φ, as

$$\left(\frac{\partial}{\partial z} + \frac{1}{c}\frac{\partial}{\partial t}\right)\varphi = \left(k - \frac{\omega}{c}\right) - \frac{k}{2\epsilon_0 \mathcal{E}_0}\Re(P_0), \tag{6.125}$$

and

$$\left(\frac{\partial}{\partial z} + \frac{1}{c}\frac{\partial}{\partial t}\right)\mathcal{E}_0 = -\frac{\sigma}{2\epsilon_0 c}\mathcal{E}_0 - \frac{k}{2\epsilon_0}\Im(P_0). \tag{6.126}$$

Inside the cavity, we need to assume discrete wavenumbers $k = k_m$, such that

$$k_m = m\frac{\pi}{L} = \frac{\omega_m}{c}. \tag{6.127}$$

We then focus on the temporal evolution. The phase equation (6.125) gives

$$\frac{\partial}{\partial t}\varphi = (\omega_m - \omega) - \frac{\omega}{2\epsilon_0\mathcal{E}_0}\Re(P_0). \tag{6.128}$$

Similarly, for the field amplitude, we get

$$\frac{\partial}{\partial t}\mathcal{E}_0 = -\frac{1}{2}\gamma_m\mathcal{E}_0 - \frac{\omega}{2\epsilon_0}\Im(P_0), \tag{6.129}$$

where we have included the damping coefficient of the cavity mode m, such that

$$\gamma_m = \frac{\sigma(\omega_m)}{\epsilon_0} = \frac{\omega_m}{Q}, \tag{6.130}$$

with the quality factor Q. In a perfect cavity, we would have $Q \to \infty$ and $\gamma_m \to 0$. But, in this limit, the photons would not be able to escape, and no laser beam could be produced outside. Equations (6.128) and (6.129) can be seen as the basic equations of our quasi-classical laser model.

In order to proceed forward, we need to write the polarisation **P**, in terms of the electric field, which depends on the atomic response. We have seen above that this response depends on the atomic populations, and can described by

$$P_0(t) = -\frac{|\mathcal{P}_{12}|^2}{\hbar}\frac{(\Delta\omega + i\gamma)}{(\Delta\omega)^2 + \gamma^2}(\rho_{22} - \rho_{11})\mathcal{E}_0(t). \tag{6.131}$$

The density matrix elements, ρ_{11} and ρ_{22} are given by equations (6.93) and (6.94), which can be generalised to account for a finite lifetime of the two atomic quantum levels, $1/\gamma_i$, with $i = 1, 2$. Such a decay results from spontaneous emission, and eventually includes collisional decay. We should also introduce finite excitation rates, λ_i, which a due to some pumping mechanism. This leads to

$$\frac{d\rho_{22}}{dt} = \lambda_2 - \gamma_2\rho_{22} - R(\rho_{22} - \rho_{11}), \tag{6.132}$$

and

$$\frac{d\rho_{11}}{dt} = \lambda_1 - \gamma_1\rho_{11} + R(\rho_{22} - \rho_{11}), \tag{6.133}$$

where the rate coefficient R is defined by

$$R = \frac{1}{2}|\Omega_R|^2 \frac{\gamma}{(\Delta\omega)^2 + \gamma^2}, \tag{6.134}$$

and $\gamma = \gamma_1 + \gamma_2$. This coefficient determines the rate at which the population difference varies in time, and depends on the radiation intensity. In steady state, we simply have

$$(\rho_{22} - \rho_{11}) = \frac{(\lambda_2/\gamma_2 - \lambda_1/\gamma_1)}{1 + 2\gamma R/\gamma_1\gamma_2} \equiv \frac{N_{\text{inv}}}{1 + R/R_s}. \tag{6.135}$$

The quantity N_{inv} is the population difference in the absence of radiation. If this population difference remains constant, which is certainly true in the steady-state regime, we get, after replacing equation (6.135) in (6.131), the following phase and amplitude equations

$$\frac{\partial}{\partial t}\varphi = (\omega_m - \omega) + \frac{A\Delta\omega}{2\gamma[1 + (B/A)(\epsilon_0 V/2\hbar\omega)\mathcal{E}_0^2]}, \tag{6.136}$$

and

$$\frac{\partial}{\partial t}\mathcal{E}_0 = -\frac{1}{2}\gamma_m \mathcal{E}_0 + \frac{A\mathcal{E}_0}{2[1 + (B/A)(\epsilon_0 V/2\hbar\omega)\mathcal{E}_0^2]}, \tag{6.137}$$

where V is the volume of the cavity, and the new quantities are

$$A = |\mathcal{P}_{12}|^2 \frac{\gamma\omega}{\epsilon_0\hbar} \frac{N_{\text{inv}}}{(\Delta\omega)^2 + \gamma^2}, \tag{6.138}$$

and

$$\frac{B}{A} = \frac{|\mathcal{P}_{12}|^2}{\hbar^2} \frac{\gamma^2}{\gamma_1\gamma_2} \frac{(2\hbar\omega/\epsilon_0 V)}{(\Delta\omega)^2 + \gamma^2}. \tag{6.139}$$

At this point, it is useful to introduce the photon number $n(\omega)$, defined as

$$n(\omega) \equiv \frac{W(\omega)}{\hbar\omega} V = \frac{\epsilon_0}{2} \frac{\mathcal{E}_0^2}{\hbar\omega} V, \tag{6.140}$$

where $W(\omega)$ is the electromagnetic energy density. The laser equations become

$$\frac{\partial}{\partial t}\varphi = (\omega_m - \omega) + \frac{A\Delta\omega}{2\gamma(1 + Bn/A)}, \tag{6.141}$$

and

$$\frac{\partial n}{\partial t} = -\gamma_m n + \frac{An}{2(1 + Bn/A)}. \tag{6.142}$$

These are the quasi-classical laser equations in their final form. Let us study the laser instability near the threshold. For low intensities, the term Bn is small, and we can expand the denominator. We get

$$\frac{\partial}{\partial t}\varphi = (\omega_m - \omega) + \frac{\Delta\omega}{2\gamma}(A - Bn), \tag{6.143}$$

and

$$\frac{\partial n}{\partial t} = (A - \gamma_m)n - Bn^2. \tag{6.144}$$

The laser threshold is defined by the condition, $dn/dt \geqslant 0$, in a regime where we have $n \simeq 0$. From here we can see that the threshold condition can be defined as

$$A > \gamma_m. \tag{6.145}$$

This is not surprising and simply states that the gain should be larger than the losses, which is equivalent to saying that the inversion of population $\mathcal{N}_{\text{inv}}$ should be larger than some critical quantity $\mathcal{N}_{\text{crit}}$, defined by

$$\mathcal{N}_{\text{crit}} = \frac{\epsilon_0 \hbar \gamma_m}{|\mathcal{P}_{12}|^2 \gamma\omega}[(\Delta\omega)^2 + \gamma^2]. \tag{6.146}$$

At exact resonance, this reduces to

$$\mathcal{N}_{\text{crit}} = \frac{\epsilon_0 \hbar \gamma_m}{|\mathcal{P}_{12}|^2 \omega}\gamma = \frac{\epsilon_0 \hbar}{|\mathcal{P}_{12}|^2}\frac{\gamma}{Q}. \tag{6.147}$$

From the laser equation (6.142) we can also see that steady state is attained for two different values of the photon number: (i) $n = 0$, which is trivial, and (ii) $n = n_0$, where

$$n_0 = \frac{A - \gamma_m}{B}. \tag{6.148}$$

Inserting this steady-state number of photons in equation (6.141) we obtain a constant phase, corresponding to the photon frequency

$$\omega = \frac{\omega_m + S\omega}{1 + S}, \tag{6.149}$$

where S is a stabilising factor defined as

$$S = \frac{A - Bn_0}{2\gamma} = \frac{\gamma_m}{2\gamma}. \tag{6.150}$$

We conclude that, in steady state, the laser frequency ω is shifted from the cavity frequency ω_m by a factor that depends on the cavity losses. Only for low cavity losses, such that $\gamma_m \ll 2\gamma$, we have $\omega \simeq \omega_m$. This frequency shift, or line shift, is called the *mode pulling effect*, and is usually quite small.

6.9 Quantum laser theory

We have shown above that the laser instability can be described by a semi-classical theory, where light is a classical field and only the atoms are quantised. This simplified theoretical approach is able to account for the laser threshold, the laser growth rate, and the steady-state solutions. Such a semi-classical model can also be extended to pulsed laser regimes. However, it ignores spontaneous emission and quantum correlations, which can be important near the threshold. It is therefore useful to consider a full quantum description of the laser, as shown next.

We first need to consider the Hamiltonian operator pertinent for the laser system. As seen before, it contains three terms corresponding to the radiation field, $\hat{H}_f$ the atoms, $\hat{H}_a$ and their interactions, $\hat{H}_{\rm int}$. In this model we also need to consider the atom coupling with the non-lasing modes of radiation, which are important for spontaneous emission. The atomic states can also be coupled with the environment, due to collisions with electrons or with other atoms, and eventually with acoustic oscillations of the medium. All these additional ingredients can be described as a thermal bath, and included in three more Hamiltonian terms, the bath Hamiltonian $\hat{H}_B$, the atom-bath and the field-bath interaction Hamiltonians, $\hat{H}_{\rm aB}$ and $\hat{H}_{\rm fB}$. The total Hamiltonian will that be

$$\hat{H} = \hat{H}_f + \hat{H}_a + \hat{H}_{\rm int} + (\hat{H}_B + \hat{H}_{\rm aB} + \hat{H}_{\rm fB}). \tag{6.151}$$

If we are just considering one cavity mode, the field Hamiltonian reduces to

$$\hat{H}_f = \hbar\omega\left(\hat{a}^\dagger\hat{a} + \frac{1}{2}\right), \tag{6.152}$$

with the usual bosonic commutation relations

$$[\hat{a}, \hat{a}^\dagger] \equiv \hat{a}\hat{a}^\dagger - \hat{a}^\dagger\hat{a} = 1. \tag{6.153}$$

The Hamiltonian of a single two-level atom can be written in the form

$$\hat{H}_a = E_1\hat{b}_1^\dagger\hat{b}_1 + E_2\hat{b}_2^\dagger\hat{b}_2 = \sum_{j=1,2} E_j\hat{b}_j^\dagger\hat{b}_j. \tag{6.154}$$

This operator simplifies if we take the zero of the energy scale coincident with the lower energy level. We then write $E_1 = 0$ and $E_2 = \hbar\omega_a$, where ω_a is the atomic transition frequency. The operator becomes

$$\hat{H}_a = \hbar\omega_a\hat{b}_2^\dagger\hat{b}_2, \tag{6.155}$$

where the fermionic anti-commutation relations hold

$$\{\hat{b}_2, \hat{b}_2^\dagger\} \equiv \hat{b}_2\hat{b}_2^\dagger + \hat{b}_2^\dagger\hat{b}_2 = 1. \tag{6.156}$$

Furthermore, we have previously shown that the atom–field interaction Hamiltonian can be written as

$$\hat{H}_{\rm int} = \hbar g\left(\hat{b}_1^\dagger\hat{b}_2 + \hat{b}_2^\dagger\hat{b}_1\right)\left(\hat{a} + \hat{a}^\dagger\right). \tag{6.157}$$

As we have seen, the term $\hat{b}_1^{\dagger}\hat{b}_2\hat{a}^{\dagger}$ represents creation of a photon and destruction of an electron state in the upper level $|\,2\rangle$, with creation of an electron state in the lower level $|\,1\rangle$. The other terms were already discussed and have similar meanings. Introducing the rotating-wave approximation, we neglect the two terms violating energy conservation. As a result, we get

$$\hat{H}_{\text{int}} = \hbar g\left(\hat{b}_1^{\dagger}\hat{b}_2\hat{a}^{\dagger} + \hat{b}_2^{\dagger}\hat{b}_1\hat{a}\right). \tag{6.158}$$

In a laser cavity, we have not just one, but a large number of atoms N, which implies a straightforward generalisation of the atom and the interaction Hamiltonians, which become

$$\hat{H}_a = \hbar\omega_0 \sum_{i=1}^{N} \hat{b}_{2i}^{\dagger}\hat{b}_{2i}, \tag{6.159}$$

and

$$\hat{H}_{\text{int}} = \hbar g \sum_{i=1}^{N}\left(\hat{b}_{1i}^{\dagger}\hat{b}_{2i}\hat{a}^{\dagger} + \hat{b}_{2i}^{\dagger}\hat{b}_{1i}\hat{a}\right). \tag{6.160}$$

Let us now consider the evolution equations for the field operators. We know that, in the Heisenberg picture, an arbitrary operator $\hat{A}$ evolves in time according to the equation

$$\frac{d\hat{A}}{dt} = \frac{\partial\hat{A}}{\partial t} + \frac{i}{\hbar}[\hat{H}, \hat{A}]. \tag{6.161}$$

If $\hat{A}$ is replaced by the destruction field operator $\hat{a}$, and is not explicitly dependent on time, we have, for $\hat{H} = \hat{H}_f$, the following evolution equation

$$\frac{d\hat{a}}{dt} = \frac{i}{\hbar}(\hat{H}_f\hat{a} - \hat{a}\hat{H}_f) = i\omega(\hat{a}^{\dagger}\hat{a}\hat{a} - \hat{a}\hat{a}^{\dagger}\hat{a}). \tag{6.162}$$

Using the commutation relation (6.153), this reduces to

$$\frac{d\hat{a}}{dt} = i\omega(\hat{a}^{\dagger}\hat{a}\hat{a} - (1 + \hat{a}^{\dagger}\hat{a})\hat{a}) = -i\omega\hat{a}. \tag{6.163}$$

Let us consider now the interaction of the radiation field mode with a thermal bath. We replace $\hat{H}_f$ by $(\hat{H}_f + \hat{H}_B + H_{fB})$, and get

$$\frac{d\hat{a}}{dt} = \frac{i}{\hbar}\left[\left(\hat{H}_f + \hat{H}_B + \hat{H}_{fB}\right), \hat{a}\right]. \tag{6.164}$$

This can also be written as

$$\frac{d\hat{a}}{dt} = -i\omega\hat{a} - \gamma_c\hat{a} + \hat{F}(t), \tag{6.165}$$

where γ_c is the damping rate of the field mode in the cavity, and the operator $\hat{F}(t)$ represents a fluctuating force with the following statistical properties

$$<\hat{F}(t)> = <\hat{F}^{\dagger}(t)> = 0, \quad <\hat{F}^{\dagger}(t)\hat{F}(t')> = 2\gamma_c n_0 \delta(t - t'), \tag{6.166}$$

and

$$<\hat{F}(t)\hat{F}^{\dagger}(t')> = 2\gamma_c(n_0 + 1)\delta(t - t'), \tag{6.167}$$

where n_0 is the mean photon number for a thermal bath at temperature T. For lasers in the optical domain, and at room temperature, we can assume that $n_0 \simeq 0$.

Now, let us consider a similar coupling between the atoms and the thermal bath. For that purpose, it is useful to introduce the atom difference operator

$$\hat{d} = \hat{b}_2^{\dagger}\hat{b}_2 - \hat{b}_1^{\dagger}\hat{b}_1. \tag{6.168}$$

The corresponding evolution equation is

$$\frac{d}{dt}\hat{d} = \frac{i}{\hbar}[\hat{H}_a, \hat{d}]. \tag{6.169}$$

Using equation (6.159), we get

$$\frac{d}{dt}\hat{d} = 0. \tag{6.170}$$

It is also useful to consider two other auxiliary operators, $\hat{\alpha}$ and $\hat{\alpha}^{\dagger}$, such that

$$\hat{\alpha} = \hat{b}_1 + \hat{b}_2, \quad \hat{\alpha}^{\dagger} = \hat{b}_2^{\dagger}\hat{b}_1, \tag{6.171}$$

which evolve as

$$\frac{d}{dt}\hat{\alpha} = -i\omega_a\hat{\alpha}, \quad \frac{d}{dt}\hat{\alpha}^{\dagger} = i\omega_a\hat{\alpha}^{\dagger}. \tag{6.172}$$

We are now in conditions to describe the atoms in a thermal bath. As we have seen in the semi-classical theory, the atom occupation numbers N_1 and N_2 were described by the balance equations

$$\frac{dN_2}{dt} = w_{21}N_1 - w_{12}N_2, \quad \frac{dN_1}{dt} = -w_{21}N_1 + w_{12}N_2, \tag{6.173}$$

where w_{21} was the pumping transition rate, and w_{12} the spontaneous decay rate. In a quantum description, we can use $N_j = \left\langle \hat{b}_j^{\dagger}\hat{b}_j \right\rangle$, and these equations are replaced by similar ones describing the atom number operators $\hat{N}_j \equiv \hat{b}_j^{\dagger}\hat{b}_j$, as

$$\frac{d\hat{N}_2}{dt} = w_{21}\hat{N}_1 - w_{12}\hat{N}_2 + \hat{\Gamma}_2(t), \quad \frac{d\hat{N}_1}{dt} = -w_{21}\hat{N}_1 + w_{12}\hat{N}_2 + \hat{\Gamma}_1(t). \tag{6.174}$$

Here, $\hat{\Gamma}_j$ are new fluctuation force operators, similar to $\hat{F}(t)$, and having similar statistical properties

$$\left\langle \hat{\Gamma}_j(t) \right\rangle = 0, \quad \left\langle \hat{\Gamma}_j(t)\hat{\Gamma}_k(t') \right\rangle = \hat{G}_{jk}\delta(t - t'), \tag{6.175}$$

where $\hat{G}_{jk}$ is a correlation operator, to be specified. Subtracting equations (6.174), we obtain an evolution equation for the difference operator defined in equation (6.168), of the form

$$\frac{d}{dt}\hat{d} = 2(w_{21}\hat{N}_1 - w_{12}\hat{N}_2) + \hat{\Gamma}, \tag{6.176}$$

where $\hat{\Gamma} = \hat{\Gamma}_2 - \hat{\Gamma}_1$. This can also be written as

$$\frac{d}{dt}\hat{d} = \gamma_\nu(\hat{d}_0 - \hat{d}) + \hat{\Gamma}, \tag{6.177}$$

where we have used

$$\gamma_\nu = w_{12} + w_{21}, \quad \hat{d}_0 = \frac{w_{21} - w_{12}}{w_{12} + w_{21}}. \tag{6.178}$$

Comparing this with the semi-classical laser model, we see that $\hat{d}_0$ corresponds to the unsaturated inversion operator, and that $\gamma_\nu = 1/\tau$, is the inverse of the relaxation time. Similarly, we can rewrite the evolution equations for the operators $\hat{\alpha}$ and $\hat{\alpha}^\dagger$, (6.172) in a more complete form, as

$$\frac{d}{dt}\hat{\alpha} = -i\omega_a\hat{\alpha} - \gamma\hat{\alpha} + \hat{\Gamma}_-(t), \quad \frac{d}{dt}\hat{\alpha}^\dagger = i\omega_a\hat{\alpha}^\dagger - \gamma\hat{\alpha}^\dagger + \hat{\Gamma}_+(t). \tag{6.179}$$

Finally, we consider the contribution of $\hat{H}_{\text{int}}$ to the field mode operator $\hat{a}$: for a single atom this is described by

$$\left(\frac{d\hat{a}}{dt}\right)_{\text{int}} = \frac{i}{\hbar}[\hat{H}_{\text{int}}, \hat{a}] = g[(\hat{a}\hat{\alpha}^\dagger\hat{a} + \hat{a}^\dagger\hat{\alpha}), \hat{a}] = -ig\hat{\alpha}. \tag{6.180}$$

Generalising to N atoms, we obtain

$$\left(\frac{d\hat{a}}{dt}\right)_{\text{int}} = -ig\sum_{i=1}^{N}\hat{\alpha}_i. \tag{6.181}$$

Adding this to equation (6.165), we obtain the contribution of the total Hamiltonian $\hat{H}$ defined in (6.151), leading to the final mode operator equation

$$\frac{d\hat{a}}{dt} = -(i\omega + \gamma_c)\hat{a} - ig\sum_{i=1}^{N}\hat{\alpha}_i + \hat{F}(t). \tag{6.182}$$

Similarly, we can calculate the contributions of H_{int} to the evolution of the atom operator $\hat{\alpha}$, using

$$\left(\frac{d\hat{\alpha}}{dt}\right)_{\text{int}} = \frac{i}{\hbar}[\hat{H}_{\text{int}}, \hat{\alpha}] = ig\hat{a}\hat{d}. \tag{6.183}$$

For any of the N atoms, the contributions from all the Hamiltonian terms are described by

$$\frac{d\hat{\alpha}_j}{dt} = -(i\omega_a + \gamma)\hat{\alpha}_j + ig\hat{a}\hat{d}_j + \hat{\Gamma}_{j-}(t), \tag{6.184}$$

for $j = 1, 2, \ldots, N$. We get a similar expression of $\hat{\alpha}_j^\dagger$, as

$$\frac{d\hat{\alpha}_j^\dagger}{dt} = +(i\omega_a - \gamma)\hat{\alpha}_j^\dagger - ig\hat{a}\hat{d}_j + \hat{\Gamma}_{j+}(t). \tag{6.185}$$

It also follows, for the difference operator

$$\frac{d}{dt}\hat{d}_j = \gamma_v(\hat{d}_0 - \hat{d}_j) + 2ig(\hat{\alpha}_j a^\dagger - \hat{\alpha}_j^\dagger a) + \Gamma_{jd}(t). \tag{6.186}$$

Following a procedure similar to that of the semi-classical laser theory, we eliminate the operators $\hat{\alpha}_j$ from equation (6.182), and obtain a closed equation for the field operator $\hat{a}$, of the form

$$\frac{d\hat{a}}{dt} = -(i\omega - \hat{G})a - \hat{C}\hat{a}^\dagger\hat{a}\hat{a} + \tilde{F}(t), \tag{6.187}$$

where the coupling $\hat{C}$ and gain $\hat{G}$ operators, are

$$\hat{C} = \frac{4g^4}{\gamma_v\gamma^2}N\hat{d}_0, \quad \hat{G} = -\gamma_c + \frac{g^2}{\gamma}N\hat{d}_0, \tag{6.188}$$

and the total fluctuation force operator is

$$\tilde{F}(t) = F(t) - i\frac{g}{\gamma}\sum_{j=1}^{N}\hat{\Gamma}_{j-}(t). \tag{6.189}$$

Formally, equation (6.187) is very similar to that of the semi-classical theory, with two main qualitative differences. First, the field amplitude is replaced by the field operator $\hat{a}$. Second, an additional fluctuation term $\tilde{F}(t)$ is included, which has no equivalent in the semi-classical model. This new term results from the field and atom interactions with the thermal bath. Therefore, the quantum laser equation takes the form of an operator Langevin equation.

It is now useful to discuss the qualitative features of the laser field below threshold. In this case, we have $\hat{G} < 0$. We also have a very small mean value for the field number operator $\hat{a}^\dagger\hat{a}$, which means that the nonlinear term $\hat{C}\hat{a}^\dagger\hat{a}\hat{a}$ can be neglected. The resulting approximate equation, valid below threshold, is

$$\frac{d\hat{a}}{dt} = -\Omega\hat{a} + \tilde{F}, \quad \Omega = \omega - i|\hat{G}|. \tag{6.190}$$

This can be formally integrated to give

$$\hat{a}(t) = \hat{a}(0)e^{-i\Omega t} + \int_0^t \tilde{F}(t')e^{i\Omega(t-t')}dt'. \tag{6.191}$$

The first term in this solution is the *ballistic term*, depending on the initial field conditions, and the second one is the force term due to the existence of fluctuations. It is appropriate to represent $\tilde{F}(t)$ as the sum of a succession of instantaneous perturbations, occurring at random times t_k, with random amplitudes and phases, as

$$\tilde{F}(t) = \sum_k \hat{f}_k e^{i\theta_k} \delta(t - t_k). \tag{6.192}$$

Inserting this in the above formal solution, and neglecting the ballistic term, which corresponds to take $\hat{a}(0) = 0$, we obtain

$$\hat{a}(t) = \sum_k \hat{f}_k \exp(-i\omega(t - t_k) - |\hat{G}|t + i\theta_k). \tag{6.193}$$

This shows the existence of successive damped oscillations, starting at random intervals, with a kind of intermittent behaviour. This solution also shows that the mean value of the field operator is equal to zero, $<\hat{a}(t)> = 0$. Let us now look at its coherence properties, using

$$\langle \hat{a}^\dagger(t)\hat{a}(t')\rangle = e^{i\Omega(t-t')}\langle \hat{a}^\dagger(t')\hat{a}(t')\rangle = e^{i\Omega(t-t')}\langle \hat{n}(t')\rangle, \tag{6.194}$$

where $\langle \hat{n}(t)\rangle$ is the average photon number. Its explicit calculation is not performed here, but can be written in the form

$$\langle \hat{n}\rangle = \frac{\gamma_c}{|G|}(n_{\text{th}} + n_{\text{sp}}), \tag{6.195}$$

where n_{th} represents the mean number of photons of the laser field mode with frequency ω, as it would be in thermal equilibrium, and n_{sp} is the mean number of spontaneously emitted photons.

When inversion of population is equal to zero, we have $Nd_0 = 0$, and $|G| = \gamma_c$. In this case, the effect of stimulated emission and absorption exactly compensate, and we get $\langle \hat{n}\rangle = n_{\text{th}} + n_{\text{sp}}$. For positive inversion, we have $|G| < \gamma_c$, and the number of photons increases, $\langle \hat{n}\rangle > (n_{\text{th}} + n_{\text{sp}})$, even below threshold. Finally, at threshold, $|G| \to 0$ as the amplification factor $\gamma_c/|G|$ tends to infinity, and the number of photons becomes larger than $(n_{\text{th}} + n_{\text{sp}})$, growing well above noise. However, indefinite growth of $\langle \hat{n}\rangle$ is prevented by the nonlinear term $\hat{C}\hat{a}^\dagger\hat{a}\hat{a}$, which becomes non-negligible. As a result, the mean number of photons will remain finite at threshold, as it should. Finally, above threshold, the field coherence properties will dramatically change, and the uncorrelated damped signal is replaced by a strongly correlated field with only a negligible fluctuation level.

In summary, we have described the formation of a collective quantum state of light, that we call laser. Our description was based on three different laser theories.

The first and simplest one is a phenomenological theory, based on balance equations, which is able to demonstrate the occurrence of laser instability, and give an estimate of the threshold conditions. The second one is a semi-classical theory where only the atomic levels are quantised, which includes a description of the field phase, and phase coherence. Finally, the third and most sophisticated one, is the quantum theory where both field and atoms are quantised, and includes the influence of spontaneous emission. We have seen that, for laser operation when a large number of photons is present in the atomic medium, spontaneous emission is negligible except near the threshold. However, as we will see later, spontaneous emission can play a dominant role in the formation of collective quantum states of light, leading to superradiance.

References

[1] Schawlow A L and Townes C H 1958 Infrared and optical masers *Phys. Rev.* **112** 1940–9

[2] Basov N G and Prokhorov A M 1955 Possible methods for obtaining active molecules for a molecular oscillator *Sov. Phys. JETP* **1** 184

[3] Perot A and Fabry C 1899 On the application of interference phenomena to the solution of various problems of spectroscopy and metrology *Astrophys. J.* **16** 87–115

[4] Maiman T 1960 Stimulated optical radiation in ruby *Nature* **187** 493–4

[5] Javan A, Bennet W R and Herriott D R 1961 Population inversion and continuous optical maser oscillation in a gas discharge containing a He-Ne mixture *Phys. Rev. Lett.* **6** 106–10

[6] Hall R N, Fenner G E, Kingsley J D, Soltys T J and Carlson R O 1962 Coherent light emission from GaAs junctions *Phys. Rev. Lett.* **9** 366–8

[7] Patel C K N 1964 Continuous-wave laser action on vibrational-rotational transitions of CO2 *Phys. Rev.* **136** A1187–93

[8] Siegmann A E 1986 *Lasers* (Sausalito, CA: University Science Books)

[9] Milonni P W and Eberly J H 1988 *Lasers* (New York: Wiley)

[10] Young M 1992 *Optics and Lasers* 4th edn (Berlin: Springer)

[11] Svelto O 2010 *Principles of Lasers* 5th edn (New York: Springer)

[12] Haken H 1964 A nonlinear theory of laser noise and coherence *Z. Phys.* **181** 96–124

[13] Scully M O and Lamb W E 1967 Quantum theory of an optical maser. I. General theory *Phys. Rev.* **159** 208–26

[14] Lax M and Louisell W H 1969 Quantum noise. XII. Density-operator treatment of field and population fluctuations *Phys. Rev. Lett.* **185** 568–91

[15] Einstein A 1917 The quantum theory of radiation *Phys. Z.* **18** 121

[16] Kastler A 1950 Quelques suggestions concernant la production optique et la détection optique d'une inégalité de population des niveaux de quantification spatiale des atomes. Application à l'expérience de Stern et Gerlach et à la résonance magnétique *J. Phys. Radium* **11** 255–65

[17] Happer W 1972 Optical pumping *Rev. Mod. Phys.* **44** 169–249

[18] Fox A G and Li T 1961 Resonant modes in a maser interferometer *Bell Syst. Tech. J.* **40** 453–88

[19] Barnett S M and Radmore P M 1988 Quantum theory of cavity quasimodes *Opt. Commun.* **68** 364–8

[20] Gea-Banacloche J, Lu N, Pedrotti L M, Prasad S, Scully M O and Wódkiewicz K 1990 Treatment of the spectrum of squeezing based on the modes of the universe. I. Theory and physical picture *Phys. Rev.* A **41** 369–80

[21] Leung P T, Liu S Y and Young K 1994 Completeness and orthogonality of quasinormal modes in leaky optical cavities *Phys. Rev.* A **49** 3057–67

[22] Ell R *et al* 2001 Generation of 5-fs pulses and octave-spanning spectra directly from a Ti: sapphire laser *Opt. Lett.* **26** 373–5

IOP Publishing

The Quantum Nature of Light
From photon states to quantum fluids of light
J T Mendonça

Chapter 7

Bose–Einstein condensates

This concept is a direct consequence of the work published by Bose and Einstein [1, 2] in the early 20s of the last century. First, Bose was able to derive Planck's law of radiation from the postulate that photons are identical particles. Second, Einstein extended his approach to an ideal gas of identical atoms, and predicted that they could condensate at the lowest energy level in the low temperature limit. This is valid for bosons, that is, for atoms with integer spin. If the atoms are fermions, with semi-integer spin, condensation will not occur because it is forbidden by Pauli's exclusion principle. After many decades, the effect of condensation was finally observed experimentally in 1995, using ultracold vapours of alkali atoms, such as rubidium 85 [3], sodium 23 [4] and lithium 7 [5]. These results started a tremendous theoretical and experimental revolution which is still going on. A number of books can be found, describing the physics of condensates [6–8]. Although the work of Bose and Einstein assumed an ideal gas of identical particles, the experiments using dilute gases showed that weak interactions between the atoms, due to close collisions at nearly zero energy, are still present. They actually play an important part in the superfluid behaviour of condensates.

In this chapter, we consider some of the basic properties of a Bose–Einstein condensate (BEC). Our discussion includes the mean-field theory, Gross–Pitaevskii equation, Bogoliubov theory and bogolon dispersion. We also consider the case of photon condensates, first in a diffusive dye solution contained in an optical cavity, and then in unbounded plasmas. We finally discuss the laser-BEC transition and study the similarities and differences between these two collective quantum states of light.

A gas of weakly interacting particles with integer spin (bosons) is a condensate when most of the particles in a given volume accumulate at their lowest energy state. Obviously, because of the exclusion principle, they cannot be fermions. Condensation occurs below a given temperature, called *critical temperature*, T_c. This means that in the process of condensation, some cooling mechanism must be in action. Another important property of the condensates is that all the wavefunctions describing every individual particle are phase-locked, through their own mean-field, and attain a collective state. This also implies the existence of some weak-interaction between bosons, so that the system can be thermalised and attain a mean-field state.

Finally, it should be noted that a finite critical temperature is always associated with a finite boson mass. Therefore, the particles (atoms or photons) need a finite mass in order to condensate. Massless particles, with $T_c = 0$, cannot condensate. This is particularly important for photons, because plane wave photons in vacuum have no mass. However, photons acquire a finite mass in specific physical conditions, such as in a cavity, or in a plasma. This photon mass was first noticed by Anderson in his theory of superconductivity [9], and indirectly inspired the concept of Higgs fields [10]. Further extension of the plasma analysis revealed the existence of a photon charge [11].

In general terms, the occurrence of Bose–Einstein condensation implies the conjugation of three different factors: (i) a cooling mechanism, which brings the gas temperature down to criticality, (ii) a finite critical temperature, which is associated to a finite mass and defines a phase transition, and (iii) weak interactions between individual particles, which build up the mean-field and create the collective state. Bose–Einstein condensation of photons in unbounded plasmas was first proposed by Zel'dovich in 1969, in an astrophysical context [12]. Condensation of photons in optical cavities was first proposed in 2000 [13]. But the experimental breakthrough came in 2010, with the first observation of photon condensation [14]. Condensation is actually a very general phenomenon, which can also take place in the presence of a classical wave spectrum [15, 16].

7.1 Basic concepts

7.1.1 Critical temperature

We start with the defining of *critical temperature*, one of the basic requirements for condensation. For this purpose, we consider a system containing a large number of bosons, for instance a dilute gas of identical atoms in some bosonic state. The quantum state of the system will be described by the wave operators

$$\hat{\psi}(\mathbf{r}, t) = \sum_{\mathbf{p}} \psi_{\mathbf{p}}(\mathbf{r}, t)\hat{a}_{\mathbf{p}}, \quad \hat{\psi}^{\dagger}(\mathbf{r}, t) = \sum_{\mathbf{p}} \psi_{\mathbf{p}}^{*}(\mathbf{r}, t)\hat{a}_{\mathbf{p}}^{\dagger}, \tag{7.1}$$

where $\hat{a}_{\mathbf{p}}$ and $\hat{a}_{\mathbf{p}}^{\dagger}$ are the destruction and creation operators of the states $\mathbf{p}$. The simplest example is that of atoms moving in free space. The form function $\psi_{\mathbf{p}}(\mathbf{r}, t)$ associated with each state is then given by

$$\psi_{\mathbf{p}}(\mathbf{r}, t) = \exp\left[(i/\hbar)(\mathbf{p} \cdot \mathbf{r} - \epsilon_{\mathbf{p}} t)\right], \tag{7.2}$$

where $\mathbf{p}$ is the particle momentum, and $\epsilon_{\mathbf{p}} = \mathbf{p}^2/2M$ the energy, for atoms with mass M. In a gas at temperature T, the mode occupation number $N_{\mathbf{p}} = \left\langle \hat{a}_{\mathbf{p}}^{\dagger}\hat{a}_{\mathbf{p}} \right\rangle$ is determined by the Bose–Einstein distribution, as

$$N_{\mathbf{p}} = \frac{1}{\exp\left[(\epsilon_{\mathbf{p}} - \mu)/T\right] - 1}, \tag{7.3}$$

where T is in energy units, and μ is the *chemical potential*. We know that, for an ideal gas, the chemical potential can only take negative values, $\mu \leqslant 0$. The number of particles in the unit volume, $n = N/V$, is obtained by integration over all the possible momentum states $\mathbf{p}$. Assuming isotropy, we get

$$n = \frac{2\pi}{\lambda_T} \int_0^\infty \frac{\sqrt{z}\, dz}{\exp(z - \mu/T) - 1}, \tag{7.4}$$

with $z = \epsilon_{\mathbf{p}}/T$ and $\lambda_T = h/\sqrt{2\pi M T}$. When we lower the temperature of the gas, the maximum possible value of the chemical potential, $\mu = 0$, is attained at some critical temperature $T = T_c$. In this case, the above integral is approximately equal to $1.3\sqrt{\pi}$. We then get a relation between critical temperature and gas density, as

$$T_c \simeq 3.3 \frac{\hbar^2}{M} n^{2/3}. \tag{7.5}$$

We can see that the critical temperature depends on the particle mass M and increases with density, as $n^{2/3}$. In the case of a gas trapped in an harmonic potential the proportionality is different, $T_c \propto n^{1/3}$. For a temperature lower than critical, $T < T_c$, a finite number N_0 of atoms will condensate in the lowest energy level, $\epsilon_{\mathbf{p}} = 0$. Its number will increase for a decreasing temperature, until we attain total condensation, $N_0 = N$, at the absolute zero $T = 0$. The scaling law is

$$\frac{N_0}{N} = 1 - \left(\frac{T}{T_c}\right)^\alpha, \tag{7.6}$$

valid for $T \leqslant T_c$. The exponent in this expression is equal to $\alpha = 3/2$ for a free gas, and $\alpha = 3$ for a trapped gas (figures 7.1 and 7.2).

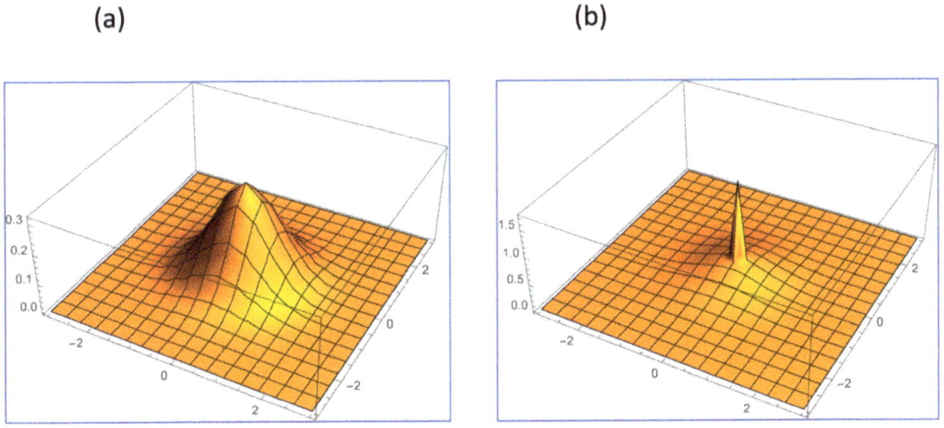

Figure 7.1. Bose–Einstein condensation of a dilute gas: particle probability distribution in velocity space (v_x, v_y) for (a) a thermal gas ($T > T_c$); and (b) a condensed gas ($T < T_c$).

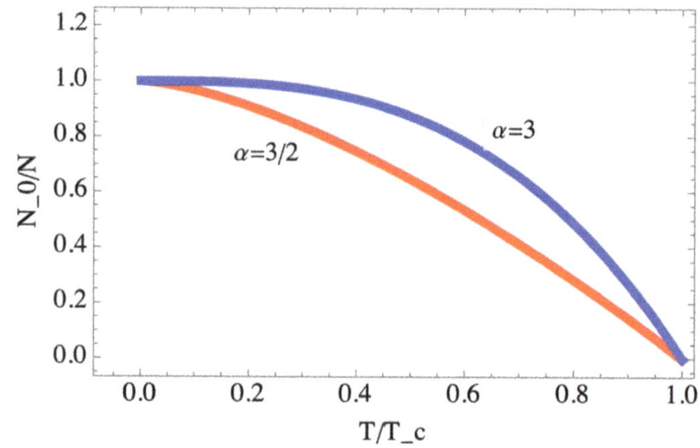

Figure 7.2. Bose–Einstein condensation: fraction of particles in the lowest energy state N_0/N, as a function of the relative temperature T/Tc, in the condensed region, for free ($\alpha = 3/2$) and trapped ($\alpha = 3$) configurations.

7.1.2 Mean-field description

It is important to notice that, once we have condensation, the fraction of condensed gas can be described by a single collective wavefunction. This important property shows that a condensed fluid is a kind of collective quantum object. In order to understand this collective state, we consider the Heisenberg equation for the boson field operator $\hat{\psi}(\mathbf{r}, t)$, defined in equation (7.1), which can be written as

$$\frac{\partial}{\partial t}\hat{\psi}(\mathbf{r}, t) = \frac{1}{i\hbar}[\hat{\psi}(\mathbf{r}, t), \hat{H}], \tag{7.7}$$

with the Hamiltonian operator

$$\hat{H} = \hat{H}_{\text{int}} + \int d\mathbf{r}\hat{\psi}^{\dagger}(\mathbf{r}, t)\left[\frac{p^2}{2M} + V_e(\mathbf{r}, t)\right]\hat{\psi}(\mathbf{r}, t). \tag{7.8}$$

Here, $V_e(\mathbf{r}, t)$ represents some external potential, and H_{int} is the interaction Hamiltonian, of the form

$$\hat{H}_{\text{int}} = \frac{1}{2}\int d\mathbf{r}\int d\mathbf{r}'\hat{\psi}^{\dagger}(\mathbf{r}, t)\hat{\psi}(\mathbf{r}', t)V(\mathbf{r} - \mathbf{r}')\hat{\psi}(\mathbf{r}, t)\hat{\psi}(\mathbf{r}', t), \tag{7.9}$$

where $V(\mathbf{r} - \mathbf{r}')$ is a generic potential describing inter-particle interactions. For instance, in a dilute gas, this is the atom–atom collision potential and, for atoms nearly at rest, can be simply defined as

$$V(\mathbf{r} - \mathbf{r}') = g\delta(\mathbf{r} - \mathbf{r}'), \quad g = 4\pi\hbar^3\frac{a}{M}, \tag{7.10}$$

where g represents the coupling parameter, and a is the s-wave scattering length, associated with low energy atomic collisions. In a photon gas, these local particle interactions should be replaced by weak non-local interactions between different photon modes, but the description would be formally similar, as shown later.

At this point, we should focus on the case where a large fraction of bosons is already condensed in the lowest energy state, as illustrated in figure 7.1(b). The quantum state of the gas can then be described by the sum of its condensed part, represented by a complex function $\Phi(\mathbf{r}, t)$, and its fluctuation part, represented by the operator $\delta\hat{\psi}(\mathbf{r}, t)$, such that

$$\hat{\psi}(\mathbf{r}, t) = \Phi(\mathbf{r}, t) + \delta\hat{\psi}(\mathbf{r}, t), \tag{7.11}$$

where the function $\Phi(\mathbf{r}, t)$ is defined as the average of the boson operator,

$$\Phi(\mathbf{r}, t) = \langle\hat{\psi}(\mathbf{r}, t)\rangle, \quad \langle\delta\hat{\psi}(\mathbf{r}, t)\rangle = 0. \tag{7.12}$$

Replacing this in equation (7.7), and using equations (7.9)–(7.12), we can then derive an evolution equation for the condensed part, of the form

$$\frac{\partial}{\partial t}\Phi(\mathbf{r}, t) = \left[\frac{p^2}{2M} + V_e(\mathbf{r}, t) + g \mid \Phi(\mathbf{r}, t) \mid^2\right]\Phi(\mathbf{r}, t). \tag{7.13}$$

This is the celebrated *Gross–Pitaevskii equation* [17, 18], or simply GP equation, which describes the mean-field properties of the condensate. This equation is valid when the fraction of condensed particles is nearly equal to one, $N_0/N \sim 1$, and the fluctuations represented by the operator $\delta\hat{\psi}(\mathbf{r}, t)$ are negligible. This is formally a nonlinear Schrödinger equation, where the nonlinear term results from inter-particle interactions. The average function $\Phi(\mathbf{r}, t)$, is sometimes called the *order parameter*, or in alternative, the *condensate wavefunction*, and describes the collective behaviour of the condensed gas. Quantum fluid equations can actually be derived from the GP equation, using the Madelung transformation [19], such that

$$\Phi(\mathbf{r}, t) = \sqrt{n(\mathbf{r}, t)}\exp\left[i\varphi(\mathbf{r}, t)\right], \tag{7.14}$$

where the fluid density n, and velocity $\mathbf{v}$, are defined as

$$n = \mid \Phi(\mathbf{r}, t) \mid^2, \quad \mathbf{v} = \frac{\hbar}{2M}\nabla\varphi(\mathbf{r}, t). \tag{7.15}$$

Replacing this in equation (7.13), and equating to zero the real and imaginary parts separately, we obtain two fluid equations, representing density and momentum conservation, as

$$\frac{\partial n}{\partial t} + \nabla \cdot (n\mathbf{v}) = 0, \quad \frac{\partial \mathbf{v}}{\partial t} + \mathbf{v} \cdot \nabla\mathbf{v} = -\frac{1}{M}\nabla(V_e + gn + V_B). \tag{7.16}$$

We can see that the motion of the condensed gas results from three different forces depending on the external potential V_e, the collision (or interaction) potential gn, and the quantum potential, also commonly referred as the *Bohm potential* [20], defined by

$$V_B = -\frac{\hbar^2}{2M}\frac{\nabla^2\sqrt{n}}{\sqrt{n}}. \tag{7.17}$$

Many interesting, and sometimes intriguing, properties of the Bose–Einstein condensates can be derived from these quantum fluid equations. One of them is associated with the elementary excitations of the fluid, which are described next.

7.1.3 Elementary excitations

Density perturbations of the condensate can be described by the quantity $\tilde{n} = n - n_0$, where n_0 represents a given equilibrium density. Assuming that the condensate is at rest, we can use $\mathbf{v}_0 = 0$ as the equilibrium velocity. Replacing this in equations (7.16), and linearising with respect to perturbations, we obtain an equation for the perturbed density, as

$$\frac{\partial^2 \tilde{n}}{\partial t^2} - \frac{n_0}{M}\nabla^2\left(g\tilde{n} - \frac{\hbar^2}{4M}\frac{\nabla^2 \tilde{n}}{n_0}\right) = 0, \tag{7.18}$$

where we have neglected the influence of the external potential V_e. For perturbations with frequency ω and wavevector $\mathbf{k}$, evolving in space and time as $\exp(i\mathbf{k} \cdot \mathbf{r} - i\omega t)$, we get the following dispersion relation

$$\omega^2 = c_s^2 k^2 + \frac{\hbar^2 k^4}{4M^2}, \quad c_s^2 = \frac{g n_0}{M}, \tag{7.19}$$

where the quantity c_s is usually called the *Bogoliubov speed.* We can see that, for long wavelengths where the first term dominates, these modes look like sound waves, with $\omega \simeq c_s k$, where c_s can be seen as the sound speed of the medium. In contrast, for short wavelengths where the second term dominates, they behave like free particles with mass M, satisfying $\epsilon_\mathbf{p} \simeq (p^2/2M)$ where $\epsilon_\mathbf{p} = \hbar\omega$ is the energy and $\mathbf{p} = \hbar\mathbf{k}$. These modes can be seen as the sound waves in the condensed fluid, and for that reason they could be called *phonons.* In alternative, they are also mentioned as *Bogoliubov excitations*, and sometimes *bogolons.*

A more sophisticated description of these modes can be made, in order to include kinetic effects, such as those associated with Landau damping. This effect was first formulated by Landau in the context of plasma physics, and was extended to Bose–Einstein condensates and other fields. Here we give a formulation based on the wave-kinetic equation for the condensate. We start from the GP equation (7.13), which can be written in the form

$$i\hbar\frac{\partial \Phi}{\partial t} = H_{GP}\Phi, \quad H_{GP} = -\frac{\hbar^2}{2m}\nabla^2 + V_0(\mathbf{r}) + g\,|\Phi(\mathbf{r}, t)|^2. \tag{7.20}$$

H_{GP} is the GP Hamiltonian, and $V_0(\mathbf{r})$ is the confining potential. We also use the coupling constant $g = 4\pi\hbar^2 a/m$ which describes the short-range atomic collisions, and the scattering length a. Introducing the Wigner function, already defined in chapter 2, as

$$W(\mathbf{q}, \mathbf{r}, t) = \int \Phi^*(\mathbf{r} - \mathbf{s}/2, t)\Phi(\mathbf{r} + \mathbf{s}/2, t)\exp(i\mathbf{q} \cdot \mathbf{s})d\mathbf{s}, \tag{7.21}$$

and applying the standard Wigner–Moyal procedure, we can derive from equation (7.20) the wave-kinetic equation

$$i\hbar\left(\frac{\partial}{\partial t} + \mathbf{v}_q \cdot \nabla\right)W = \int V_k(t)\Delta W \exp(i\mathbf{k}\cdot\mathbf{r})\frac{d\mathbf{k}}{(2\pi)^3}, \tag{7.22}$$

where $\mathbf{v}_q = \hbar\mathbf{q}/m$ is the atom velocity, and ΔW is defined by

$$\Delta W = W^- - W^+, \qquad W^\pm \equiv W(\mathbf{q} \pm \mathbf{k}/2, \mathbf{r}, t). \tag{7.23}$$

The quantity $V_k(t)$ in equation (7.22) is the space Fourier transform of the total potential, according to

$$V(\mathbf{r}, t) = \int V_k(t)\exp(i\mathbf{k}\cdot\mathbf{r})\,\frac{d\mathbf{k}}{(2\pi)^3}, \tag{7.24}$$

where $V(\mathbf{r}, t) = V_0 + gn$. The atom density n can be expressed in terms of the condensate quasi-probability, as

$$n(\mathbf{r}, t) \equiv |\psi(\mathbf{r}, t)|^2 = \int W(\mathbf{q}, \mathbf{r}, t)\frac{d\mathbf{q}}{(2\pi)^3}. \tag{7.25}$$

In order to discuss the elementary excitations in the condensate, we assume that the quasi-probability can be divided in two distinct parts, $W = W_0 + \tilde{W}$. Here, W_0 is the equilibrium distribution, and $|\tilde{W}| \ll |W_0|$, describes the elementary excitations of the medium. Neglecting non-uniformity and boundaries, we can make the simple assumption of a plane wave perturbation, defined by

$$\tilde{W}(\mathbf{q}, \mathbf{r}, t) = \tilde{W}_k(\mathbf{q})\exp(i\mathbf{k}\cdot\mathbf{r} - i\omega t), \tag{7.26}$$

where ω is the mode frequency. The corresponding density perturbation will be $\tilde{n}(\mathbf{r}, t) = \tilde{n}_k \exp(i\mathbf{k}\cdot\mathbf{r} - i\omega t)$, as above. Linearising equation (7.22) with respect to the perturbations, we get

$$\tilde{W}_k = g\frac{\Delta W_0}{\hbar(\omega - \mathbf{k}\cdot\mathbf{v}_q)}\tilde{n}_k. \tag{7.27}$$

Integrating over the atom momentum states $\mathbf{q}$, and using equation (7.25), we arrive at the kinetic dispersion relation

$$1 - g\int \frac{\Delta W_0}{\hbar(\omega - \mathbf{k}\cdot\mathbf{v}_q)}\frac{d\mathbf{q}}{(2\pi)^3} = 0. \tag{7.28}$$

This can also be written in the form

$$1 - g\int \frac{W_0(\mathbf{q})}{\hbar}\left[\frac{1}{(\omega_- - \mathbf{k}\cdot\mathbf{v}_q)} - \frac{1}{(\omega_+ - \mathbf{k}\cdot\mathbf{v}_q)}\right]\frac{d\mathbf{q}}{(2\pi)^3} = 0, \tag{7.29}$$

where $\omega_\pm = \omega \pm \hbar k^2/2M$. In order to understand the meaning of this kinetic dispersion, let us first consider the zero-temperature limit. In this case, we can use the equilibrium value of $W_0(\mathbf{q}) = (2\pi)^3 n_0 \delta(\mathbf{q} - \mathbf{q}_0)$, where n_0 is the unperturbed density. Here $\mathbf{q}_0$ defines a constant drift velocity $\mathbf{v}_0 = \hbar \mathbf{q}_0 / M$, which is zero for a condensate at rest. Equation (7.29) becomes

$$1 - g\frac{n_0}{\hbar}\left[\frac{1}{(\omega_- - \mathbf{k}\cdot\mathbf{v}_0)} - \frac{1}{(\omega_+ - \mathbf{k}\cdot\mathbf{v}_0)}\right](2\pi)^3 = 0. \tag{7.30}$$

Rearranging terms and using the Bogoliubov speed, $c_s = \sqrt{g n_0 / M}$, this can also be written as

$$(\omega - \mathbf{k}\cdot\mathbf{v}_0)^2 = k^2 c_s^2 + \frac{\hbar^2 k^4}{4m^2}. \tag{7.31}$$

For $\mathbf{q}_0 = 0$, this reduces to our previous dispersion relation (7.19). It is also useful to consider the relation between the phase velocity $v_\phi = \omega / k$ and the group velocity $v_g = \partial\omega / \partial k$. For a condensate at rest, we have

$$v_\phi v_g = c_s^2 + k^2 \frac{\hbar^2}{M^2}. \tag{7.32}$$

The contributions of the collective behaviour and single particle behaviour are clearly visible here. In the limit of long wavelengths, we simply have $v_\phi v_g = c_s^2$, the square of the Bogoliubov speed, showing that these sound waves are surely collective oscillations of the medium. In contrast, for short wavelengths, quantum dispersion is relevant and eventually dominant. It represents the free streaming of atoms with momentum k.

Let us now consider finite temperature effects. Going back to equation (7.29), it can be rewritten as

$$1 - \frac{g}{\hbar k}\int G_0(u)\left[\frac{1}{(u - \omega_+/k)} - \frac{1}{(u - \omega_-/q)}\right]\frac{du}{2\pi} = 0. \tag{7.33}$$

Here, u and q represent the atom velocity and the atom momentum components parallel to propagation, as defined by

$$\mathbf{v}_q = u\frac{\mathbf{k}}{k} + \mathbf{v}_\perp, \quad \mathbf{q} = q\frac{\mathbf{k}}{k} + \mathbf{q}_\perp. \tag{7.34}$$

The new function $G_0(q)$ is the reduced distribution

$$G_0(q) = \int W_0(q, \mathbf{q}_\perp)\frac{d\mathbf{q}_\perp}{(2\pi)^2}. \tag{7.35}$$

And the integrals in equation (7.33) take the form

$$\int \frac{G_0(u)}{(u - v_\pm)} du = \mathcal{P}_{\rm pr}\int \frac{G_0(u)}{(u - v_\pm)} du + i\pi G_0(v_\pm), \tag{7.36}$$

where $v_\pm = \omega_\pm / k$. The symbol $\mathcal{P}_{\rm pr}$ represents the principal part of the integral, in the Cauchy sense. Using this in equation (7.33), we obtain an expression of the form $\epsilon(\omega, k) = 0$. Splitting this into real and imaginary parts, as $\epsilon = \epsilon_r + i\epsilon_i$, and assuming a real value of k, the mode frequency becomes complex, $\omega = \omega_r + i\gamma$. Assuming $|\gamma| \ll \omega_r$, the wave frequency ω_r and damping rate γ will satisfy

$$\epsilon_r(\omega_r, k) = 0, \quad \gamma = -\frac{\epsilon_i(\omega_r, k)}{(\partial\epsilon_r/\partial\omega)_{\omega_r}}. \tag{7.37}$$

Temperature effects are usually negligible in what concerns ω_r. We are then allowed to use $G_0(u) = 2\pi G_0 \delta(u)$ in the first of these equations, which then reduces to

$$\epsilon_r(\omega_r, k) = 1 - \frac{gk^2}{m}\frac{G_0}{\omega^2 - \hbar^2 k^4/4M^2} = 0. \tag{7.38}$$

Writing $G_0 = n_0$ this reduces again to the dispersion relation (7.19). As for the damping rate, we get

$$\gamma = \frac{gkc_s^2}{4\hbar\omega_r}[G_0(v_+) - G_0(v_-)]. \tag{7.39}$$

This is the Landau damping of bogolons, which is a kinetic damping without dissipation. In thermal equilibrium, $G_0(v_+) < G_0(v_-)$, and Landau damping is negative. However, inversion of population can eventually occur in a disturbed condensate, with $G_0(v_+) > G_0(v_-)$, and $\gamma > 0$. In this case, the condensate is kinetically unstable and bogolons can eventually be created from noise. In the semi-classical limit, such that $|k| \ll |q|$, we can develop the quantities $G_0(v_\pm)$ around $v = \omega/k$, and Landau damping takes a more familiar form. Equation (7.41) becomes

$$\gamma \simeq \frac{k^3 c_s^4}{4 n_0 \omega_r}\left(\frac{\partial G_0}{\partial v}\right)_{v=\omega/k}. \tag{7.40}$$

For a condensate in equilibrium at a finite temperature T, we can use the Bose–Einstein distribution

$$G_0(v) = 2\pi n_0 \{e^{[E(v)-\mu]\beta} - 1\}^{-1}, \tag{7.41}$$

where $E(v) = Mv^2/2 = \hbar^2 q^2/2M$, $\beta = 1/T$. The damping rate γ becomes

$$\gamma \simeq -\frac{k^3 c_s^4}{4\omega_r}\beta e^{[E(\omega/k)-\mu]\beta}\{e^{[E(\omega/k)-\mu]\beta} - 1\}^{-2}. \tag{7.42}$$

The damping rate γ is always negative, for any possible value of the phase velocity ω/k. This is a property of thermal equilibrium. However, out of equilibrium situations can eventually occur where the condensate becomes unstable. As an example, we can consider two interacting condensates moving at different velocities v_j, for $j = 1, 2$, such that

$$G_0(v) = \sum_{j=1,2} \frac{2\pi n_j}{\exp\left\{\left[E(v - v_j) - \mu\right]\beta_j\right\} - 1}, \tag{7.43}$$

where n_j are the densities and $\beta_j = 1/T_j$. In this case, γ can eventually change sign, leading to unstable regions of bogolon frequency. Generalisation of this discussion to include long-range dipolar interactions, quantum fluctuations and finite energy collisions can be found in [21]. This is relevant to the discussion of the possible existence of *rotons*, which are bogolons with a negative group velocity.

7.1.4 Vortices

The Bogoliubov oscillations, or bogolons, discussed above are elementary perturbations of the condensate, and can be considered as sound waves propagating in a quantum fluid. These perturbations are essentially linear structures, with amplitudes that are negligible compared with the mean density n_0 of the medium. In contrast, nonlinear perturbations can be excited in a condensate, such as vortices, with perturbations of the order of n_0. Vortices become unstable in rotating media, and can be defined in a two-dimensional geometry, which is essentially the plane perpendicular to the rotation axis. A huge amount of literature exists on these nonlinear objects (see the reviews [22, 23]). A vortex is a nonlinear solution of the GP equation, of form

$$\Phi(\mathbf{r}, t) = \phi(r)\exp(-i\mu t/\hbar + il\theta), \tag{7.44}$$

where we have used cylindrical variables $\mathbf{r} = (r, \theta, z)$, and uniformity along the z-axis was assumed. Here, μ is the chemical potential and l is a positive or negative integer defining the topological charge of the vortex. Replacing this in equation (7.13), and neglecting the external potential, we obtain

$$\mu\phi = -\frac{\hbar^2}{2M}\left[\frac{1}{r}\frac{d}{dr}\left(r\frac{d\phi}{dr}\right) + \frac{l^2}{r^2}\phi\right] + g\phi^3. \tag{7.45}$$

Using the normalised vortex amplitude, $\chi = \phi/\sqrt{n_0}$, this is transformed into

$$\chi = -\frac{1}{\rho}\frac{d}{d\rho}\left(\rho\frac{d\chi}{d\rho}\right) + \frac{l^2}{\rho^2}\chi + \phi^3, \tag{7.46}$$

with the normalised radial variable $\rho = r/\xi$, where $\xi = \hbar/\sqrt{2Mgn_0}$ is the *healing length*, defining the characteristic size of the vortex. A typical vortex solution is represented in figure 7.3. For, $l = 1$, it is approximately given by

$$\chi(\rho) = \frac{\rho}{\sqrt{2 + \rho^2}}. \tag{7.47}$$

Of particular interest is the energy density associated with vortex solutions. For the unit length along z, this is determined by

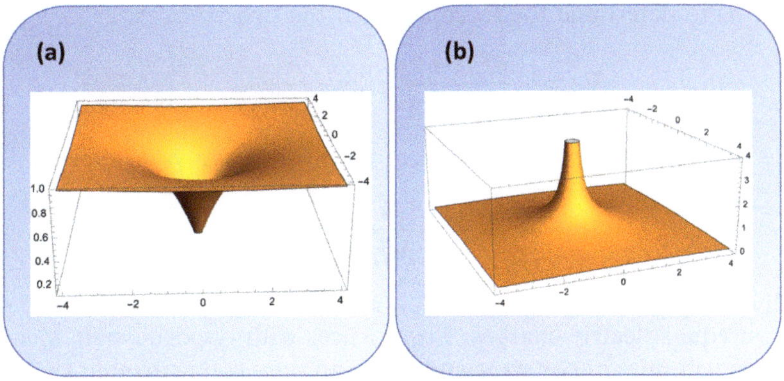

Figure 7.3. Vortex in a BEC: we represent the single charge vortex $l = \pm 1$, located at $(x = 0, y = 0)$: (a) normalised density $n(\rho)/n_0 = \chi(\rho)$, and (b) absolute value of the velocity around the vortex, $\mathbf{v} = l(1/\rho)\mathbf{e}_\theta$.

$$E_v = 2\pi n_0 \mu \int \left[\left(\frac{d\chi}{d\rho} \right)^2 + \frac{l^2}{r^2}\chi^2 + \frac{1}{2}\chi^4 \right] d\rho. \tag{7.48}$$

The dominant contribution comes from the kinetic energy associated with fluid rotation around the singularity. Integrating over a distance $R \gg \xi$, and using the approximate solution (7.47), we get

$$E_v = \pi \frac{\hbar^2 n_0}{M} l^2 \ln\left(\frac{R}{\xi} \right). \tag{7.49}$$

It is also useful to consider vortex–vortex interactions. For this purpose we assume two vortices with topological charges l_j, for $j = 1, 2$, located at positions $\mathbf{r}_j$, defined in the perpendicular plane. They will move with velocities

$$\mathbf{v}_j(\mathbf{r}) = l_j \frac{\hbar}{M} \mathbf{e}_z \times \frac{(\mathbf{r} - \mathbf{r}_j)}{\left|\mathbf{r} - \mathbf{r}_j\right|^2}. \tag{7.50}$$

Noting that $\mathbf{r}/r^2 \equiv -\nabla \ln(1/r) \equiv \nabla(\ln r)$, this can also be written as

$$\mathbf{v}_j(\mathbf{r}) = l_j \frac{\hbar}{M} \nabla\left[\ln(\mathbf{r} - \mathbf{r}_j)\right]. \tag{7.51}$$

The interaction energy between the two vortices can then be defined as

$$E_{12} = M n_0 \int (\mathbf{v}_1 \cdot \mathbf{v}_2) d\mathbf{r}. \tag{7.52}$$

Using equations (7.51), this gives, after integration by parts

$$E_{12} = 2\pi l_1 l_2 \frac{\hbar^2 n_0}{M} \ln\left(\frac{\xi}{|\mathbf{r}_1 - \mathbf{r}_2|} \right). \tag{7.53}$$

This allows us to determine the force between the two vortices, as

$$\mathbf{F}_{12} = -\frac{\partial E_{12}}{\partial r_{12}} \frac{\mathbf{r}_{12}}{|r_{12}|} = 2\pi l_1 l_2 \frac{\hbar^2 n_0}{M r_{12}}, \tag{7.54}$$

where $\mathbf{r}_{12} = (\mathbf{r}_1 - \mathbf{r}_2)$ is the vector distance. It will then evolve in time as

$$\frac{d\mathbf{r}_{12}}{dt} = (l_1 + l_2)\frac{\hbar}{M r_{12}^2}(\mathbf{e}_z \times \mathbf{r}_{12}). \tag{7.55}$$

For equally charged vortices, $l_1 = l_2 = l$, it will increase with time, as they repel each other as two equal electric charges. For vortices with opposite sign, $l_1 = l$, $l_2 = -l$, their distance will remain constant and they can only rotate around their centre-of-mass, with an angular frequency $\Omega_l = 2\hbar l / M r_{12}^2$. They will then form a bound object with total topological charge equal to zero, $l_1 + l_2 = 0$.

7.1.5 BEC in lower dimensions

We have seen above that vortices are defined in a plane perpendicular to the rotation axis of the condensate. In practice, this means that they will maintain their structure for an arbitrary length along z and can, in principle, be excited in a pancake condensate with a nearly negligible axial depth. How far can we go in the reduction of this third dimensional size?

The answer to that question is relevant to the case of photons, as shown later, but is difficult and complex, due to the change in character of the correlations between the atoms, when we make the transition between three to two or even one dimensions (2D and 1D). It has indeed been proven theoretically that, the absence of long-range correlations in a 2D geometry prevents condensation. This is known as the *Bogoliubov–Hohenberg theorem* (see [25] for a review and for pertinent references). However, this is not a problem from the experimental point of view, because in the absence of condensation another phase transition is possible in 2D, which leads to the development of superfluidity.

This is the so-called *BKT transition*, name derived from the work of Berezinskii [26] and Kosterlitz and Thouless [27], in the early 70s. A critical temperature $T_{\rm KT}$ for this new phase transition, which is in reality a topological one, can be defined. Above this critical transition vortices and antivortices are essentially free structures with opposite topological charges. But, below $T_{\rm KT}$, bound pairs of vortex–antivortex tend to form, because they have lower energies than free vortices. If, above this temperature, the vortex fluid is similar to a vortex plasma, where vortices play a role similar to positive and negative electric charges, and are attracted to each other by a Coulomb-type of potential, below $T_{\rm KT}$ the medium becomes a kind of insulator where no free charges can exist [28].

The value of this transition temperature can be estimated using simple thermodynamic arguments, as follows. The free energy of a single vortex is $\kappa \ln(R/\xi)$, where R is the system size, ξ the vortex size, and $\kappa \simeq (\pi\hbar^2/M)n_0$, as shown in equation (7.49). In general, the equilibrium density will depend on the temperature, $n_0 = n_0(T)$. Assuming that $R \gg \xi$, the number of possible vortices can be estimated as $(R/\xi)^2$. We define the Boltzmann entropy of the vortex fluid as $S_v = k_B \ln W_v$, where $W_v \simeq (R/\xi)^2$ is the number of states. Therefore $S_v = 2k_B \ln(R/\xi) - \text{const}$. In the thermodynamic limit $(T/\xi \to \infty)$, the constant can be forgotten. The total free energy is then

$$F_v = E_v - TS_v = \left[\frac{\pi\hbar^2 n_0(T)}{M} - 2T\right] \ln\left(\frac{R}{\xi}\right). \tag{7.56}$$

Vortices can only exist for $F > 0$, which defines a critical temperature $T = T_{\mathrm{KT}}$ for vortex formation, which can be estimated as

$$T_{\mathrm{KT}} = \frac{\pi}{2}\frac{\hbar^2}{M} n_0(T_{\mathrm{KT}}). \tag{7.57}$$

This gives a simple estimate of the critical temperature for the Kosterlitz–Thouless transition.

7.2 Photon condensation

7.2.1 Basic processes

As we have seen, one of basic requirements for Bose–Einstein condensation is the existence of a finite mass. More precisely, three different requirements can be defined: finite mass, weak interactions and cooling mechanism. First, the bosons should have a finite mass M, which is a necessary condition for the existence of a critical temperature T_c. Second, the bosons should be a weakly interacting gas. They need to collide or interact with each other, otherwise they will not attain thermal equilibrium. Third, the gas of bosons should be in contact with a cold source, such that the temperature could be lowered below its critical value. In the absence of weak interactions, this cold source could be used to thermalise the gas. We have seen that, in Bose–Einstein condensation of a diluted gas, weak interactions are guaranteed by atom–atom collisions at low energy, and the cooling mechanism is provided by laser cooling and evaporation.

Let us see how these three conditions can be achieved in a photon gas. First, we have the problem of the photon mass. According to current views, photons propagating in vacuum have no mass. But the question of a non-zero limit of the photon mass was discussed by many people through the years, as it would have an impact on our description of the electromagnetic interactions [24]. So, the absence of a photon mass in vacuum seems to prevent any hope of achieving photon condensation. However, photons acquire mass in a bounded medium, such as when they are confined in optical cavities or propagate in waveguides. The origin of this mass comes from the existence of physical boundaries. But a finite photon mass can also occur in the absence of boundaries, for media possessing free charges, such as in a plasma, a configuration that will be discussed latter.

In order to understand the origin of photon mass in a cavity, we take the simple example of photon propagation in the empty region between two parallel mirrors.

Assuming that the axis of the cavity is the z-axis, the dispersion relation of a photon with frequency ω is determined by

$$\omega = c(k_\perp^2 + k_z^2)^{1/2}, \quad k_\perp = \left(k_x^2 + k_y^2\right)^{1/2}. \tag{7.58}$$

The corresponding photon momentum is $\mathbf{p} = \hbar\mathbf{k}$. Cavity modes are quantised along the z-direction, such that

$$p_z = \hbar k_z = \hbar\frac{\nu\pi}{L} \gg p_\perp = \left(p_x^2 + p_y^2\right)^{1/2}, \tag{7.59}$$

where ν is a quantum number and L is the length of the cavity. From this, we then get the photon energy $\epsilon(\mathbf{p}) = \hbar\omega$, as

$$\epsilon(p_\perp) \simeq m_{\rm ph}c^2 + \frac{p_\perp^2}{2m_{\rm ph}}, \quad m_{\rm ph} = \hbar\frac{\pi}{Lc} \simeq \hbar\frac{\omega_0}{c^2}, \tag{7.60}$$

where ω_0 is the photon frequency with the lowest possible energy. The photon mass $m_{\rm ph}$ is finite and depends on the cavity length. The photon energy relates to the transverse momentum $p_\perp$ by an expression formally identical to the energy of a non-relativistic particle. Therefore, the quantised parallel momentum p_z determines its mass. Radiation trapped in a cavity can then be seen as a gas of two-dimensional massive bosons. Photon condensation will correspond to an accumulation of most photons in its ground state.

This elementary discussion has a clear physical meaning. It explains the existence of a finite photon mass and indicates what we expect in a condensed photon gas. But it is not completely correct, for a number of reasons. Photons do not 'propagate' inside a cavity. They are elementary excitations of the field mode occupying the entire cavity, with a transverse structure characterised by the perpendicular wave-number $k_\perp$. For a small curved cavity, with length much smaller than the radius of curvature R of the cavity mirror, the photon energy equation can be approximately written as

$$\epsilon(k_\perp) \simeq m_{\rm ph}c^2 + \frac{\hbar^2 k_\perp^2}{2m_{\rm ph}} + \frac{1}{2}m_{\rm ph}\Omega^2 r^2, \quad \Omega^2 = \frac{2c^2}{LR}. \tag{7.61}$$

This is similar to the energy of an atom trapped in an harmonic potential with effective frequency Ω. Let us now consider the second ingredient necessary for condensation, which is the interaction between photons. This interaction will be responsible for gas thermalisation, assuming that the photon lifetime inside the cavity is much longer than the thermalisation time scale. Photon interactions can be achieved by nonlinear processes occurring in the cavity, which implies that the cavity should not be empty, but filled with some nonlinear optical medium.

Experiments proving the existence of photon condensation were realised in microcavities filed with a dye liquid. The heavy dye molecules provide the cooling reservoir, and photon scattering with the dye molecules also provides thermalisation, two ingredients needed for condensation. Given the existence of photon losses, a

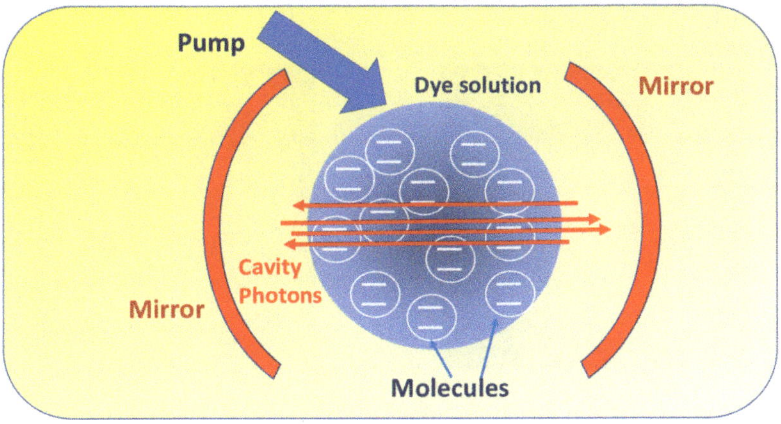

Figure 7.4. Photon condensation in a cavity filled with dye molecules: (a) experimental scheme; (b) phase transition, showing a sharp increase in the number of ground state photons, above a given threshold.

laser pump beam is injected from the outside, coupling to the cavity modes through scattering. This is illustrated in figure 7.4.

Cooling is provided by the scattering properties of dye molecules. At a given temperature T, their emission and absorption coefficients satisfy the so-called *Kennard–Stepanov relation* [29, 30], $R_{\rm em}(\omega)/R_{\rm abs}(\omega) \simeq \exp(-\hbar\omega/T)$. This property favours scattering of lower frequencies by the molecules, thus pushing the photon spectrum towards the lowest frequency, $\omega_0 = m_{\rm ph}c^2/\hbar$. The physical reason is the excitation of vibrational modes in the molecules, which dissipate as internal phonons in the dye liquid. Photon condensation process can then be described as a decaying process similar to Brillouin scattering, where an incident photon excites a phonon and another photon of lower energy [31]. Photon condensation can be achieved at room temperature ($T \sim 300K$), above a certain pump laser intensity, when the number of photons in the cavity exceeds some critical value.

A simple description of a condensed photon gas can be made using the following arguments [32]. Let us assume that, in equilibrium at a some temperature below critical $T \leqslant T_c$, the Bose–Einstein distribution is valid, and we can define the photon energy distribution as

$$W(\mathbf{r}, \mathbf{k}) = \frac{1}{\exp\left(\epsilon(\mathbf{k})/T - \mu/T - 1\right)}. \tag{7.62}$$

But, in a cavity, we should consider a two-dimensional distribution instead, which would be given by

$$W(\mathbf{r}_\perp, \mathbf{k}_\perp) = \int dz \int \frac{dk_z}{2\pi} W(\mathbf{r}, \mathbf{k}). \tag{7.63}$$

Considering the above definition for the longitudinal wavenumber k_z, equation (7.58), we get

$$W(\mathbf{r}_\perp, \mathbf{k}_\perp) = \frac{\nu}{2} \frac{1}{\exp\left(\epsilon(\mathbf{k}_\perp)/T - \mu/T - 1\right)}. \tag{7.64}$$

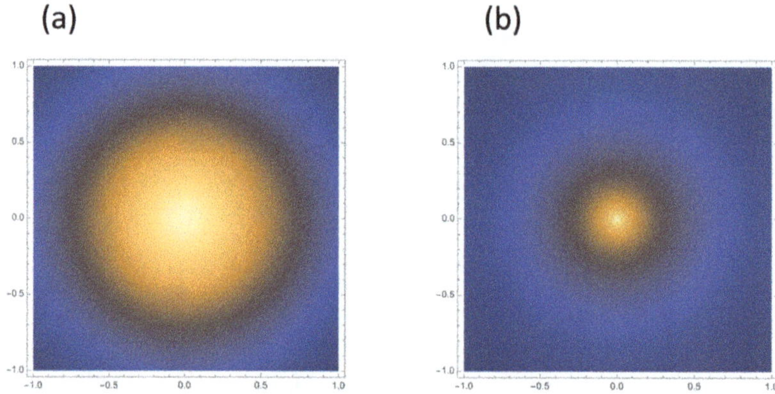

Figure 7.5. Normalised intensity distribution inside the cavity: (a) before condensation, at $\mu < m_{\rm ph}c^2$; and (b) at condensation $\mu = m_{\rm ph}c^2$, as determined by the photon number distribution of equation (7.66).

Integrating over the perpendicular momentum, we get the radial distribution of the photon number as

$$N_{\rm ph}(r) = \int dz \int \frac{dk_z}{2\pi} W(\mathbf{r}_\perp, \mathbf{k}_\perp). \tag{7.65}$$

Using equation (7.61), we finally arrive at the expression

$$N_{\rm ph}(r) = \frac{\nu}{2\pi}\frac{m_{\rm ph}T}{\hbar^2} \ln\left\{1 - \exp\left[\mu/T - m_{\rm ph}c^2/T - m_{\rm ph}\Omega^2 r^2/2T\right]\right\}^{-1}. \tag{7.66}$$

This provides the radial intensity of a condensed photon gas in the cavity, as illustrated in figure 7.5, for a chemical potential below and at the critical value $\mu = m_{\rm ph}c^2$. This result is in qualitative agreement with the observed photon condensation [14]. Note also that equation (7.62) allows the calculation of photon number at criticality, and therefore determines the critical temperature T_c, as

$$N_c = \int d\mathbf{r} \int \frac{d\mathbf{k}}{(2\pi)^3} W(\mathbf{r}, \mathbf{k}) = \int d\mathbf{r}_\perp \int \frac{d\mathbf{k}_\perp}{(2\pi)^2} W(\mathbf{r}, \mathbf{k}) \tag{7.67}$$

calculated at $\mu = m_{\rm ph}c^2$. We then get

$$N_c = \frac{\nu\pi^2}{6}\left(\frac{T_c}{\hbar\Omega}\right)^2. \tag{7.68}$$

This shows that the critical temperature is proportional to the square-root of the radiation intensity. Increasing the external pump, we can therefore arrive at an intensity such that the temperature of the dye liquid will become equal or smaller than the critical temperature, $T \leqslant T_c$, therefore allowing photon condensation at room temperature, as observed.

7.2.2 Temporal evolution

The temporal evolution of the radiation intensity inside the cavity can be described by a detailed balance equation of the photon number. In such an equation, the various processes of photon gain and loss are taken into account, as confirmed by a rigorous quantum analysis. The number of photons in the ground cavity state $N_{ph} = N(\omega)$ is determined by the balance between gain and losses [33], written in the usual form

$$\frac{d}{dt}N(\omega) = M_{\uparrow}[A(\omega) + B_{21}(\omega)N(\omega)] - M_{\downarrow}B_{12}(\omega)N(\omega) - \Gamma N(\omega), \tag{7.69}$$

where $A(\omega)$, $B_{21}(\omega)$ and $B_{12}(\omega)$ are the Einstein coefficients, determining the spontaneous emission, induced decay and radiation absorption, respectively. Here, we have also used the molecular numbers in the upper and lower electronic states, $M_{\uparrow}$ and $M_{\downarrow}$, and the cavity loss rate $\Gamma(\omega)$. It is known that the Einstein coefficients satisfy the relation $A(\omega)/B_{21}(\omega) = g(\omega)$, where $g(\omega)$ is the photon mode degeneracy. We have also mentioned that, for a dye liquid in thermal equilibrium at temperature T, the Kennard–Stepanov scale $B_{21}(\omega)/B_{12}(\omega) \propto \exp(-\hbar\omega/T)$ should be use.

The balance equation can be written in another form. Introducing the fraction of excited molecules $f = M_{\uparrow}/M_{dye}$, where M_{dye} is the total number of dye molecules in the cavity, and using the simplified notation $n = N(\omega)$, we then get [34]

$$\frac{dn}{dt} = Ef(n + 1) - An(1 - f) - \kappa n, \tag{7.70}$$

where E, A and κ are the new emission, absorption and decay coefficients. This has to be completed with another balance equation, describing the temporal evolution of the number of excited molecules. It depends on the external pumping and loss rates $\Gamma_{\uparrow}$, and $\Gamma_{\downarrow}$, as

$$\frac{df}{dt} = -Ef(n + 1) + An(1 - f) + \Gamma_{\uparrow}(1 - f) - \Gamma_{\downarrow}f. \tag{7.71}$$

In equilibrium, $d/dt = 0$, these equations lead to the molecule and photon number steady-state values, $f = f_{eq}$ and $n = n_{eq}$, such that

$$f_{eq} = \frac{(\kappa + A)n_{eq}}{E(n_{eq} + 1) + An_{eq}}, \quad n_{eq} = \frac{\Gamma_{\uparrow}}{\kappa}\left[(1 - f_{eq}) + \frac{\Gamma_{\downarrow}}{\Gamma_{\uparrow}}f_{eq}\right]. \tag{7.72}$$

The evolution of the photon density towards equilibrium is illustrated in figure 7.6, for two different values of the pumping rate coefficient $\Gamma_{\uparrow}$. We can see that, for a low pump intensity, the average number of photons in the lower cavity mode stays close to zero. In contrast, for a large pumping rate, stimulated emission dominates over the spontaneous process, and the number of photons tends to a large saturation value $n_{eq} \gg 1$. This is a signature of photon condensation inside the cavity.

The above balance equations can also be derived from a microscopic quantum model, describing the details of the collective photon-matter processes taking place

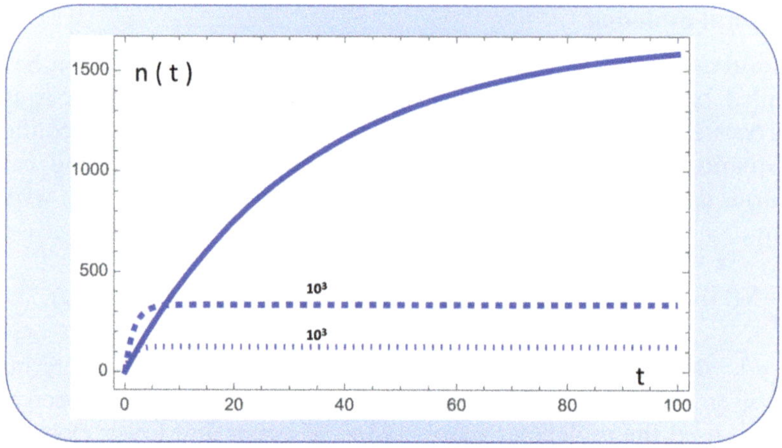

Figure 7.6. Transition towards equilibrium, described by of equations (7.70)–(7.72): (a) absence of photon condensation, for low pumping rates, such that $f = f_{eq} = 0.1$ (dot-dashed) and 0.2 (dashed curve), where $n \leqslant 1$; (b) condensation, for high pumping rate and $f = f_{eq} = 0.5$, with $n \gg 1$.

in the cavity. For this purpose, we consider the system of M_{dye} dye molecules interacting with the photon modes inside the spherical cavity. Photons are described by creation and annihilation operators $a_\nu^\dagger$ and a_ν for each cavity mode ν, and the two-level molecules are described by the Pauli matrices σ_i, for $i = 1, \ldots, M_{dye}$. The energy difference between the two relevant electronic levels is Δ, and each of them is a ladder of rovibrational sublevels. The internal low energy transitions between these sublevels can be seen as internal phonons, described by operators $b_i^\dagger$ and b_i. The total Hamiltonian of the system is a sum of three terms,

$$H = H_{ph} + H_{dye} + H_{int}, \tag{7.73}$$

where the photon, molecular and interaction terms are determined (in $\hbar = 1$ units) by

$$H_{ph} = \sum_\nu \omega_\nu a_\nu^\dagger a_\nu, \tag{7.74}$$

where ω_ν are the mode frequencies,

$$H_{dye} = \sum_i \left\{ \frac{\Delta}{2}\sigma_i^z + \Omega[b_i^\dagger b_i + \sqrt{S}\sigma_i^z(b_i + b_i^\dagger)] \right\}, \tag{7.75}$$

where Δ is the energy difference between the two electronic levels, Ω is the vibrational mode spacing, and S the coupling strength with the internal phonons, and finally

$$H_{int} = g\sum_{i,\nu}(a_\nu\sigma_i^+ + a_\nu^\dagger\sigma_i^-), \tag{7.76}$$

where g is the coupling coefficient between the cavity modes and the molecules. From here, we can establish the master equation describing the density operator

ρ for both photons and molecules [35]. Defining the mean photon number for the lower energy mode $\langle n\rangle$ and the number of excited molecules $\langle M\rangle$, as

$$\langle n\rangle = \langle a^{\dagger}(t)a(t)\rangle, \quad \langle M\rangle = \sum_i \langle \sigma_i^{+}(t)\sigma_i^{-}(t)\rangle, \tag{7.77}$$

we obtain the evolution equations

$$\frac{d\langle n\rangle}{dt} = Ef(\langle M\rangle + \langle nM\rangle) - A(\langle n\rangle M_{\rm dye} - \langle nM\rangle) - \kappa\langle n\rangle, \tag{7.78}$$

and

$$\begin{aligned}\frac{d\langle M\rangle}{dt} = &-E(\langle nM\rangle + \langle M\rangle) + A(\langle n\rangle M_{\rm dye} - \langle nM\rangle) + \Gamma_{\uparrow}(M_{\rm dye} - \langle M\rangle)\\ &- \Gamma_{\downarrow}\langle M\rangle.\end{aligned} \tag{7.79}$$

In the semi-classical limit, the mean values of n and M can be factorised, as $\langle nM\rangle \simeq \langle n\rangle\langle M\rangle$. Furthermore, defining $n \equiv \langle n\rangle$, and $f = \langle M\rangle / M_{\rm dye}$, we are reduced to the above balance equations (7.70) and (7.71).

7.3 Condensation in plasma

7.3.1 Compton cooling

Let us now discuss photon condensation in free space, independent of any cavity boundaries, as it eventually occurs in a plasma. Here, the photon modes are not discrete, but form a continuum, as described by the respective wavevectors $\mathbf{k}$. Photon condensation can be established through their interaction with the electrons, via Compton scattering. In order to describe such a process we use the photon balance equation, which is nothing but a quantum Boltzmann equation for the photon spectral distribution $N(\mathbf{k})$. It takes the form

$$\frac{\partial N(\mathbf{k})}{\partial t} = c\int d\mathbf{p}\int dW\{f_e(\mathbf{p}')N(\mathbf{k}')[1 + N(\mathbf{k})] - f_e(\mathbf{p})N(\mathbf{k})[1 + N(\mathbf{k}')]\}. \tag{7.80}$$

Here, $f_e(\mathbf{p})$ represents the electron distribution function, and dW the differential transition probability from a photon state $\mathbf{k}'$ to another photon state $\mathbf{k}$, through Compton scattering. This scattering process satisfies the momentum conservation relation, $\mathbf{p} + \hbar\mathbf{k} = \mathbf{p}' + \hbar\mathbf{k}'$. Notice that this equation ignores any inelastic photon absorption or emission process, and only considers momentum and energy exchanges between the plasma and radiation resulting from single-particle Compton scattering.

In the non-relativistic case, when $T_e \ll m_e c^2$, and assuming that the energy transfer due to scattering is small compared with the electron thermal energy, $\Delta \equiv \hbar(\omega - \omega')/T_e \ll 1$, we can use the Maxwellian distribution

$$f_e(\mathbf{p}) = \frac{n_{e0}}{(2\pi m_e T_e)^{3/2}}\exp\left(-\frac{p^2}{2m_e T_e}\right), \tag{7.81}$$

and expand the distributions $f_e(\mathbf{p}')$ and $N(\mathbf{k}')$ around the values $\mathbf{p}$ and $\mathbf{k}$. This allows us to reduce the quantum Boltzmann equation (7.80), to an expression of the form

$$\frac{\partial N}{\partial t} = [N' + N(1 + N)]\mathcal{I}_1(\Delta) + [N'' + (2N' + N)(1 + N)]\mathcal{I}_2(\Delta^2), \tag{7.82}$$

where we have used the notation $N \equiv N(x, t)$, and $N' = \partial N/\partial x$, and introduced the dimensionless variable $x = \hbar\omega/T_e$. The two integrals in this expression are defined by

$$\mathcal{I}_1(\Delta) = c\int d\mathbf{p}\int dW \Delta f_e(\mathbf{p}), \quad \mathcal{I}_2(\Delta^2) = \frac{c}{2}\int d\mathbf{p}\int dW \Delta^2 f_e(\mathbf{p}). \tag{7.83}$$

We use $dW = d\Omega(d\sigma/d\Omega)_C$, where $d\Omega$ is element of the solid angle, and take the Compton differential cross section in the non-relativistic limit, as

$$\left(\frac{d\sigma}{d\Omega}\right)_C \simeq \frac{3}{16\pi}(1 + \cos^2\theta)\sigma_T, \tag{7.84}$$

where θ is the scattering angle, $\sigma_T = 8\pi r_e^2/3$ is the Thomson cross section, and $r_e = e^3/(4\pi\epsilon_0 m_e c^2)$ is the classical radius of the electron. Taking equation (7.81) into account, we obtain

$$\mathcal{I}_1(\Delta) = (4T_e - \hbar\delta)\frac{\hbar\delta}{m_e c}\frac{n_{e0}}{T_e}\sigma_T, \; \mathcal{I}_2(\Delta^2) = \delta^2 T_e \frac{n_{e0}}{T_e}\sigma_T, \tag{7.85}$$

where $\delta^2 = (\omega^2 - \omega_p^2)$, and ω_p is the plasma frequency. Finally, introducing the Compton time τ, and the new variable κ, such that

$$\tau = \frac{n_{e0}T_e}{m_e c^2}\sigma_T t, \quad \kappa = \frac{\hbar\delta}{T_e} = \frac{\delta}{\omega}x, \tag{7.86}$$

we reduce the photon kinetic equation (7.82) to a much simpler form [37]

$$\frac{\partial N}{\partial \tau} = \frac{f(\kappa)}{\kappa^2}\frac{\partial}{\partial \kappa}\left\{\kappa^4\left[\frac{\partial N}{\partial \kappa}f(\kappa) + N(1 + N)\right]\right\}, \tag{7.87}$$

where $f(\kappa)$ is the Jacobian function, defined as

$$f(\kappa) = \frac{\kappa}{\sqrt{\kappa^2 + (\hbar\omega_p/T_e)^2}}. \tag{7.88}$$

This equation takes into account the specific plasma dispersion properties, namely the existence of a cut-off $\omega = \omega_p$, below which photon modes cannot exist. Steady-state solutions satisfy the condition

$$\frac{dN}{N^2 + N} = -\frac{d\kappa}{\kappa}\sqrt{\kappa^2 + (\hbar\omega_p/T_e)^2}, \tag{7.89}$$

and can be written explicitly as (figure 7.7)

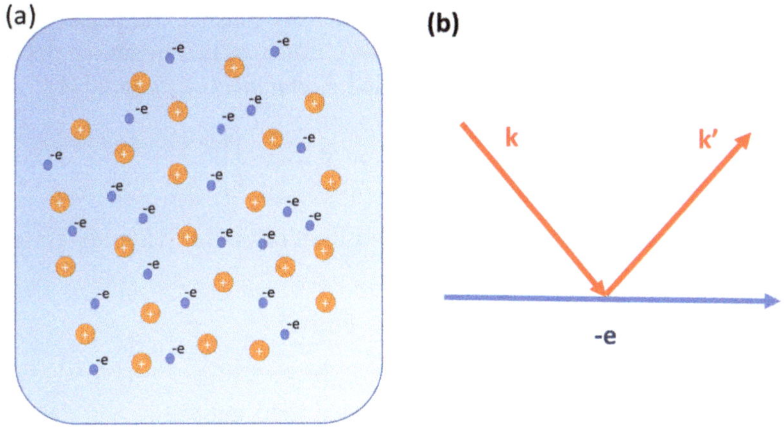

Figure 7.7. Photon condensation in plasma: (a) plasma as a collection of electrons and ions with global charge neutrality; (b) Compton scattering as a cooling mechanism.

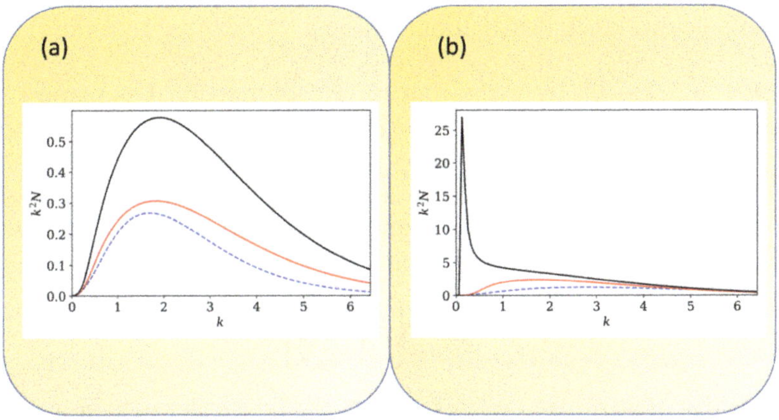

Figure 7.8. Photon condensation in plasma, as described by equation (7.87): (a) condensation is absent in hot plasmas, when the electron temperature is larger than the initial photon temperature, $T_{\rm ph} = 0.8T_e$; (b) condensation occurs in cold plasmas, when $T_{\rm ph} = 1.5T_e$. Photon spectrum: Initial conditions (dashed blue), intermediate state (in red) and final state (in black). (Courtesy of J L Figueiredo.)

$$N(\kappa) = \frac{1}{ze^{\sqrt{\kappa^2+a^2}-\Theta(a)} - 1}, \quad \Theta(\kappa) = a\ Tanh^{-1}\left(\frac{1}{a}\sqrt{\kappa^2 + a^2}\right), \tag{7.90}$$

where $a = \hbar\omega_p/T_e$, and $z = e^{-\kappa_0}$ is the fugacity. Notice that $\kappa^2 + a^2 = \hbar\omega/T_e$. We can study the temporal evolution towards condensation, if we assume an initial Planckian distribution, $N(\tau = 0) = [\exp(\hbar\omega/T_{\rm ph}) - 1]^{-1}$. For an electron temperature lower than the initial photon temperature, $T_e < T_{\rm ph}$, we observe condensation with an accumulation of photons at the cut-off frequency, $\omega \sim \omega_p$. This clearly shows that the electrons are the cold source which provides photon condensation, as illustrated in figure 7.8.

The analysis is simplified if we assume a low density plasma and neglect the influence of the electron plasma cut-off. We get $f(\kappa) \to 1$, and $\kappa = x$. The above photon kinetic equation (7.87) is then reduced to the so-called *Kompaneets equation* [36]

$$\frac{\partial N}{\partial \tau} = \frac{1}{x^2}\frac{\partial}{\partial x}\left\{x^4\left(\frac{\partial N}{\partial x} + N + N^2\right)\right\}. \tag{7.91}$$

The solutions of this equation were discussed in detail in [38]. Steady-state solutions are determined by the Riccati equation

$$\frac{dN}{dx} + N^2 + N = \frac{\text{const.}}{x^4}. \tag{7.92}$$

Assuming that the constant is equal to zero, we then get the

$$N = \frac{1}{ze^x - 1}, \qquad z = \text{const.} \tag{7.93}$$

where z is a constant. We can take $z = e^{-\mu}$, where μ is the dimensionless chemical potential, and get a Planck distribution. This can be reduced to a Bose distribution when $\mu = 0$.

Zel'dovich and Levich were the first to realise that the Kompaneets equation (7.91) predicts photon condensation for $T_{\text{ph}} > T_e$ [12]. They further simplified the equation by assuming a large occupation number, $N \gg 1$. This is certainly valid in the low frequency region, when condensation is taking place. In this case, we can also drop the second term on the r.h.s., and reduce it to a diffusion equation, of the form

$$\frac{\partial N}{\partial \tau} = \frac{1}{x^2}\frac{\partial}{\partial x}\left\{x^4\left(\frac{\partial N}{\partial x}\right)\right\}. \tag{7.94}$$

Assuming further that $N^2 \gg |\partial N/\partial x|$, we get the first-order equation

$$\frac{\partial N}{\partial \tau} = \frac{1}{x^2}\frac{\partial}{\partial x}(x^4 N^2). \tag{7.95}$$

Multiplying this by x^2, and using $f = x^2 N$, we obtain

$$\frac{\partial f}{\partial \tau} - 2f\frac{\partial f}{\partial x} = 0. \tag{7.96}$$

Integration by the method of characteristics leads to $x + 2\tau f = F(f)$. Therefore the solution takes the form

$$x = F(f) - 2\tau f. \tag{7.97}$$

If we start with a given function $f = f(x, \tau = 0)$, this curve will move in time along the characteristics, and will attain the point $x = 0$ at $\tau = F(f)/2f$. Noting that the (dimensionless) frequency x cannot be negative, we conclude that an accumulation of photons will occur in time at $x = 0$. Photon condensation will then correspond to the formation of a *shock front* at zero frequency. The critical time for the formation of a Bose condensate can then be estimated as

$$\tau \geqslant \tau_C = \frac{T_e}{2T_{\rm ph}}. \tag{7.98}$$

Invariant solutions of the Kompaneets equation (7.91) can also be found for $N \gg 1$, of the form [38]

$$N(x, \tau) = \frac{1}{x} \frac{1}{[(1 - \mu)e^{2\tau}]}, \tag{7.99}$$

where $C > 0$ is a constant. A singularity occurs at a critical time, $\tau^* = -\ln\sqrt{1 - \mu}$. For $\mu = T_e/T_{\rm ph}$, this is larger than the estimated value of equation (7.98).

7.3.2 Photon interactions

We have seen above that coupling with electrons via Compton scattering could lead to photon condensation. Therefore, electron–photon coupling was considered the dominant process. But, photons also interact weakly with each other, due to the nonlinear plasma response, thus creating an alternative channel for radiation transport (figure 7.9). In the strong radiation regime, this could eventually dominate over the Compton process and lead to condensation without electron cooling [39]. But, even in the weak radiation regime where Compton processes eventually dominate, these nonlinear photon interactions allow for photon thermalisation, and for the collective behaviour of the condensed medium. In order to understand the nonlinear photon processes, we start with the wave equation in a plasma, which can be written as

$$\left(\nabla^2 - \frac{1}{c^2}\frac{\partial^2}{\partial t^2}\right)\mathbf{a} = \frac{\omega_p^2}{\gamma c^2}\mathbf{a}, \tag{7.100}$$

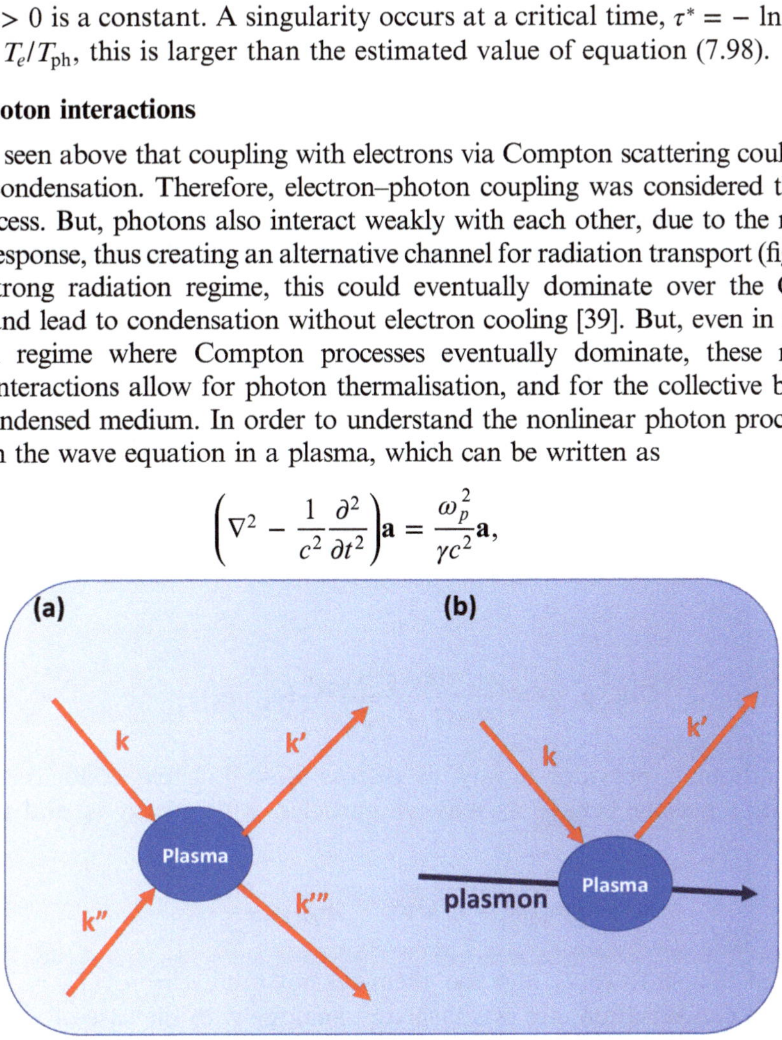

Figure 7.9. Nonlinear photon processes in a plasma: (a) photon–photon interactions, mediated by four-wave mixing; (b) photon–plasmon interactions.

where $\mathbf{a} = e\mathbf{A}/m_e c$ is the normalised potential, ω_p the plasma frequency, and $\gamma = \sqrt{1 + a^2}$ the relativistic gamma-factor. Using a space Fourier decomposition, of the form

$$\mathbf{a} \equiv \mathbf{a}(\mathbf{r}, t) = \sum_{\mathbf{k}} \mathbf{a}_k \exp(i\mathbf{k} \cdot \mathbf{r}), \tag{7.101}$$

and expanding γ in power series, using the unit polarisation vector $\mathbf{e}_k = \mathbf{a}_k/|\mathbf{a}_k|$, we get to the lowest order a nonlinear equation of the form [37]

$$\left(\frac{\partial^2}{\partial t^2} + \omega_{\mathbf{k}}^2\right) a_{\mathbf{k}} = - \sum_{\mathbf{k}', \mathbf{k}''} C(\mathbf{k}, \mathbf{k}', \mathbf{k}'') a_{\mathbf{k}'} a_{\mathbf{k}''} a_{\mathbf{k}'''}, \tag{7.102}$$

where the mode frequency is determined by $\omega_{\mathbf{k}}^2 \simeq \omega_p^2 + k^2/c^2$, and $\mathbf{k}''' = \mathbf{k} + \mathbf{k}'' - \mathbf{k}'$. Photon coupling is provided by the factor

$$C(\mathbf{k}, \mathbf{k}', \mathbf{k}'') = \frac{\omega_p^2}{2c^2} (\mathbf{e}_{\mathbf{k}'} \cdot \mathbf{e}_{\mathbf{k}''})(\mathbf{e}_{\mathbf{k}}^* \cdot \mathbf{e}_{\mathbf{k}'''}), \tag{7.103}$$

equation (7.102) describes a collection of coupled harmonic oscillators. After quantisation, the photon states associated with these oscillators are described by an Hamiltonian of the form

$$\hat{H} = \sum_{\mathbf{k}} \hbar\omega_{\mathbf{k}} \left(\hat{a}_{\mathbf{k}}^\dagger \hat{a}_{\mathbf{k}} + \frac{1}{2} \right) + \hat{H}_{\text{int}}, \tag{7.104}$$

where the interaction Hamiltonian $\hat{H}_{\text{int}}$ describes the nonlinear photon–photon interactions, and is determined by

$$\hat{H}_{\text{int}} = \frac{1}{2} \sum_{\mathbf{k},\mathbf{p},\mathbf{q}} C(\mathbf{k}, \mathbf{p}, \mathbf{q}) \hat{a}_{\mathbf{k}+\mathbf{p}}^\dagger \hat{a}_{\mathbf{q}-\mathbf{p}}^\dagger \hat{a}_{\mathbf{k}} \hat{a}_{\mathbf{q}}, \tag{7.105}$$

with

$$C(\mathbf{k}, \mathbf{p}, \mathbf{q}) = \frac{m_e^2}{e\epsilon_0} \frac{(\mathbf{e}_{\mathbf{k}+\mathbf{p}} \cdot \mathbf{e}_{\mathbf{q}-\mathbf{p}})}{\left(1 + \omega_{\mathbf{k}}^2/\omega_p^2\right)} \left(\mathbf{e}_{\mathbf{k}}^* \cdot \mathbf{e}_{\mathbf{q}}\right). \tag{7.106}$$

This Hamiltonian provides a way to discuss Bose–Einstein condensation. The photons in a plasma behave as massive particles, with energy $\epsilon_{\mathbf{k}}$ and mass m_{ph} determined by

$$\epsilon_{\mathbf{k}} = \sqrt{m_{\text{ph}}^2 c^4 + \hbar^2 k^2 c^4}, \quad m_{\text{ph}} = \frac{\hbar\omega_p}{c^2}. \tag{7.107}$$

The critical temperature T_c and the chemical potential $\mu_c = \mu_{\text{ph}}(T_{\text{ph}} = T_c)$, corresponding to condensation, are established in analogy with the case of atoms as

$$T_c = \pi \frac{\hbar^2}{m_{\text{ph}}} n_e^{2/3}, \quad \mu_c = m_{\text{ph}} c^2. \tag{7.108}$$

Notice that condensation can be achieved by Compton cooling, as described, and also eventually by photon–photon interactions [39]. In any case, even if Compton cooling is dominant, nonlinear photon coupling is still needed to guarantee the existence of Bogoliubov oscillations in the photon gas. This is relevant for the possible occurrence of superfluidity, as seen later. In order to establish the photon Bogoliubov mode dispersion, we start from the photon number operator

$$\hat{N} = \sum_{\mathbf{k}} \hat{a}_{\mathbf{k}}^{\dagger}\hat{a}_{\mathbf{k}} \simeq N_0 + \sum_{\mathbf{k}\neq 0} \hat{a}_{\mathbf{k}}^{\dagger}\hat{a}_{\mathbf{k}}, \tag{7.109}$$

where N_0 is the number of condensed photons in the lowest energy state, $\mathbf{k} = 0$. Using this in the Hamiltonian (7.104)–(7.105), and assuming $N_0 \gg 1$, we get

$$\hat{H} = \hat{H}_0 + \hat{H}_1, \tag{7.110}$$

where the condensed part of the Hamiltonian is determined by

$$\hat{H}_0 = (\hbar\omega_p - \mu_{\mathrm{ph}})N_0 + \frac{1}{2}C(0, 0, 0)N_0^2, \tag{7.111}$$

and the remaining part, which is associated with the non-condensed photons, is given by

$$\hat{H}_1 = \sum_{\mathbf{k}\neq 0}\left[(\epsilon_{\mathbf{k}} + g_{\mathbf{k}}N_0)\hat{a}_{\mathbf{k}}^{\dagger}\hat{a}_{\mathbf{k}} + \frac{1}{2}g_{\mathbf{k}}N_0(\hat{a}_{\mathbf{k}}^{\dagger}\hat{a}_{-\mathbf{k}}^{\dagger}\hat{a}_{-\mathbf{k}}\hat{a}_{\mathbf{k}})\right], \tag{7.112}$$

where the quantity $g_{\mathbf{k}} = C(\mathbf{k}, 0, 0)/N_0$ is the photon interaction parameter. This Hamiltonian description can be simplified by the introduction of a *Bogoliubov transformation*, such that

$$\hat{\alpha}_{\mathbf{k}} = u_{\mathbf{k}}\hat{a}_{\mathbf{k}} + v_{\mathbf{k}}\hat{a}_{-\mathbf{k}}^{\dagger}, \quad \hat{\alpha}_{-\mathbf{k}}^{\dagger} = u_{\mathbf{k}}\hat{a}_{-\mathbf{k}}\dagger + v_{\mathbf{k}}\hat{a}_{\mathbf{k}}, \tag{7.113}$$

where the coefficients $u_{\mathbf{k}}$ and $u_{\mathbf{k}}$ satisfy the condition $|u_{\mathbf{k}}|^2 - |v_{\mathbf{k}}|^2 = 1$. Writing the total Hamiltonian (7.109) in terms of the new operators, we are reduced to a diagonalised form

$$\hat{H} = E_0 + \sum_{\mathbf{k}\neq 0} \omega_{\mathbf{k}}\hat{\alpha}_{\mathbf{k}}^{\dagger}\hat{\alpha}_{\mathbf{k}}, \tag{7.114}$$

where the first term, E_0, corresponds to the energy of the photon condensate, and the non-condense part is described as a sum over a number of new modes, with momentum $\mathbf{k}$ and frequency $\omega_{\mathbf{k}}$, satisfying the dispersion relation

$$\hbar\omega_{\mathbf{k}} = \sqrt{\epsilon_{\mathbf{k}}(\epsilon_{\mathbf{k}} + 2g_{\mathbf{k}}N_0)}\,. \tag{7.115}$$

These modes are not the usual photons, as can easily be realised, but correspond to low frequency oscillations of the photon gas. These are the photon *Bogoliubov oscillations* or photon *bogolons*, analogous to those of atomic condensates. Noting that we are in the limit of very low wavenumbers, $k^2 \to 0$, the mode energy can be expanded as

$$\epsilon_{\mathbf{k}} = \hbar\sqrt{\omega_p^2 + k^2c^2} - \mu_{\text{ph}} \simeq \frac{\hbar^2 k^2}{2m_{\text{ph}}}. \tag{7.116}$$

This finally leads to

$$\omega_{\mathbf{k}}^2 = k^2 C_B^2 + \frac{\hbar^2 k^4}{4m_{\text{ph}}^2}, \quad C_B = \sqrt{\frac{gN_0}{m_{\text{ph}}}}, \tag{7.117}$$

where C_B is the Bogoliubov speed of the photon gas, and $g \simeq g_{q=0}$. This describes low frequency oscillations of the photon density inside the plasma, and is completely distinct from the usual plasma oscillations, plasmons and phonons. Possible intersection between such bogolon modes and the plasmons can eventually occur, for wavenumbers of the order of

$$k^2 \simeq 2\hbar m_{\text{ph}}\omega_p = 2\frac{\omega_p^2}{\hbar c^2}. \tag{7.118}$$

This could eventually lead to additional photon cooling, due to the conversion of non-condensed photons into plasmons, and later dissipation of plasmons by electron Landau damping. In order to illustrate this intersection, we represent in figure 7.10 the bogolon and plasmon dispersion curves, represented respectively by

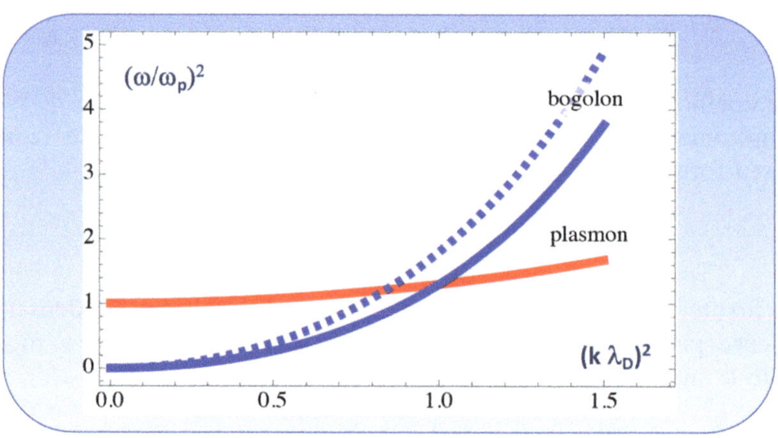

Figure 7.10. Plasmon-bogolon intersection: normalised dispersion relations (7.119) are represented, with plasmon dispersion (in red) and bogolon dispersion (in blue). We have used $\nu = 1$, $\lambda_D^2 = 0.1$ and $\beta^2 = 0.3$. The intersection between the two modes occurs at $k^2\lambda_D^2 = 1$, already in the region of strong Landau damping. Another bogolon dispersion curve, with $\nu^2 = 1.5$ is also shown (blue dashed), showing possible intersection at the undamped region $k^2\lambda_D^2 < 1$.

$$X^2 \equiv \frac{\omega^2}{\omega_p^2} = \nu^2\lambda_D^2 + \beta^2 k^4, \qquad Y^2 \equiv \frac{\omega^2}{\omega_p^2} = 1 + k^2\lambda_D^2, \tag{7.119}$$

where $\nu = C_B/S_e$ is the ratio between the Bogoliubov speed and the electron thermal velocity $S_e = \sqrt{3T_e/m_e}$, the quantity $\lambda_D = S_e/\omega_p^2$ is the electron Debye length, and $\beta = \hbar/(2m_{\rm ph}\omega_p)$ is the quantum factor.

7.3.3 Photon–plasmon coupling

We have seen the complexity of photon behaviour in a plasma, due to wave and particle processes that can take place simultaneously. Another source of complexity is due to the possible interaction of the photons with plasma density perturbations, in particular, with plasmons. We have seen that Compton scattering gives access to cooling and to condensation. But, in our description, the electron distribution function $f_e(\mathbf{p})$ appearing in equation (7.80) was assumed constant. This is not true in general, and an electron kinetic equation should be coupled to the photon kinetic equation. This means that we should use the Vlasov equation for the electron distribution function f_e. Formally, this is a single-particle Liouville equation, valid in a broad range of plasma conditions, which can be written as

$$\left(\frac{\partial}{\partial t} + \frac{\mathbf{p}}{m_e} \cdot \nabla + \mathbf{F}_e \cdot \frac{\partial}{\partial \mathbf{p}}\right) f_e = 0, \tag{7.120}$$

where $\mathbf{p} = m_e\mathbf{v}$ is the electron momentum, and the force $\mathbf{F}_e$ acting on the electrons is

$$\mathbf{F} = -\nabla\phi - \frac{\hbar e^2}{2m_e}\nabla \int \frac{d\mathbf{k}}{(2\pi)^3}\omega_{\mathbf{k}} N_{\mathbf{k}}(\mathbf{r}, t). \tag{7.121}$$

The first term is the electrostatic force, associated with electron plasma waves (plasmons), and is determined by the Poisson equation

$$\nabla^2\phi = \frac{e}{\epsilon_0}\left[\int f_e(\mathbf{r}, \mathbf{p}, t)d\mathbf{p} - n_0\right]. \tag{7.122}$$

Here n_0 is the equilibrium density. The second term in equation (7.121) is the radiation force or ponderomotive force term. As we can see, inhomogeneities in the photon distribution will induce a force, which pushes the electrons out of the high intensity regions. This is at the origin of the concept of *photon charge* [11]. The medium is a mixture of two interacting fluids, the electron and photon fluids, as described by the coupled equations (7.80) and (7.120). Other two-fluids models have been discussed in the context of superfluidity. The influence of the electrostatic force, associated with the plasmon potential ϕ, on the Compton cooling process has been considered in [40]. As for the ponderomotive force term, it could be at the origin of suprathermal acceleration of particles by photons [41]. They both would require a more careful analysis.

7.4 Polariton condensation

We now turn to the concept of exciton–polariton, analyse the polariton equations and establish the analogy between polariton condensates and atomic and photonic condensates. Semiconductor microcavities can be seen as an interface between photons and condensed matter. See figure 7.11(a) for an illustration. A planar semiconductor plate, which is a kind of *quantum well*, is located in the middle of a microcavity delimited by two Bragg mirrors. The photon modes inside the cavity are strongly coupled to the excitonic transitions that can occur in the semiconductor, and this coupling leads to the creation of new modes, involving the excitation of both light and matter. These new modes are quasi-particles of a mixed origin, usually called *polaritons*. They are bosons with a very small mass, and can condensate in the lowest quantum level of the cavity at room temperature.

The field of exciton–polaritons in these planar microcavities has received considerable attention in recent years [42]. This is mainly due to the fact that this field is a two-dimensional bosonic field, which can suffer phase transitions analogous to Bose–Einstein condensation [43]. Such a condensation was first observed in 2006 by Karprzak and collaborators [44]. The main difference with respect to the usual condensation of dilute atom vapour is that we are now dealing with a strongly dissipated system, where the exciton losses are compensated by an external pump. In this sense it is closer to condensation of photons in a dye cavity. Usually, the external pump is fixed, and occupies a well-defined region of the planar cavity. But a moving source could eventually be considered. The coupled photonic and excitonic fields, represented by $\phi(\mathbf{r}, t)$ and $\chi(\mathbf{r}, t)$ respectively, can be described by the following system of wave equations [42]

$$i\hbar\frac{\partial\phi^\sigma}{\partial t} = -\left[\frac{\hbar^2}{2m_\phi}\nabla^2 + \frac{i\hbar}{2\tau_\phi}\right]\phi^\sigma + \frac{\hbar}{2}\Omega_R\chi^\sigma + H_{\mathrm{eff}}\phi^{-\sigma} + P^\sigma, \tag{7.123}$$

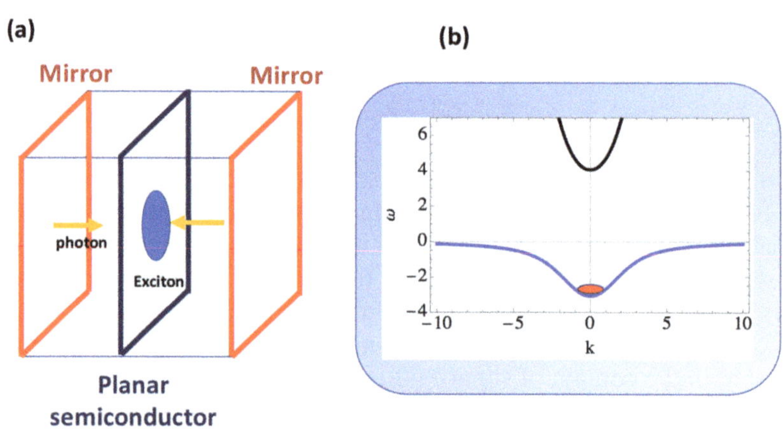

Figure 7.11. Polaritons in a semiconductor microcavity: (a) structure of a planar microcavity; (b) dispersion relation of exciton–polaritons in the cavity, as described by equations (7.129), in arbitrary units. The two dispersion branches are shown, as well as the location of the polariton condensation.

$$i\hbar\frac{\partial\chi^\sigma}{\partial t} = -\left[\frac{\hbar^2}{2m_\chi}\nabla^2 + \frac{i\hbar}{2\tau_\chi}\right]\chi^\sigma + \frac{\hbar}{2}\Omega_R\phi^\sigma + (\alpha_1 \mid \chi^\sigma \mid^2 + \alpha_2 \mid \chi^{-\sigma} \mid^2)\chi^\sigma. \quad (7.124)$$

In these equations $\sigma = \pm 1$ represents the two possible polarisation states, $m_\phi \ll m_e$ and $m_\chi \leqslant m_e$, are the cavity photon and the quantum well photon masses (and, for comparison, m_e the electron mass). Typical values are $m_\phi = 5 \times 10^{-5} m_e$, and $m_\chi = 0.4 m_e$. The quantities τ_ϕ and τ_χ are the lifetimes of cavity photons and excitons, with typical values $\tau_\phi = 10$ ps and $\tau_\chi = 400$ ps.

In equation (7.123), $P^\sigma \equiv P^\sigma(\mathbf{r}, t)$ is the photonic pump intensity, resulting from an external source, and tuned in quasi-resonance with respect to the lower polariton branch, as specified bellow. The interaction constants α_1 and α_2 in equation (7.124) describe the exciton–exciton interactions, with $|\alpha_2| \ll |\alpha_1|$. Typically, we have $\alpha_1 = 0.02 meV/\mu m^2$ and $\alpha_2 = -0.2\alpha_1$. Equation (7.123) also shows the in-plane effective magnetic field H_{eff}, coupling the two spin components $\sigma = \pm 1$, as due to the photon TE–TM split. It is defined as

$$H_{\text{eff}} = \beta\left(\frac{\partial}{\partial x} \mp i\frac{\partial}{\partial y}\right)^2, \quad \beta = \frac{\hbar^2}{4}\left(\frac{1}{m_\phi^{\text{TM}}} - \frac{1}{m_\phi^{\text{TE}}}\right), \quad (7.125)$$

where $m_\phi^{\text{TE}} = m_\phi$ and $m_\phi^{\text{TM}} = 0.95 m_\phi$ and the masses of the photonic TE and TM modes. We start with the derivation of the appropriate dispersion relations for the photon–exciton coupled field. In a first-order approximation, we neglect the photon–exciton coupling, the effective magnetic field coupling, the spin-coupling in the exciton–exciton interaction terms, as well as the optical pump. The resulting equations are

$$i\hbar\frac{\partial\phi^\sigma}{\partial t} = -\left[\frac{\hbar^2}{2m_\phi}\nabla^2 + \frac{i\hbar}{2\tau_\phi}\right]\phi^\sigma, \quad (7.126)$$

$$i\hbar\frac{\partial\chi^\sigma}{\partial t} = -\left[\frac{\hbar^2}{2m_\chi}\nabla^2 + \frac{i\hbar}{2\tau_\chi}\right]\phi^\sigma + (\alpha_1 \mid \chi^\sigma \mid^2)\chi^\sigma. \quad (7.127)$$

We can then find solutions of the form

$$\phi^\sigma(\mathbf{r}, t) = \phi_0^\sigma \exp(i\mathbf{k}\cdot\mathbf{r} - i\omega_\phi t), \quad \chi^\sigma(\mathbf{r}, t) = \chi_0^\sigma \exp(i\mathbf{k}\cdot\mathbf{r} - i\omega_\chi t). \quad (7.128)$$

Replacing this in equations (7.126)–(7.127), we obtain the dispersion relations

$$\omega_\phi = \frac{\hbar k^2}{2m_\phi} + i\gamma_\phi, \quad \omega_\chi = \frac{\hbar k^2}{2m_\chi} + \alpha_1 \left|\chi_0^\sigma\right|^2 + i\gamma_\chi. \quad (7.129)$$

This defines two independent frequency (or energy) branches, for the upper (photonic) states and the lower (excitonic) states, where a nonlinear correction for the lower branch is retained, as well as two damping coefficients $\gamma_\phi = 1/2\tau_\phi$ and $\gamma_\chi = 1/2\tau_\chi$. For a given value of the wavenumber k, we have $\omega_\phi \gg \omega_\chi$, as a result of the mass difference $m_\phi \ll m_\chi$. As a refinement to this simple description, we can reintroduce coupling between the photon and the exciton fields, corresponding to the terms containing the coupling frequency Ω_R in equations (7.123)–(7.124), but still ignoring the quantities $H_{\text{eff}} \simeq 0$, $P^\sigma = 0$ and $\alpha_2 = 0$. Using solutions of the form

$$(\phi^\sigma, \chi^\sigma)(\mathbf{r}, t) = \left(\phi_0^\sigma, \chi_0^\sigma\right)\exp(i\mathbf{k}\cdot\mathbf{r} - i\omega t), \tag{7.130}$$

we can then derive the coupled dispersion equation

$$(\omega - \omega_\phi)(\omega - \omega_\chi) = \frac{\Omega_R^2}{4}, \tag{7.131}$$

where ω_ϕ and ω_χ are determined by equation (7.129). Solving for ω, we get two solutions $\omega_\pm$, such that

$$\omega_\pm = \frac{1}{2}\left[(\omega_\phi + \omega_\chi) \pm \sqrt{(\omega_\phi + \omega_\chi)^2 + \Omega_R^2}\right]. \tag{7.132}$$

This corresponds to the two well known ω_+ polariton and ω_- exciton branches, which are represented in figure 7.11. In the linear and non-dissipative limit, they would simply reduce to the parabolic dispersion curves $\omega_\phi = \hbar k^2/2m_\phi$, and $\omega_\chi = \hbar k^2/2m_\chi$. In typical experiments, the cavity modes are excited by a pump laser beam. Excess of energy is dissipated through the emission of phonons, exciton scattering and other loss mechanisms, leading to the appearance of a population of low energy polaritons, which progressively accumulate in the lowest polariton level, as indicated in the figure. Occurrence of a Bose–Einstein condensate of polaritons is observed above a threshold for the pump intensity, in a way similar to that of photon condensation in a dye cavity.

7.5 BEC–laser transition

Similarities between a photon BEC and a laser are quite striking. They both need a pump field to excite the upper level population of atoms or molecules, and both lead to the coherent excitation of a single photon mode. This becomes more obvious when we compare the equations governing the photon number in these two processes. A simple nonlinear equation can be written for the laser, equation (6.144), in the form

$$\frac{\partial n}{\partial t} = (A - \gamma_c)n - Bn^2, \tag{7.133}$$

where $A = D_0 w$ represents the oscillator strength of the radiative transition, γ gives the cavity losses, and $B = 2D_0 w^2 \tau$ the decay rate. We have seen in section 6.8 that, for sufficiently high inversion of population, $D > \gamma_c/w$, stimulated emission dominates over the cavity losses and laser instability takes place. The number of photons of the resonant cavity mode will grow in time, until nonlinear saturation is reached. A similar, although not identical process occurs in a photon BEC, as shown by equation (7.70) of section 7.3, written here in a slight different way, to facilitate comparison

$$\frac{dn}{dt} = B\left[M_e(n+1) - M_g n \exp \hbar\delta/T\right] - \kappa n, \tag{7.134}$$

where B is the Einstein emission coefficient, M_e and M_g are the mean number of molecules in the excited and ground states. The exponential factor results from the Kennard–Stepanov relation, where $\delta = (\omega - \omega_s)$ is the detuning between the photon mode frequency ω and the zero-phonon line of the molecule, and T is the dye bath temperature. Given that the total number of molecules is constant, $M = M_e + M_g$, the fraction of excited molecules $f = M_e/M$ will grow until it attains an equilibrium value, as indicated by equation (7.72). In the limit of negligible losses $\kappa \to 0$, we can see from (7.134) that a Bose–Einstein distribution will be attained, such that

$$\left(\frac{n+1}{n}\right) = \frac{(1 - f_{\text{eq}})}{f_{\text{eq}}} \exp\left[\frac{\hbar(\omega - \omega_0)}{T}\right]. \tag{7.135}$$

This discussion shows the striking differences, despite the similarities, between a photon BEC and a laser. First, the BEC state is an equilibrium state, at a given temperature, which is the temperature of the dye solution (the room temperature). This state depends on both spontaneous and stimulated emission, as indicated by the presence of the $(n + 1)$ term in this expression.

In contrast, spontaneous emission is ignored in the laser, which is a state out-of-equilibrium, where light is concentrated in a single mode of the cavity, not necessarily coincident with the lowest cavity frequency. The laser therefore requires a strong pump, necessary to obtain an inversion of population $D_0 > 0$, which is not required by photon BECs. For that reason, photon condensation is observed at much lower pump intensity and lower frequency [33, 34, 45, 46]. Similar effects had previously been observed in polariton condensates [47]. But the important fact is that the same experimental configuration is able to support both regimes, and that we can observe a transition from laser action to condensation, by decreasing the pump power and making other small adjustments in the cavity system.

7.6 Photon kinetics

The process of photon condensation can be approached from a more general point of view, when photons propagate in an arbitrary medium. The starting point of this approach is the wave-kinetic equation of photons, which is discussed now in detail. This will then coupled with the equations for the particles in the medium, neutral atoms, molecules or electrons.

As a starting point for any description of collective photon processes, which include Bose–Einstein condensation, we should start with a photon kinetic equation, which can be derived in quite general conditions (see [48, 49]). We consider a system of atoms and photons, described by the evolution equation

$$\frac{\partial \rho_{\text{tot}}}{\partial t} = \frac{1}{i\hbar}[H, \rho_{\text{tot}}], \tag{7.136}$$

where we define the total Hamiltonian H in the usual way, as

$$H = H_0 + H_{\text{int}}, \quad H_0 = H_a + H_f, \tag{7.137}$$

where, according to the usual conventions, H_{int} represents the interaction between atoms and radiation, and H_0 is the unperturbed field containing the atomic part H_a and the field part H_f described by

$$H_f = \sum_{\mathbf{k}} \hbar\omega_{\mathbf{k}}\left(a_{\mathbf{k}}^\dagger a_{\mathbf{k}} + \frac{1}{2}\right). \tag{7.138}$$

It is useful to move onto the interaction picture, where two new operators are used, as defined by

$$\rho = e^{iH_0t/\hbar}\rho_{\text{tot}}e^{-iH_0t/\hbar}, \quad V = e^{iH_0t/\hbar}H_{\text{int}}e^{-iH_0t/\hbar}. \tag{7.139}$$

In this picture, equation (7.136) can be rewritten as

$$\frac{\partial\rho}{\partial t} = \frac{1}{i\hbar}[V, \rho]. \tag{7.140}$$

A formal integration of this equation leads to

$$\rho(t) = \rho(0) + \frac{1}{i\hbar}\int_0^t [V(t-\tau), \rho(t-\tau)]d\tau. \tag{7.141}$$

For each photon mode $\mathbf{k}$, we define a photon number operator $N_{\mathbf{k}} = a_{\mathbf{k}}^\dagger a_{\mathbf{k}}$. It is also useful to introduce a local photon number operator $N_{\mathbf{k}}(\mathbf{r})$, defined in a symmetrised form as

$$N_{\mathbf{k}}(\mathbf{r}) = \int a^\dagger(\mathbf{k} - \mathbf{q}/2)\, a(\mathbf{k} + \mathbf{q}/2)e^{i\mathbf{q}\cdot\mathbf{r}}\frac{d\mathbf{q}}{(2\pi)^3}. \tag{7.142}$$

In alternative, this could be written as a sum, using

$$N_{\mathbf{k}}(\mathbf{r}) = \sum_{\mathbf{k}} a^\dagger(\mathbf{k} - \mathbf{q}/2)a(\mathbf{k} + \mathbf{q}/2)e^{i\mathbf{q}\cdot\mathbf{r}}. \tag{7.143}$$

In the interaction picture, we should be able to use, in the same way as in equations (7.139), a time-dependent operator $N(\mathbf{r}, \mathbf{k}, t)$ defined by

$$N(\mathbf{r}, \mathbf{k}, t) = e^{iH_0t/\hbar}N_{\mathbf{k}}(\mathbf{r})e^{-iH_0t/\hbar}. \tag{7.144}$$

These equations are valid for a given photon polarisation. For unpolarised light, we should introduce a sum over the two possible states of polarisation, σ. Noting that $H_0 = H_a + H_f$ and that the atom Hamiltonian commutes with $N_{\mathbf{k}}(\mathbf{r})$ in equation (7.144), we can easily conclude that

$$N(\mathbf{r}, \mathbf{k}, t) = \int a^\dagger(\mathbf{k} - \mathbf{q}/2)\, a(\mathbf{k} + \mathbf{q}/2)\, e^{i(\mathbf{q}\cdot\mathbf{r} - \Delta\omega t)}\frac{d\mathbf{q}}{(2\pi)^3}, \tag{7.145}$$

where $\Delta\omega = (\omega_+ - \omega_-)$, with $\omega_\pm = \omega(\mathbf{k} \pm \mathbf{q}/2)$. Here, we have assumed that $\omega_\mathbf{k}$ is a well-defined function of the photon wavevector $\mathbf{k}$, as determined by an appropriate dispersion relation. In vacuum, we simply have $\omega_\mathbf{k} = kc$, and $\omega_\pm = | \mathbf{k} \pm \mathbf{q}/2 | c$. At this point, it should be noticed that, in a medium, the description of photon states is only meaningful when the wavelength is much shorter than the photon mean free path. This means that we should take $|\mathbf{k}| \gg |\mathbf{q}|$ in the above integral. As a result, we are led to the conclusion that the photon number operator satisfies the conservation law

$$\left(\frac{\partial}{\partial t} + \mathbf{v}_\mathbf{k} \cdot \nabla\right) N(\mathbf{r}, \mathbf{k}, t) = 0, \tag{7.146}$$

where

$$\mathbf{v}_\mathbf{k} = \frac{\partial \omega_\mathbf{k}}{\partial \mathbf{k}} = \lim_{|\mathbf{q}| \to 0}\left(\frac{\Delta\omega}{\mathbf{q}}\right) \cdot \frac{\mathbf{k}}{|\mathbf{k}|}, \tag{7.147}$$

is the photon group velocity. Let us now define the *photon Wigner function* as the mean value of the symmetrised photon number operator, such that

$$W(\mathbf{r}, \mathbf{k}, t) = Tr.\{\rho(t) N(\mathbf{r}, \mathbf{k}, t)\}, \tag{7.148}$$

where the density operator $\rho(t)$ evolves according to equation (7.140). Applying the operator $(\partial/\partial t + \mathbf{v}_\mathbf{k} \cdot \nabla)$ to the definition of the photon Wigner function, as stated in equation (7.145), and noting that ρ is not explicitly dependent on $\mathbf{r}$, we obtain

$$\left(\frac{\partial}{\partial t} + \mathbf{v}_\mathbf{k} \cdot \nabla\right) W(\mathbf{r}, \mathbf{k}, t) = Tr.\left\{\frac{\partial \rho}{\partial t} N(\mathbf{r}, \mathbf{k}, t)\right\}. \tag{7.149}$$

Using equation (7.140), this is equivalent to

$$\left(\frac{\partial}{\partial t} + \mathbf{v}_\mathbf{k} \cdot \nabla\right) W(\mathbf{r}, \mathbf{k}, t) = \frac{1}{i\hbar} Tr.\{[V(t), \rho(t)] N(\mathbf{r}, \mathbf{k}, t)\}. \tag{7.150}$$

Finally, using the property of trace invariance under a cyclic permutation of the operators, and using a more compact notation, we obtain

$$\left(\frac{\partial}{\partial t} + \mathbf{v}_\mathbf{k} \cdot \nabla\right) W = \frac{1}{i\hbar} Tr.\{[N, V]\rho\}, \tag{7.151}$$

where N and W are functions of $(\mathbf{r}, \mathbf{k}, t)$. This is an important starting point, from where we can attempt to derive a closed equation for the photon Wigner function W. For such purpose, we use the formal solution of equation (7.141), and assume that the initial state $\rho(0)$ is diagonal in the basis of the eigenvectors of the number operator N. This means that the first term in equation (7.141) has no contribution to the evolution described by equation (7.13). We then get

$$\left(\frac{\partial}{\partial t} + \mathbf{v}_\mathbf{k} \cdot \nabla\right) W = -\frac{1}{\hbar^2} \int_0^t Tr.\{[N(\mathbf{r}, \mathbf{k}, t), V(t)] V(t - \tau) \rho(t - \tau)\} d\tau. \tag{7.152}$$

This can be transformed into a closed equation for W after some simplifying assumptions. First, we assume that radiation is *weakly coupled* to the atoms in the medium, such that the total density operator can be factorised as

$$\rho \simeq \rho_A \, \rho_F. \tag{7.153}$$

Second, that the system is *Markovian*, such that

$$\rho(t - \tau) \simeq \rho(t), \tag{7.154}$$

and we can extend the integral in equation (7.152) to infinity. Third, that the interaction potential $V(t)$ results from dipole interactions with the atoms, and is described by

$$V(t) = \sum_a \mathbf{p}_a \cdot \mathcal{E}(\mathbf{r}_a, t), \tag{7.155}$$

where $\mathbf{r} = \mathbf{r}_a$ is the position of each atom a, and $\mathbf{p}_a$ the corresponding dipole operator. Finally, the electric field operator is given by

$$\mathcal{E}(\mathbf{r}, t) = i \int \sqrt{\frac{\omega_\mathbf{k}}{2\epsilon_0}} \, a_\mathbf{k} \exp(i\mathbf{k} \cdot \mathbf{r}) d\mathbf{k} + h.\,c. \tag{7.156}$$

We are then led to an expression of the form

$$\left(\frac{\partial}{\partial t} + \mathbf{v}_\mathbf{k} \cdot \nabla\right) W = \int d\mathbf{k}' \int d\mathbf{k}'' K'(\mathbf{r}, t; \mathbf{k}, \mathbf{k}', \mathbf{k}'') \langle a^\dagger_{\mathbf{k}'} a_{\mathbf{k}''} \rangle + \cdots, \tag{7.157}$$

plus a similar term containing $\langle a_{\mathbf{k}'} a^\dagger_{\mathbf{k}''} \rangle$. Other terms of the form $\langle aa \rangle$ and $\langle a^\dagger a^\dagger \rangle$ are neglected when we use the rotating wave approximation. Now, using the identity

$$\langle a_{\mathbf{k}'} a^\dagger_{\mathbf{k}''} \rangle = \delta(\mathbf{k}' - \mathbf{k}'') + \langle a^\dagger_{\mathbf{k}'} a_{\mathbf{k}''} \rangle, \tag{7.158}$$

we can see that a term independent of the radiation field operators will emerge from the missing term in equation (7.157), and represents the contribution from spontaneous emission. As a result, this equation will take the form

$$\left(\frac{\partial}{\partial t} + \mathbf{v}_\mathbf{k} \cdot \nabla\right) W = S(\mathbf{r}, \mathbf{k}, t) + \int d\mathbf{k}' \int d\mathbf{k}'' K'(\mathbf{r}, t; \mathbf{k}, \mathbf{k}', \mathbf{k}'') \langle a^\dagger_{\mathbf{k}'} a_{\mathbf{k}''} \rangle, \tag{7.159}$$

where the source term $S(\mathbf{r}, \mathbf{k}, t)$ represents spontaneous emission. Finally, we should notice that the last term in this equation can be written in terms of the Wigner function. Using equations (7.142) and (7.145), we get

$$\langle a^\dagger_{\mathbf{k}'} a_{\mathbf{k}''} \rangle = \int W(\mathbf{r}, (\mathbf{k}' + \mathbf{k}'')/2, t) \, e^{i(\mathbf{k}' - \mathbf{k}'') \cdot \mathbf{r})} d\mathbf{r}. \tag{7.160}$$

We can then obtain a close form for the Wigner equation, as

$$\left(\frac{\partial}{\partial t} + \mathbf{v}_\mathbf{k} \cdot \nabla\right) W(\mathbf{r}, \mathbf{k}, t) = S(\mathbf{r}, \mathbf{k}, t) + \int d\mathbf{r}' \int d\mathbf{k}' K(\mathbf{r}', t; \mathbf{k}, \mathbf{k}') W(\mathbf{r}', \mathbf{k}', t), \tag{7.161}$$

where the new coupling function $K(\mathbf{r}', t; \mathbf{k}, \mathbf{k}', \mathbf{k}")$ is defined by an integral of the previous one, over the spectrum $\mathbf{k}"$. The specific form of this function depends on the medium we consider. This could be due to photon scattering by dye molecules, as described by the Kennard–Stepanov relation in section 7.2, or to Compton scattering as in the case of section 7.3.

In alternative to this formal derivation, we can use a phenomenological description of the optical medium where, instead of a collective ensemble of atoms, photon propagation in the medium is described a generic dispersion relation. This is a simple approach, is shown in the appendix.

References

[1] Bose S N 1924 Plancks Gesetz und Lichtquantenhypothese *Z Phys.* **26** 178

[2] Einstein A 1925 Quantentheorie des einatomigen idealen Gases *Sitzber. Kgl. Preuss. Akad. Wiss.* **1** 3

[3] Anderson M H, Ensher J R, Matthews M R, Wieman C E and Cornell E A 1995 Observation of Bose-Einstein condensation in a dilute atomic vapor *Science* **269** 198

[4] Davis K B, Mewes M-O, Andrews M R, van Druten N J, Durfee D S, Kurn D M and Ketterle W 1995 Bose-Einstein condensation in a gas of sodium atoms *Phys. Rev. Lett.* **75** 3969

[5] Bradley C C, Sackett C A, Tollett J J and Hulet R G 1995 Evidence of Bose-Einstein condensation in an atomic gas with attractive interactions? *Phys. Rev. Lett.* **75** 1687

[6] Pethick C J and Smith H 2002 *Bose-Einstein Condensation in Dilute Gases* (Cambridge: Cambridge University Press)

[7] Pitaevskii L P and Stringari A 2003 *Bose-Einstein Condensation* (Oxford: Clarendon)

[8] Leggett A J 2006 *Quantum Liquids* (Oxford: Oxford University Press)

[9] Anderson P W 1963 Plasmons, gauge invariance, and mass *Phys. Rev.* **130** 439

[10] Green D 2013 The search for and discovery of the Higgs boson at CMC *Int. J. Mod. Phys.* A **28** 1330003

[11] Mendonça J T, Martins A M, Guerreiro A 2000 Field quantization in a plasma: photon mass and charge *Phys. Rev.* E **62** 298

[12] Zel'dovich Y B and Levich E V 1969 Bose condensation and shock waves in photon spectra *Sov. Phys. JETP* **28** 1287

[13] Chiao R Y 2000 Bogoliubov dispersion relation for a photon fluid: is this a superfluid? *Opt. Commun.* **179** 157

[14] Klaers J, Schmitt J, Vewinger F and Weitz M 2010 Bose-Einstein condensation of photons in an optical microcavity *Nature* **468** 545

[15] Zakharov V E and Nazarenko S V 2005 Dynamics of the Bose-Einstein condensation *Physica* D **201** 203–11

[16] Connaughton C, Josserand C, Picozzi A, Pomeau Y and Rica S 2005 Condensation of classical nonlinear waves *Phys. Rev. Lett.* **95** 263901

[17] Gross E P 1961 Structure of a quantized vortex in boson systems *Nuovo Cimento* **20** 454–77

[18] Pitaevskii L P 1961 Vortex lines in an imperfect Bose gas *Sov. Phys. JETP* **13** 451–4

[19] Madelung E 1926 Eine anschauliche Deutung der Gleichung von Schrodinger *Naturwissenschaften* **14** 1004

[20] Bohm D 1952 A suggested interpretation of the quantum theory in terms of "Hidden" Variables. I *Phys. Rev.* **85** 166–79
Bohm D 1952 A suggested interpretation of the quantum theory in terms of "Hidden" Variables. II *Phys. Rev.* **85** 180–93
[21] Mendonça J T, Terças H and Gammal A 2018 Quantum Landau damping of dipolar Bose-Einstein condensates *Phys. Rev.* A **97** 063610
[22] Sonin E B 1987 Vortex oscillations and hydrodynamics of rotating superfluids *Rev. Mod. Phys.* **59** 87–155
[23] Fetter A L 2009 Rotating trapped Bose-Einstein condensates *Rev. Mod. Phys.* **81** 647–91
[24] Goldhaber A S and Nieto M M 2010 Photon and graviton mass limits *Rev. Mod. Phys.* **82** 939
[25] Posazhennikova A 2006 Weakly interacting, dilute Bose gases in 2D *Rev. Mod. Phys.* **78** 1111–34
[26] Berezinskii V L 1971 Destruction of long-range order in one-dimensional and two-dimensional systems having a continuous symmetry group *Sov. Phys. JETP* **32** 493–500
Berezinskii V L 1972 **34** 610–16
[27] Kosterlitz J M and Thouless D J 1973 Ordering, metastability and phase transitions in two-dimensional systems *J. Phys. C: Solid State Phys.* **6** 1181–203
[28] Minnhagen P 1987 The two-dimensional Coulomb gas, vortex unbinding and superfluid-superconducting films *Rev. Mod. Phys.* **59** 1001–66
[29] Kennard E H 1926 On the interaction of radiation with matter and on fluorescent exciting power *Phys. Rev.* **28** 672
[30] Stepanov B I and Kazachenko L P 1971 Universal relationship between absorption and emission spectra taking the solvent effect into account *J. Appl. Spectr.* **14** 596
[31] Snoke D W and Girvin S M 2013 Dynamics of phase coherence onset in Bose condensate of photons by incoherent phonon emission *J. Low Temp. Phys.* **171** 1
[32] Kruchkov A 2014 Bose-Einstein condensation of light in a cavity *Phys. Rev.* A **89** 033862
[33] Schmitt J, Damm T, Dung D, Vewinger F, Klaers J and Weitz M 2015 Thermalization kinetics of light: from laser dynamics to equilibrium condensation of photons *Phys. Rev.* A **92** 011602(R)
[34] Keeling J and Kirton P 2016 Spatial dynamics, thermalization, and gain clamping in a photon condensate *Phys. Rev.* A **93** 013829
[35] Walker B T, Rodrigues J D, Dhar H S, Oulton R F, Mintert F and Nyman R A 2020 Non-stationary statistics and formation jitter in transient photon condensation *Nat. Commun.* **11** 1390
[36] Kompaneets A S 1957 The establishment of thermal equilibrium between quanta and electrons *Sov. Phys. JETP* **4** 730
[37] Mendonça J T and Terças H 2017 Bose-Einstein condensation of photons in a plasma *Phys. Rev.* A **95** 063611
[38] Ibragimov N H 2010 Time-dependent exact solutions of the nonlinear Kompaneets equation *J. Phys. A: Math. Theor.* **43** 502001
[39] Tsintsadze L N 2004 Bose-Einstein condensation and intermediate state of the photon gas *Phys. Plasmas* **11** 855–8
[40] Erochenkova G and Chandre C 2015 Photon plasma-wave interaction via Compton scattering *J. Phys. A: Math. Theor.* **49** 495502
Erochenkova G and Chandre C 2016 **49** 089501

[41] Mendonça J T 2001 *Theory of Photon Acceleration* (Bristol: Institute of Physics Publishing)
[42] Shelykh I A, Kavokin A V, Rubo Yu G, Liew T C H and Malpuech G 2010 Polariton polarization-sensitive phenomena in planar semiconductor microcavities *Semicond. Sci. Technol.* **25** 013001
[43] Malpuech G, Rubo Y G, Laussy F P, Bigenwald P and Kavokin A V 2003 Polariton laser: thermodynamics and quantum kinetic theory *Semicond. Sci. Technol.* **18** S395–404
[44] Karprzak J *et al* 2006 Bose-Einstein condensation of exciton polaritons *Nature* **447** 409–14
[45] Hesten H J, Nyman R A and Mintert F 2018 Decondensation in nonequilibrium photonic condensates: when less is more *Phys. Rev. Lett.* **120** 040601
[46] Kirton P and Keeling J 2013 Nonequilibrium model of photon condensation *Phys. Rev. Lett.* **111** 100404
[47] Deng H, Weihs G, Snoke D, Bloch J and Yamamoto Y 2003 Polariton lasing vs. photon lasing in a semiconductor microcavity *Proc. Natl Acad. Sci.* **100** 15318–23
[48] Cooper J and Zoller P 1984 Radiative transfer equations in broad-band, time-varying fields *Astrophys. J.* **277** 813–9
[49] Rosato J 2010 The photon picture of radiation transport *Am. J. Phys* **78** 851–7

IOP Publishing

The Quantum Nature of Light
From photon states to quantum fluids of light
J T Mendonça

Chapter 8

Collective atomic emission

We have seen previously that a large ensemble of identical atoms can be collectively coupled with a single photon state by the process of stimulated emission, and this is the basis of the laser concept. Here we consider the collective spontaneous decay of an ensemble of atoms, when stimulated emission is not present, a process called *superradiance*. This was discovered by Dicke [1], for atoms contained in a small volume, with size of the order of the radiation wavelength. The concept was generalised later to the case of much larger volumes [2]. Historically, this can be considered as a precursor of the laser, although based on collective spontaneous emission. In its simplest version, this can be described as a kind of *superfluorescence*, where the spontaneous radiative decay of excited atoms significantly differs from the usual fluorescence of an uncorrelated collection of atoms.

The concept of superradiance was studied and refined by many researchers over the years, as described in the reviews [3–5]. It inspired similar concepts and models, as those of collective Rayleigh scattering, where in contrast with the superradiance models, the atoms can move from their original positions, due to the electromagnetic forces. As a result, they can eventually bunch, increasing in this way the cooperative emission process. This was first proposed by Bonifacio and collaborators [6] (see also [7]), as the atomic analogue of the *free-electron-laser* (FEL), and is sometimes called *collective atomic recoil laser* (CARL).

This chapter includes a short description of cyclotron superradiance, first proposed by O'Neil [9], where the atoms are replaced by electrons in a magnetic field. This is based on collective cyclotron radiation, and can be used as an effective cooling mechanism for electrons or positrons in a cavity, which is relevant for antimatter experiments [10, 11]. Recent extensions of superradiance show that intense radiation shock fronts can be produced, for specific particle (electron or atom) configurations [12, 13].

It should be noted that, in recent years, another type of superradiance has been considered in the literature. This is based on wave scattering by a rotating obstacle, with no obvious connection with Dicke's superradiance. This *superradiant scattering* by rotating bodies was first introduced in the astrophysical context by Zel'dovich [14], and proposed as a means to excite low-frequency emission of radiation by black holes [15, 16] (see for a review [17]). It has been revived in the context of nonlinear optics and explored in optical analogue experiments of astrophysical processes [18]. In this case, the black hole could be replaced by an optical vortex as the rotating scatterer.

8.1 Superradiance

Let us first consider the basic concepts associated with superradiance, and define its basic geometry. For that purpose, we assume a large number of N_a identical atoms, occupying a volume V, interacting with the electromagnetic field. In analogy with single-atom spontaneous decay, we assume that the atoms only have two internal states, the ground state and the excited state, with energies E_g and E_e, described by the state vectors $|\, g\rangle$ and $|\, e\rangle$. We assume that, initially, all the atoms are excited in the upper energy level. The collective spontaneous emission of photons with frequency ω_0, nearly equal to the resonant atomic frequency $\omega_a = (E_e - E_g)/\hbar$, can significantly differ from the spontaneous decay of individual atoms, described in chapter 4. See figure 8.1 for an illustration. The system is described by the total Hamiltonian

$$\hat{H} = \hat{H}_0 + \hat{H}', \quad \hat{H}_0 = \hat{H}_a + \hat{H}_f, \tag{8.1}$$

where the unperturbed terms are associated with the N_a atoms and the field, as

$$\hat{H}_a = \frac{1}{2}\hbar\omega_a\hat{S}_z, \quad \hat{S}_z = \sum_{j=1}^{N_a}\hat{\sigma}_{zj}, \quad \hat{H}_f = \sum_{\mathbf{k}}\hbar\omega_{\mathbf{k}}\left(\hat{a}_{\mathbf{k}}^{\dagger}\hat{a}_{\mathbf{k}} + \frac{1}{2}\right). \tag{8.2}$$

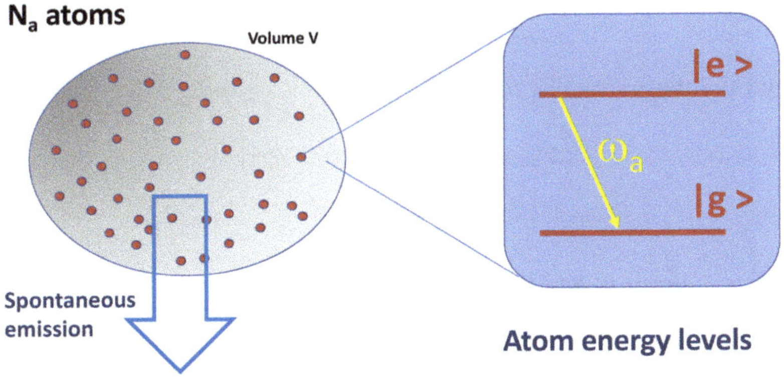

Figure 8.1. Geometry of superradiance: a large number $N_a \gg 1$ of identical two-level atoms in a volume V. If, at time $t = 0$, they are all in the excited state $|\, e\rangle$, they can coherently decay in time, emitting a burst of radiation. This deviates from the spontaneous decay of an uncorrelated collection of individual atoms.

The interaction term is given by

$$\hat{H}' = -\sum_{j=1}^{N_a}\left\{\hat{\mathcal{E}}^+(\mathbf{r}_j) + \hat{\mathcal{E}}^-(\mathbf{r}_j)\right\}\cdot\vec{\sigma}_{aj}, \tag{8.3}$$

where the electric field operators $\hat{\mathcal{E}}^{\pm}$ are defined at the location of the different atoms, $\mathbf{r} = \mathbf{r}_j$, and the vector operator of each atom, $\vec{\sigma}_{aj}$, is defined by

$$\vec{\sigma}_{aj} = \mathcal{P}\mathbf{e}_a\left(\hat{\sigma}_{+j} + \hat{\sigma}_{-j}\right), \tag{8.4}$$

with the electric dipole matrix element $\mathcal{P}$, and the unit vector $\mathbf{e}_a$ defining the direction of polarisation of the atomic transition. Here, we have also used the pseudo-spin operators introduced in chapter 4, defined as

$$\hat{\sigma}_{zj} = \left|e_j\right\rangle\left\langle e_j\right| - \left|g_j\right\rangle\left\langle g_j\right|, \quad \hat{\sigma}_{+j} = \left|e_j\right\rangle\left\langle g_j\right|, \quad \hat{\sigma}_{-j} = \left|g_j\right\rangle\left\langle e_j\right|. \tag{8.5}$$

Finally, the electric field operators, for positive and negative frequency parts, are

$$\hat{\mathcal{E}}^+(\mathbf{r}) = -i\sum_{\mathbf{k}}\mathbf{C}_{\mathbf{k}}a_{\mathbf{k}}e^{i\mathbf{k}\cdot\mathbf{r}},\ \hat{\mathcal{E}}^-(\mathbf{r}) = +i\sum_{\mathbf{k}}\mathbf{C}_{\mathbf{k}}a_{\mathbf{k}}^{\dagger}e^{-i\mathbf{k}\cdot\mathbf{r}}, \quad \mathbf{C}_{\mathbf{k}} = \sqrt{\frac{\hbar ck}{2\epsilon_0 V}}\mathbf{e}_{\mathbf{k}}, \tag{8.6}$$

and $\mathbf{e}_{\mathbf{k}}\cdot\mathbf{k} = 0$, for transverse waves. At this point, it is useful to define the population inversion operator $\hat{S}_z(\mathbf{r})$, appearing in equation (8.2), and the related atomic polarisation operator $\hat{\mathcal{P}}^{\pm}(\mathbf{r})$, as

$$\hat{S}_z(\mathbf{r}) = \sum_{j=1}^{N_a}\delta(\mathbf{r} - \mathbf{r}_j)\hat{\sigma}_{zj}, \quad \hat{\mathcal{P}}^{\pm}(\mathbf{r}) = \mathcal{P}\mathbf{e}_a\sum_{j=1}^{N_a}\delta(\mathbf{r} - \mathbf{r}_j)\hat{\sigma}_{\pm j}. \tag{8.7}$$

The evolution of these operators can be described by the Heisenberg equations, which are, for a generic operator $\hat{A}$, given by

$$\frac{d\hat{A}}{dt} = \frac{\partial\hat{A}}{\partial t} + \frac{1}{i\hbar}[\hat{A}, \hat{H}]. \tag{8.8}$$

Here, they take the particular form

$$\frac{d\hat{S}_z}{dt} = \frac{i}{\hbar}\left(\hat{\mathcal{E}}^+ + \mathcal{E}^-\right)(\mathcal{P}^+ - \mathcal{P}^-), \tag{8.9}$$

and

$$\frac{d\hat{\mathcal{P}}^+}{dt} = i\omega_a\mathcal{P}^+ + \frac{2i}{\hbar}\mathcal{P}^2\mathbf{e}_a\left\{\mathbf{e}_a\cdot\left(\hat{\mathcal{E}}^+ + \mathcal{E}^-\right)\right\}\hat{S}_z, \tag{8.10}$$

where $\hat{\mathcal{P}}^{\pm} \equiv \hat{\mathcal{P}}^{\pm}(\mathbf{r}, t)$ and $\hat{\mathcal{E}}^{\pm} \equiv \hat{\mathcal{E}}^{\pm}(\mathbf{r}, t)$. These Heisenberg equations should be completed with an evolution equation for the electric field, which takes the form

$$\left(\frac{\partial^2}{\partial t^2} - c^2\nabla\times\nabla\times\right)\hat{\mathcal{E}}^+ = -\frac{1}{\epsilon_0}\frac{\partial^2\hat{\mathcal{P}}^-}{\partial t^2}. \tag{8.11}$$

These are the so-called *Maxwell–Bloch equations* for the ensemble of N_a atoms interacting with the radiation field. It is now useful to consider radiation propagating along a given direction, with wavevector $\mathbf{k}_0$ and frequency $\omega_0 \simeq \omega_a$, such that $k_0 = \omega_0/c$. Defining with generality the z-axis in the propagation direction, we specify the atom and field operators as

$$\hat{\mathcal{E}}^{\pm} = \hat{E}^{\pm}(z, t)\mathbf{e}_0 \exp[\pm i(k_0 z - \omega_0 t)], \quad \hat{\mathcal{P}}^{\pm} = \hat{P}^{\pm}(z, t)\mathbf{e}_a \exp[\pm i(k_0 z - \omega_0 t)], \tag{8.12}$$

where the amplitudes $\hat{E}^{\pm}(z, t)$ and $\hat{P}^{\pm}(z, t)$ are slowly varying functions of z and t. Replacing this in equations (8.9)–(8.11), taking $\mathbf{e}_0 = \mathbf{e}_a$, and making the appropriate ordering of the operators, we get the slowly varying equations

$$\frac{\partial \hat{S}_z}{\partial t} = \frac{i}{\hbar}(\hat{P}^+\hat{E}^+ - \hat{E}^-\hat{P}^-), \quad \frac{\partial \hat{P}^+}{\partial t} = \frac{2i}{\hbar}\mathcal{P}^2\hat{E}^-\hat{S}_z, \tag{8.13}$$

and

$$\left(\frac{\partial}{\partial z} + \frac{1}{c}\frac{\partial}{\partial t}\right)\hat{E}^+ = \frac{i\omega_0}{2\epsilon_0 c}\hat{P}^-. \tag{8.14}$$

These are the amplitude equations, or the envelope equations, associated with the above Maxwell–Bloch system. Using a new temporal variable $\tau = t - z/c$, we can now rewrite these equations as

$$\frac{\partial \hat{S}_z}{\partial \tau} = \frac{i}{\hbar}(\hat{P}^+\hat{E}^+ - \hat{E}^-\hat{P}^-), \quad \frac{\partial \hat{P}^+}{\partial \tau} = \frac{2i}{\hbar}\mathcal{P}^2\hat{E}^-\hat{S}_z, \tag{8.15}$$

and

$$\frac{\partial}{\partial \tau}\hat{E}^+ = \frac{i\omega_0}{2\epsilon_0 c}\hat{P}^-. \tag{8.16}$$

This last equation can be integrated, for propagation along the z-axis, to give

$$\hat{E}^+(z, t) = \frac{i\omega_0}{2\epsilon_0 c}\int_0^z \hat{P}^-(z', t)dz'. \tag{8.17}$$

This leaves us with two equations for $\hat{S}_z$ and $\hat{P}^+$. In order to solve them, we need to rewrite the definitions in equation (8.7) in terms of the new one-dimensional model. Defining a radiation beam propagating along z with transverse waist w, we now have

$$\hat{S}_z(z) = \frac{1}{\pi w^2}\sum_j \delta(z - z_j)\hat{\sigma}_{zj}, \quad \hat{\mathcal{P}}^{\pm}(z, \tau) = \frac{\mathcal{P}}{\pi w^2}e^{\mp i\omega_0 \tau}\sum_j \delta(z - z_j)\hat{\sigma}_{\pm j}. \tag{8.18}$$

Replacing this, and equation (8.17), in the envelope equations (8.15), we get

$$\frac{d}{d\tau}\hat{\sigma}_{zj} = -\Gamma\mu\sum_{j\geqslant i}\left(\hat{\sigma}_{+i}\hat{\sigma}_{-j} + \hat{\sigma}_{-i}\hat{\sigma}_{+j}\right), \quad \frac{d}{d\tau}\hat{\sigma}_{+j} = i\omega_0\hat{\sigma}_{+j} + 2\Gamma\mu\sum_{j\geqslant i}\hat{\sigma}_{+i}\hat{\sigma}_{zj}, \tag{8.19}$$

with the new parameters

$$\Gamma = \frac{\omega_0^3 \mathcal{P}^2}{3\pi\epsilon_0 \hbar c^3}, \qquad \mu = \frac{3}{8\pi^2}\frac{\lambda_0^2}{w^2}, \tag{8.20}$$

where $\lambda_0 = 2\pi c/\omega_0$ is the wavelength, and Γ is the single-atom spontaneous decay rate. In order to proceed further, we introduce the collective operators

$$\hat{D}_\pm = \sum_j \hat{\sigma}_{\pm j}, \qquad \hat{D}_z = \sum_j \hat{\sigma}_{zj}. \tag{8.21}$$

We then get the evolution equations for these collective operators, as

$$\frac{d}{d\tau}\hat{D}_z = -\Gamma\mu\hat{D}_+\hat{D}_-, \qquad \frac{d}{d\tau}\hat{D}_+ = i\omega_0\hat{D}_+ + \Gamma\mu\hat{D}_+\hat{D}_z, \tag{8.22}$$

To simplify, we replace the operators by c-numbers. Notice first that these equations conserve the quantity D, defined as

$$D = D_z^2 + D_+D_-. \tag{8.23}$$

This conservation law allows us to use a vectorial representation, where D is the length of a vector in a three-dimensional space. If the photons initially emitted by the N_a excited atoms is small, we can assume that $D_z(0) \simeq \mathcal{N}$, and $D_+(0) \simeq 0$. Defining two angles θ and φ in this space, such that

$$D_z(\tau) = N_a \cos\theta(\tau), \qquad D_+(\tau) = N_a \sin\theta(\tau) e^{i\varphi(\tau) + i\omega_0\tau}. \tag{8.24}$$

These equations define a point on the surface of a sphere of radius $\mathcal{N}$, usually called the *Bloch sphere*, as illustrated in figure 8.2. Comparing this with equations (8.22), we obtain

$$\frac{d\varphi}{d\tau} = 0, \qquad \frac{d\theta}{d\tau} = \frac{1}{2\tau_s}\sin\theta, \tag{8.25}$$

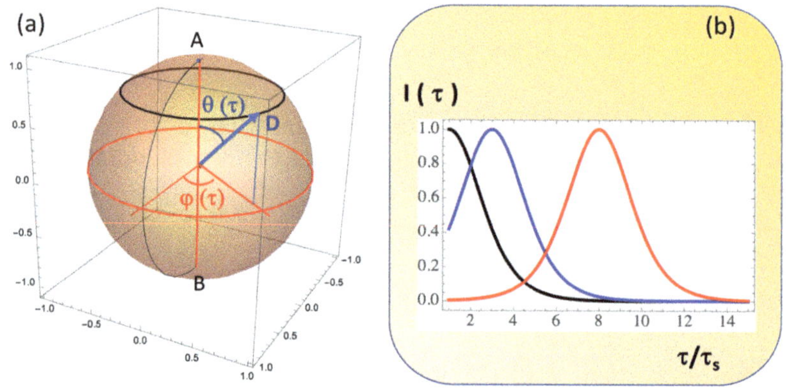

Figure 8.2. (a) Bloch sphere of superradiant decay, in normalised units, $D(\tau)/N$. The vector representing the quantum state of the system rotates around the vertical axis, with periodic $\theta(\tau)$ and constant φ; (b) radiation intensity as a function of time: in normalised units $I(\tau)/I_0$, where $I_0 = \hbar\omega_0 N_a/2\tau_s$, for increasing delay times, $\tau_d/\tau_s = 1$ in black, $\tau_d/\tau_s = 3$ in blue, and $\tau_d/\tau_s = 8$ in red.

where the quantity τ_s defines the superradiant time-scale, and is given by

$$\tau_s = \frac{1}{N_a \Gamma \mu}. \tag{8.26}$$

The solutions for θ and φ are therefore able to describe a closed trajectory on the surface of the Bloch sphere. For the assumed initial conditions $\theta(0) \equiv \theta_i \simeq 0$, equation (8.25) will describe the trajectory of a purely damped oscillator, moving from a position near the point A on the figure, to point B. However, the initial value cannot be exactly equal to zero, otherwise motion is not possible. The deviations of $\theta_i \neq 0$ with respect to the position A can be provided by quantum fluctuations.

It is then plausible to assume that the average initial angle is inversely proportional to the square-root of the number of atoms in the system, The larger this number is, the smaller the amplitude of the fluctuations

$$\langle \theta_i \rangle = \frac{2}{\sqrt{N_a}}. \tag{8.27}$$

Using this as an estimate of the initial conditions for the quantum system described by equations (8.25), with $\theta(0) = \langle \theta_i \rangle$, we can easily get

$$\tan\left(\frac{\theta(\tau)}{2}\right) = \tan\left(\frac{\langle \theta_i \rangle}{2}\right) \exp\left(-\frac{\tau}{2\tau_s}\right). \tag{8.28}$$

Defining the radiation intensity as

$$I(\tau) = -\hbar\omega_0 \frac{dD_z}{d\tau} = \hbar\omega_0 \frac{N_a}{2\tau_s} \sin^2\theta(\tau), \tag{8.29}$$

we can write more explicitly that

$$I(\tau) = \hbar\omega_0 \frac{N_a}{2\tau_s} \frac{1}{\cosh^2[(\tau - \tau_d)/2\tau_s]}, \tag{8.30}$$

where we have defined the *radiation delay time*, τ_d, as

$$\tau_d = 2\tau_s \ln\left(\frac{1}{2}\langle \theta_i \rangle\right) = \tau_s \ln N_a. \tag{8.31}$$

As we can see, the delay time for superradiant decay increases logarithmically with the number of atoms in the system, as illustrated in figure 8.2(b). For increasing values of τ_d, the radiated signal evolves from the exponential decay, typical of spontaneous decay of a collection of independent atoms, to a coherent burst of radiation. These solutions can be improved, as follows. Going back to equations (8.15), we assume that the solutions for these equations depend on both position z and time τ, not just in τ. For that purpose, we use

$$S_z = n_a \cos\theta(z, \tau), \quad P^+ = i\mathcal{P} n_a \sin\theta(z, \tau)\exp(i\varphi), \tag{8.32}$$

where $n_a = N_a/V$ is the atom density, and $V = \pi L w^2$ is the volume of the beam. Taking the derivative of P^+ with respect to time, we get

$$\frac{\partial P^+}{\partial \tau} = i\mathcal{P} n_a \cos\theta e^{i\varphi} \frac{\partial \theta}{\partial \tau} = \frac{2i}{\hbar}\mathcal{P}^2 E^- n_a \cos\theta, \tag{8.33}$$

from where we obtain an equation for the angle θ, as

$$\frac{\partial \theta}{\partial \tau} = \frac{2}{\hbar}\mathcal{P} E^- e^{i\varphi}. \tag{8.34}$$

On the other hand, from equation (8.16) we also have

$$\frac{\partial}{\partial z}E^- = \frac{\omega_0}{2\epsilon_0 c}\mathcal{P} n_a \sin\theta e^{i\varphi} = \frac{\hbar}{2\mathcal{P}} e^{i\varphi} \frac{\partial^2}{\partial z \partial \tau}\theta. \tag{8.35}$$

This leads to a closed equation for the angle θ, of the form

$$\frac{\partial^2 \theta}{\partial z \partial \tau} = \frac{\mathcal{P}^2}{\hbar} \frac{n_a \omega_0}{\epsilon_0 c} \sin\theta. \tag{8.36}$$

Using the above definition of the temporal delay time, τ_s, and introducing the characteristic length of the system $\mathcal{L}$, we can rewrite this equation in a more appropriate form, as

$$\frac{\partial^2 \theta}{\partial z \partial \tau} = \frac{1}{\mathcal{L}\tau_s} \sin\theta. \tag{8.37}$$

with

$$\mathcal{L} = \frac{\mathcal{P}^2 \tau_s}{c} \frac{N_a \omega_0}{\hbar \epsilon_0 c}. \tag{8.38}$$

This is the well-known *sine-Gordon equation.* Now, using the normalised space and time variables, η and ζ, defined as

$$\eta = \frac{z}{\mathcal{L}}, \quad \zeta = \frac{z}{\tau_s}, \tag{8.39}$$

we can write this equation in its simplest form

$$\frac{\partial^2 \theta}{\partial \eta \partial \zeta} = \sin\theta. \tag{8.40}$$

This equation is known to satisfy soliton solutions of a particular type, the so-called breather solutions, which display the intrinsic nonlinear properties of the super-radiance process.

8.2 Collective recoil emission

We now turn to a different problem, and consider the influence of atomic motion on collective radiation processes. For that purpose, we assume again that a large number N_a of identical atoms interacts with the radiation field. But now we include the atomic motion and the recoil effect induced by photon emission. For this reason, the total Hamiltonian of the system depends on both the atoms position $\mathbf{r}_j$, as previously, and the atoms momentum $\mathbf{p}_j$, and equation (8.1) is replaced by

$$\hat{H}(\mathbf{r}_j, \mathbf{p}_j) = \hat{H}_0 + \hat{H}'(\mathbf{r}_j) + \sum_{j=1}^{N_a} \frac{p_j^2}{2M}, \tag{8.41}$$

where H_0 is still defined by equation (8.2), but we now should write the interaction Hamiltonian as (figure 8.3)

$$H'(\mathbf{r}_j) = \sum_{j=1}^{N_a} h_j'(\mathbf{r}_j), \tag{8.42}$$

with

$$h_j'(\mathbf{r}_j) = -i\hbar \sum_{\mathbf{k}} \left\{ g_{\mathbf{k}} a_{\mathbf{k}}^{\dagger} \sigma_{-j} e^{-i\mathbf{k}\cdot\mathbf{r}_j} - g_{\mathbf{k}}^{*} \sigma_{+j} a_{\mathbf{k}} e^{+i\mathbf{k}\cdot\mathbf{r}_j} \right\}, \tag{8.43}$$

where the mode coupling coefficient $g_{\mathbf{k}}$ is

$$g_{\mathbf{k}} = (\mathcal{P}_{\mathrm{eg}} \cdot \mathbf{e}_{\mathbf{k}}) \sqrt{\frac{\omega_{\mathbf{k}}}{2\hbar\omega_a}}. \tag{8.44}$$

Here, $\mathcal{P}_{\mathrm{eg}}$ is the atomic dipole moment and $\mathbf{e}_{\mathbf{k}}$ is the unit polarisation direction of field mode $\mathbf{k}$. The centre-of-mass motion of the atoms is determined by the equations of motion

$$\frac{d\mathbf{r}_j}{dt} = \frac{\mathbf{p}_j}{M}, \quad \frac{d\mathbf{p}_j}{dt} = -\frac{\partial H}{\partial \mathbf{r}_j} = -\hbar \sum_{\mathbf{k}} \mathbf{k} \left\{ g_{\mathbf{k}} a_{\mathbf{k}}^{\dagger} \sigma_{-j} e^{-i\mathbf{k}\cdot\mathbf{r}_j} + h.\ c. \right\}. \tag{8.45}$$

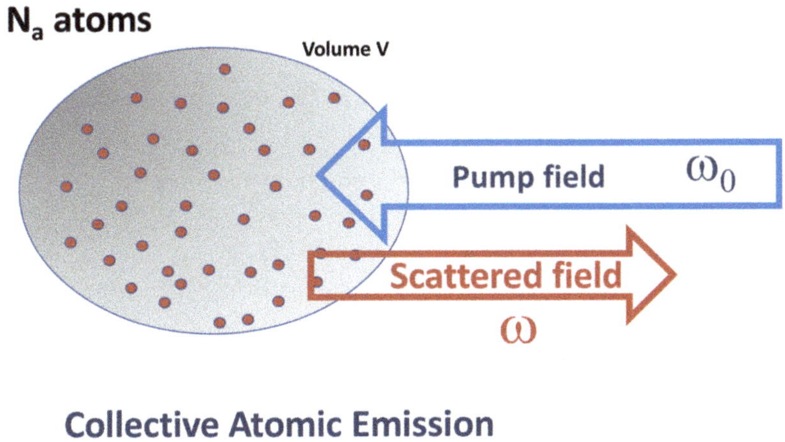

Figure 8.3. Geometry of collective Rayleigh scattering: a large number $N_a \gg 1$ of identical two-level atoms in a volume V. If, at time $t = 0$, they are all in the ground state $|\, g\rangle$. The pump field with frequency ω_0 much larger than the atomic transition frequency ω_a is collectively scattered by the atoms. This leads to the so-called 'collective atomic recoil laser' (CARL), as the atomic analogue to the free-electron laser (FEL).

On the other hand, the field operators satisfy the Heisenberg equations (Maxwell's equations), of the form

$$\frac{da_{\mathbf{k}}}{dt} = -i\omega_{\mathbf{k}} a_{\mathbf{k}} + \sum_{j=1}^{N_a} g_{\mathbf{k}} \sigma_{-j} e^{-i\mathbf{k}\cdot\mathbf{r}_j}. \tag{8.46}$$

Using the Heisenberg equations and making some simplifying assumptions, it is then possible to arrive at the coupled equations for field and atom operators. The atom equations can be written as

$$\frac{d\sigma_{-j}}{dt} = -i\omega_a \sigma_{-j} + \sum_{\mathbf{k}} g_{\mathbf{k}} \sigma_{zj} a_{\mathbf{k}} e^{i\mathbf{k}\cdot\mathbf{r}_j}, \qquad \frac{d\sigma_{zj}}{dt} = -2\sum_{\mathbf{k}} g_{\mathbf{k}} a_{\mathbf{k}}^{\dagger} e^{-i\mathbf{k}\cdot\mathbf{r}_j} + h.\ c. \tag{8.47}$$

Let us assume that the field spectrum is dominated by the existence of a large pump field, corresponding to a particular wavevector $\mathbf{k}_0$, and mode frequency $\omega_{\mathbf{k}} = \omega_0$, such that all the operators describing the system are modulated by this frequency. It is then useful to introduce new (slowly varying) operators, $\tilde{a}_{\mathbf{k}}$ and $\tilde{\sigma}_{-j}$, such that

$$a_{\mathbf{k}} = \tilde{a}_{\mathbf{k}} e^{-i\omega_0 t}, \qquad \sigma_{-j} = \tilde{\sigma}_{-j} e^{i(\mathbf{k}_0\cdot\mathbf{r}_j - \omega_0 t)}. \tag{8.48}$$

Replacing this in the previous two equations, we get

$$\frac{d\tilde{\sigma}_{-j}}{dt} = i\left(\Delta_0 - \mathbf{k}_0 \cdot \frac{\mathbf{p}_j}{M}\right)\tilde{\sigma}_{-j} + \sum_{\mathbf{k}} g_{\mathbf{k}} \sigma_{zj} \tilde{a}_{\mathbf{k}} e^{i(\mathbf{k}-\mathbf{k}_0)\cdot\mathbf{r}_j}, \tag{8.49}$$

and

$$\frac{d\sigma_{zj}}{dt} = -2\sum_{\mathbf{k}} g_{\mathbf{k}} \tilde{a}_{\mathbf{k}}^{\dagger} e^{-i(\mathbf{k}-\mathbf{k}_0)\cdot\mathbf{r}_j} + h.\ c., \tag{8.50}$$

with $\Delta_0 = \omega_0 - \omega_a$. Equation (8.46) becomes

$$\frac{d\tilde{a}_{\mathbf{k}}}{dt} = i\Delta_{\mathbf{k}} \tilde{a}_{\mathbf{k}} + \sum_{j=1}^{N_a} g_{\mathbf{k}}^{*} \tilde{\sigma}_{-j} e^{-i(\mathbf{k}-\mathbf{k}_0)\cdot\mathbf{r}_j}, \tag{8.51}$$

with $\Delta_{\mathbf{k}} = \omega_{\mathbf{k}} - \omega_0$, and the centre-of-mass equation becomes

$$\frac{d\mathbf{p}_j}{dt} = -\hbar \sum_{\mathbf{k}} \mathbf{k}\left\{g_{\mathbf{k}} \tilde{a}_{\mathbf{k}}^{\dagger} \tilde{\sigma}_{-j} e^{-i(\mathbf{k}-\mathbf{k}_0)\cdot\mathbf{r}_j} + h.\ c.\right\}. \tag{8.52}$$

Let us now simplify these equations by assuming that the atoms are initially in the ground state $|g_j\rangle$, and that the relevant field modes are all very far from the atomic transition, $\omega_{\mathbf{k}} \neq \omega_a$. In this case, we can assume that the population in the excited state remains small during the process, and their contribution can be neglected. This assumption can be stated as

$$\sigma_{zj} = -I, \qquad \frac{d\tilde{\sigma}_{-j}}{dt} \simeq 0, \tag{8.53}$$

where I is the identity operator. Using equations (8.49)–(8.50), we obtain

$$\tilde{\sigma}_{-j} \simeq \frac{-i}{(\Delta_0 - \mathbf{k}_0 \cdot \mathbf{p}_j/M)} \sum_{\mathbf{k}} g_{\mathbf{k}} \tilde{a}_{\mathbf{k}} e^{i(\mathbf{k} - \mathbf{k}_0)\cdot \mathbf{r}_j}. \tag{8.54}$$

Replacing this in equation (8.51), we get

$$\frac{d\tilde{a}_{\mathbf{k}}}{dt} = i\Delta_{\mathbf{k}}\tilde{a}_{\mathbf{k}} - \sum_{j=1}^{N_a} \frac{ig_{\mathbf{k}}^*}{(\Delta_0 - \mathbf{k}_0 \cdot \mathbf{p}_j/M)} \sum_{\mathbf{k}'} g_{\mathbf{k}'} \tilde{a}_{\mathbf{k}'} e^{i(\mathbf{k}' - \mathbf{k}_0)\cdot \mathbf{r}_j}. \tag{8.55}$$

Going back to the momentum equation (8.52), and retaining the dominant terms, which correspond to $\mathbf{k}' = \mathbf{k}_0$, we can then get

$$\frac{d\mathbf{p}_j}{dt} = -i\hbar \sum_{\mathbf{k}} (\mathbf{k} - \mathbf{k}_0) \left\{ \frac{g_{\mathbf{k}}^* g_0 \tilde{a}_0}{(\Delta_0 - \mathbf{k}_0 \cdot \mathbf{p}_j/M)} \tilde{a}_{\mathbf{k}}^{\dagger} e^{i(\mathbf{k} - \mathbf{k}_0)\cdot \mathbf{r}_j} + h.\ c. \right\}. \tag{8.56}$$

Introducing the coupling factor

$$g = \frac{g_{\mathbf{k}}^* g_0}{\Delta_0} \tilde{a}_0, \tag{8.57}$$

and neglecting the Doppler shifts in the denominator of equation (8.56), we then arrive at the coupled equations describing the field mode operators $\tilde{a}_{\mathbf{k}}$ and the atom centre-of-mass moments $\mathbf{p}_j$, as given by

$$\frac{d\mathbf{p}_j}{dt} = i\hbar g \sum_{\mathbf{k}} (\mathbf{k} - \mathbf{k}_0) \{ \tilde{a}_{\mathbf{k}} e^{i(\mathbf{k} - \mathbf{k}_0)\cdot \mathbf{r}_j} + \tilde{a}_{\mathbf{k}}^{\dagger} e^{-i(\mathbf{k} - \mathbf{k}_0)\cdot \mathbf{r}_j} \} \tag{8.58}$$

and

$$\frac{d\tilde{a}_{\mathbf{k}}}{dt} = i\Delta_{\mathbf{k}}\tilde{a}_{\mathbf{k}} - ig \sum_{j=1}^{N_a} e^{i(\mathbf{k}' - \mathbf{k}_0)\cdot \mathbf{r}_j}. \tag{8.59}$$

At this point, it should be useful to replace the operators by c-numbers and introduce new dimensionless variables defined by

$$\boldsymbol{\theta}_j = (\mathbf{k} - \mathbf{k}_0) \cdot \mathbf{r}_j, \quad \mathbf{P}_j = \frac{2\mathbf{p}_j}{|\mathbf{k} - \mathbf{k}_0|}, \quad A_{\mathbf{k}} = \frac{i}{\omega_r} g, \tag{8.60}$$

with the dimensionless time variable, $\tau = 4\omega_r t$, and the frequency $\omega_r = \hbar |\mathbf{k} - \mathbf{k}_0|^2/2^3 M$. If the amplitude of the different scattered modes $\mathbf{k}$ is small, we can assume that they interact independently with the atoms and the pump field. We can therefore treat single mode evolution. Furthermore, if we restrict our discussion to a one-dimensional geometry, we are reduced to the following equations

$$\frac{d\theta_j}{d\tau} = P_j, \quad \frac{d\theta_j}{d\tau} = -\left[A \exp(-i\theta_j) + c.\ c. \right], \tag{8.61}$$

and

$$\frac{dA}{d\tau} = i\delta A + \frac{\alpha}{N_a}\sum_{j=1}^{N_a} \exp(-i\theta_j). \tag{8.62}$$

Here, δ is the dimensionless detuning parameter, and α determines the strength of the interaction between the atoms and radiation. They are defined as

$$\delta = \frac{1}{\omega_r}(\omega_{\mathbf{k}} - \omega_0), \quad \alpha = \frac{N_a g^2}{2\omega_r^2}. \tag{8.63}$$

We now follow the approach used in [19], and introduce the microscopic density of atoms. This quantity, called *Klimontovich density*, is not a regular function but allows us to establish a precise link between a dynamical description of the particle trajectories and a statistical description of the system of particles. It is determined by sum of Dirac delta functions, as

$$N(\theta, P, \tau) = \sum_{j=1}^{N_a} \delta\big(\theta - \theta_j(\tau)\big)\, \delta\big(P - P_j(\tau)\big), \tag{8.64}$$

where the functions $\theta_j(\tau)$ and $P_j(\tau)$ represent the classical trajectories of the atoms and are given by the solutions of the centre-of-mass equations of motion (8.43). Using this density, it is then possible to rewrite the mode amplitude equation (8.62) as

$$\frac{dA}{d\tau} = i\delta A + \frac{\alpha}{N_a}\int d\theta \int dP\ N(\theta, P, \tau)\exp(-i\theta). \tag{8.65}$$

On the other hand, it can be shown using equations (8.56), that the microscopic density satisfies the so-called *Klimontovich equation* [8], which is a conservation equation in the single-particle phase space (θ, P), of the form

$$\left(\frac{\partial}{\partial\tau} + P\frac{\partial}{\partial\theta} + F\frac{\partial}{\partial P}\right)N(\theta, P, \tau) = 0, \tag{8.66}$$

where F is determined by

$$F = -\big[A\exp(-i\theta_j) + c.\ c.\big]. \tag{8.67}$$

This equation provides a microscopic kinetic description of the N_a atoms system in the presence of radiation. The only obvious problem is that the Klimontovich density $N(\theta, P, \tau)$ is singular. But this problem can be overcome if we take a statistical average, such that

$$W(\theta, P, \tau) = \langle N(\theta, P, \tau)\rangle. \tag{8.68}$$

This new distribution can now be described as a regular function, if we neglect the deviations $\Delta N = N(\theta, P, \tau) - W(\theta, P, \tau)$, it is described by a kinetic equation of the form

$$\frac{d}{d\tau}W(\theta, P, \tau) \equiv \left(\frac{\partial}{\partial\tau} + P\frac{\partial}{\partial\theta} + F\frac{\partial}{\partial P}\right)W(\theta, P, \tau) = 0. \tag{8.69}$$

In terms of this atomic distribution function, the amplitude equation (8.65) can now be written as

$$\frac{dA}{d\tau} = i\delta A + \frac{\alpha}{N_a}\int d\theta \int dP\ W(\theta, P, \tau)\exp(-i\theta). \tag{8.70}$$

We now study the evolution of the coupled equations (8.66) and (8.70), describing the field amplitude A and the density distribution W, and show that coherent emission of scattered radiation can eventually occur. For that purpose, we assume perturbations of these two quantities with frequency Ω, such that

$$A(\tau) = \tilde{A}\exp(-i\Omega\tau), \qquad W(\theta, P. \tau) = W_0(\theta, P) + \tilde{W}\exp(iq\theta - i\Omega\tau). \tag{8.71}$$

It is obvious that $W_0(\tau, P)$ is the unperturbed density, and that the perturbations $\tilde{W}$ describe the occurrence of atomic bunching. We should notice that the force F in equation (8.66) couples the perturbed quantities $\tilde{A}$ and $\tilde{W}$ when $q = \pm 1$, and is responsible for the bunching process. A standard perturbative analysis of the coupled equations then leads to a dispersion relation for perturbations in the medium, of the form

$$1 + \frac{\alpha}{N_a(\Omega + \delta)}\int d\theta \int dP \frac{\partial W_0/\partial P}{(\Omega - Pq)} = 0. \tag{8.72}$$

This is equivalent to write

$$1 - \frac{\alpha}{qN_a(\Omega + \delta)}\int d\theta \int dP \frac{W_0(\theta, P)}{(\Omega - Pq)^2} = 0. \tag{8.73}$$

For a cold atomic medium, with temperature $T_a \simeq 0$, we can simply use $W_0(\theta, P) = (1/2\pi)N_a\delta(P - P_0)$, where P_0 is the normalised momentum of the atomic beam. In the beam frame, we have $P_0 = 0$, and using $q = 1$, we simply have

$$\Omega^2(\Omega + \delta) = \alpha. \tag{8.74}$$

This equation can easily be solved and the unstable regimes, where the imaginary part of Ω is positive values, can be identified. In the regime of negligible detuning $\delta \to 0$, the resulting instability growth rate is $\Gamma \equiv \Im(\Omega) = \sqrt{3}\alpha^{1/3}/2$. It increases with the coupling parameter g, and with the atom number N_a, which are both contained in the definition of α. This regime has no threshold, and instability takes place for any finite value of the coupling parameter. In contrast, the existence of detuning, $\delta \neq 0$, introduces a finite threshold and reduces the instability growth rate, as illustrated in figure 8.4.

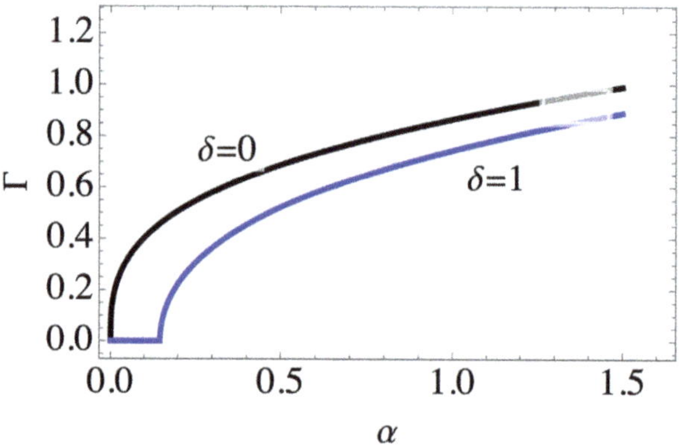

Figure 8.4. Unstable region: instability growth rate Γ for a cold atomic gas, as a function of the coupling parameter α, for $\delta \to 0$ and $\delta = 1$.

Similar results can also be found using a dynamical analysis [20]. But the interest of the present statistical approach is that we can treat the influence of the temperature of the system, T_a very easily. For instance, a thermal medium can be described

$$W_0(\theta, P) = \frac{N_a}{(2\pi)^{3/2} P_a} \exp(-P^2/2P_a^2), \tag{8.75}$$

where $P_a = MT_a/\hbar^2$. This would introduce correction terms in the dispersion (8.74), changing the thresholds and growth rates. Out-of-equilibrium distributions could equally be described by equation (8.73).

8.3 Quantum recoil

The above statistical description is valid for very low photon energies, when compared with the atomic kinetic energy, which corresponds to the condition $|p_j| \gg \hbar|\mathbf{k}|$. But, when this inequality is reversed, we need to consider the momentum exchanges between atoms and photons, and quantum recoil effects will add to the bunching process and have to be taken into account. For that purpose, we need to use the quantum description of centre-of-mass motion. This can be done with the help of the quasi-probability function

$$W(\mathbf{r}, \mathbf{p}, t) = \int \langle \psi(\mathbf{r} - \mathbf{s}/2, t) | \, \psi(\mathbf{r} + \mathbf{s}/2, t) \rangle e^{i\mathbf{p}\cdot\mathbf{s}/\hbar} d\mathbf{s}. \tag{8.76}$$

For the present purposes, it is more useful to define the one-dimensional function

$$W(\theta, P, \tau) = \int \langle \psi(\theta - \theta_s/2, t) | \, \psi(\theta + \theta_s/2, t) \rangle e^{iP\theta} d\theta_s, \tag{8.77}$$

where we use normalised variables (θ, P, τ). It is well-known that this quasi-probability is described by a *wave-kinetic equation*, of the form

$$i\left(\frac{\partial}{\partial \tau} + P\frac{\partial}{\partial \theta}\right) W = \int \tilde{V}(q, \tau) \Delta W e^{iq\theta} dq, \tag{8.78}$$

where $\tilde{V}(\theta, \tau) = \int \tilde{V}(q, \tau)\exp(iq\theta)dq$ is the Fourier transform of the potential. The auxiliary quantity ΔW is determined by

$$\Delta W \equiv [W_- - W_+], \quad W_\pm = W(P \pm q/2). \tag{8.79}$$

In order to understand the effects of a finite quantum recoil, let us assume the simple case where the potential only contains a single Fourier component, as determined by

$$V(\theta, \tau) = iA \exp(-i\theta) + c.\,c., \tag{8.80}$$

with the normalised field variable $A \equiv A(\tau)$. Now, if we perform a perturbative analysis using the quantum kinetic equation, we are led to the new dispersion relation

$$1 - \frac{\alpha}{N(\Omega + \delta)} \int d\theta \int dP \frac{W_0(\theta, P)}{[(P - \Omega)^2 - 1/4]} = 0. \tag{8.81}$$

This is a generalisation of equation (8.46) when the quantum recoil effects are included. It is useful to consider a cold atomic beam, such that $W_0(\theta, P) = (1/2\pi)N_a\delta(P - P_0)$. We are then reduced to

$$(\Omega + \delta)[(\Omega - P_0)^2 - 1/4] = \alpha. \tag{8.82}$$

In the quasi-classical limit, this new dispersion relation reduces to the above equation (8.74). The dispersion relation is represented in figure 8.5 for a cold atomic gas. Comparison with the quasi-classical result is also shown. We can see that quantum recoil effects clearly reduce the value of the instability threshold. The above one-dimensional model can be generalised to two and three-dimensions. But the corresponding analytical solutions are more difficult to find. They can be replaced by numerical simulations, as shown by [20]. In this case, the radiation instability associated with collective atomic scattering leads to the formation of atomic bunches, as illustrated in figure 8.6.

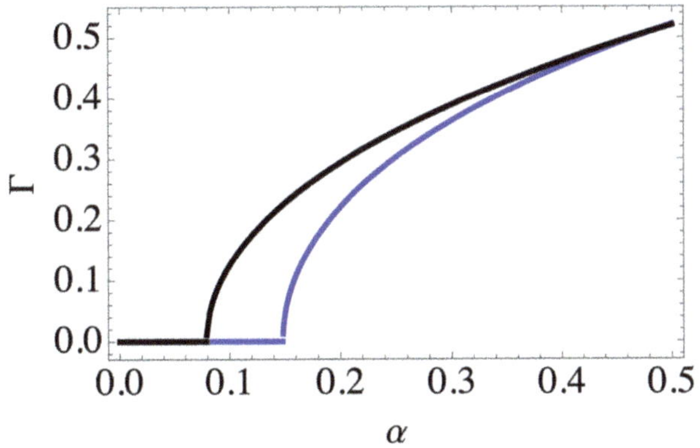

Figure 8.5. Damping rate with quantum recoil Γ as a function of α, for $\delta = 1$ (black curve). For comparison, the quasi-classical damping with no recoil is also shown (blue curve).

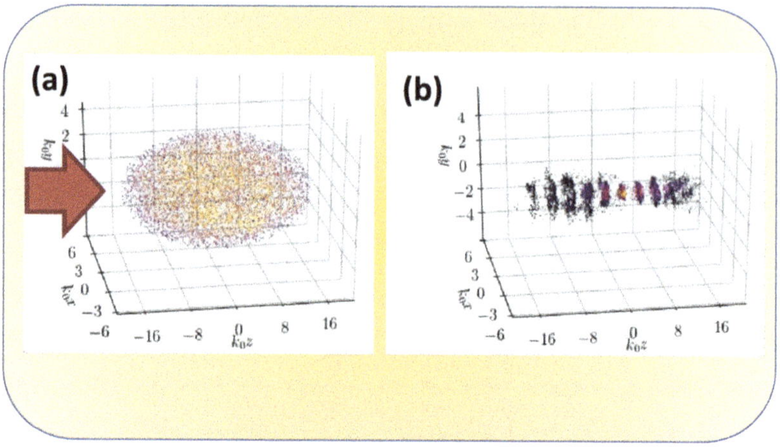

Figure 8.6. Numerical simulation of collective light scattering in 3D, using $N_a = 5000$ atoms: (a) initial atom distribution; (b) atom distribution after a simulation time $t = 0.01/\omega_r$, where $\omega_r = \hbar k_0^2/2M$ is the recoil frequency. The laser pump is propagating in the positive z-direction. (Courtesy R Ayllon.)

8.4 Cyclotron superradiance

Let us now consider superradiant emission of radiation in a microwave cavity by a non-neutral gas made of a large number N_e of electrons or positrons, confined by a static magnetic field (see figure 8.7). No atoms are present inside the cavity. Collective cyclotron emission could provide an effective cooling mechanism, which is important to attain long confinement times and perform antimatter experiments.

We use a semi-classical description, where the electromagnetic field is quantised but the electron motion is described classically. This model could be easily refined by including a quantum description of the guiding-centre motion of the charged particles, as we did with the atoms in the previous section. We start with the total Hamiltonian operator, $\hat{H} = \hat{H}_f + \hat{H}_e$, where the field term, $\hat{H}_f$, and the charged particle term (let us call it the electron term to specify), $\hat{H}_e$, take the familiar forms

$$\hat{H}_f = \sum_\nu \hbar\omega_\nu\left(\hat{a}_\nu^\dagger \hat{a}_\nu + \frac{1}{2}\right), \tag{8.83}$$

and

$$H_e = \sum_{j=1}^{N_e} \frac{1}{2m_e}\left[\mathbf{p}_j + e\mathbf{A}(\mathbf{r}_j, t)\right]^2 + V(\mathbf{r}, t), \tag{8.84}$$

where $\mathbf{r}_j$ are the particle positions, and $V(\mathbf{r}, t)$ is a generic potential that describes possible external forces used for confinement, as well as the electrostatic fields created by inter-particle interactions. The electromagnetic potential $\mathbf{A}(\mathbf{r}_j, t)$ describes the cavity field and the magnetic confinement, and can be written as

$$\mathbf{A}(\mathbf{r}, t) = \sum_\nu [\mathbf{F}_\nu(\mathbf{r})\hat{a}_\nu^\dagger + h.\, c.] + \mathbf{A}_0(\mathbf{r}). \tag{8.85}$$

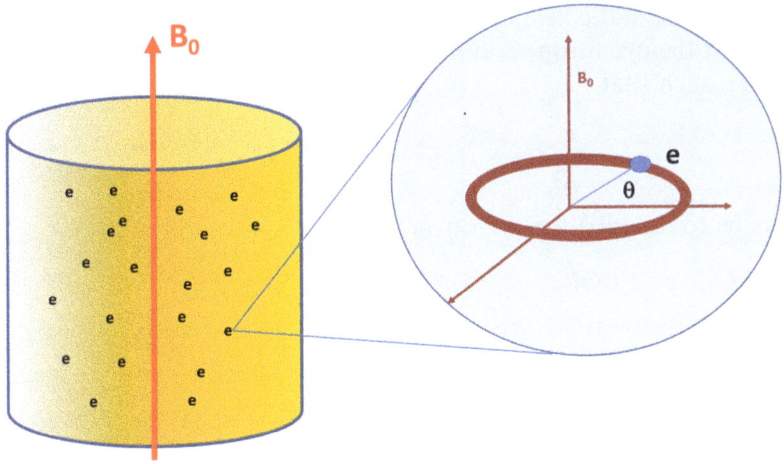

Figure 8.7. Collective cyclotron emission: a non-neutral electron (or positron) gas is confined in a microwave cavity by a static magnetic field. Unperturbed cyclotron motion: $\theta(t) = \omega_c t$.

Here, the quantity ν denotes the set of numbers identifying the cavity modes, and $\mathbf{F}_\nu(\mathbf{r})$ contains the field polarisation vector and the form function of these particular modes. In the case of free propagation, we could assume plane-wave modes, such that $\nu \equiv \mathbf{k}$ is the mode wavenumber, and we would be reduced to $\mathbf{F}_\nu(\mathbf{r}) = \mathbf{e}_\mathbf{k} \exp(i\mathbf{k} \cdot \mathbf{r})$. In equation (8.85), the static potential is such that $\nabla \times \mathbf{A}_0(\mathbf{r}) = \mathbf{B}_0$ is the confining magnetic field. The Heisenberg equations for the field operators are

$$\frac{d\hat{a}_\nu^\dagger}{dt} = -i\omega_\nu \hat{a}_\nu^\dagger + \frac{e}{2\omega_\nu} \sum_j^{N_e} \mathbf{v}_j \cdot \mathbf{F}_\nu(\mathbf{r}_j), \tag{8.86}$$

where the electron velocity is given by

$$\mathbf{v}_j = \frac{1}{m_e}\left[\mathbf{p}_j + e\mathbf{A}(\mathbf{r}_j, t)\right]. \tag{8.87}$$

As for the electron equations of motion, we assume that their guiding-centre is slowly moving as compared with the cyclotron motion, and neglect this slow motion. We also neglect contributions from possible collective oscillations, which is justified for a dilute gas, when the electron plasma frequency is much smaller than the cyclotron frequency, $\omega_p \ll \omega_c$. Therefore, each electron interacts individually with the field, and the electrostatic potential in equation (8.83) can be ignored. We then get

$$\frac{d\mathbf{v}_j}{dt} = -\omega_c(\mathbf{v}_j \times \mathbf{e}_z) + \frac{e}{m_e} \sum_\nu \left[-i\omega_\nu \mathbf{F}_\nu(\mathbf{r}_j)\hat{a}_\nu^\dagger + h.\, c.\right]. \tag{8.88}$$

Here, the static magnetic field was assumed in the z-direction, $\mathbf{B}_0 = B_0\mathbf{e}_z$, and we have used the definition of the cyclotron frequency $\omega_c = eB_0/m_e$. In a plane perpendicular to the static field, we can use polar coordinates and define the velocity associated with the cyclotron motion as $\mathbf{v} = u\mathbf{e}_\theta$. We can then define the auxiliary unit vector $\mathbf{e}_c$, such that

$$\mathbf{e}_c = \frac{1}{\sqrt{2}}(\mathbf{e}_r + i\mathbf{e}_\theta), \qquad u_j = i\sqrt{2}(\mathbf{e}_c \cdot \mathbf{v}_j). \tag{8.89}$$

This allows us to describe the cyclotron motion using

$$\frac{d\theta_j}{dt} = u_j, \qquad \theta_j = i\sqrt{2}(\mathbf{e}_c \cdot \mathbf{r}_j). \tag{8.90}$$

We can then write equation (8.88) in a much simple form

$$\frac{du_j}{dt} = -i\omega_c u_j + \left[f(\theta_j)\hat{a}_\nu^\dagger + h.\ c.\right], \tag{8.91}$$

where only a single cavity mode was retained, and a new quantity was defined as

$$f(\theta) = \frac{e}{m_e}\sqrt{2}\left[\mathbf{e}_c \cdot \mathbf{F}_\nu(\theta)\right]\omega_\nu. \tag{8.92}$$

Similarly, using this quantity and the scalar particle velocity u_j, we can rewrite the Heisenberg equation as

$$\frac{d\hat{a}_\nu^\dagger}{dt} = -i\omega_\nu\hat{a}_\nu^\dagger - \frac{1}{4\omega_\nu^2}\sum_j^{N_e}\left[if(\theta_j)u_j + h.\ c.\right]. \tag{8.93}$$

As a final step, we consider a variable transformation, into rotating-wave frame variables, replacing $(u_j,\ \hat{a}_\nu^\dagger)$ by the new quantities $(\tilde{u}_j,\ \tilde{a})$, such that

$$u_j = \tilde{u}_j \exp(i\omega_c t), \qquad \hat{a}_\nu^\dagger = \tilde{a}\exp(i\omega_c t). \tag{8.94}$$

Equations (8.91) and (8.93) can then be transformed into

$$\frac{d\tilde{u}_j}{dt} = \left[f(\theta_j)\tilde{a} + h.\ c.\right], \tag{8.95}$$

and

$$\frac{d\tilde{a}}{dt} = i\delta\tilde{a} - \frac{i}{4\omega_\nu^2}\sum_j^{N_e} f(\theta_j)\tilde{u}_j, \tag{8.96}$$

where we have neglected the oscillations and used the frequency detuning, $\delta = (\omega_c - \omega_\nu)$, assumed small. We can also define $d\tilde{\theta}_j/dt = \tilde{u}_j$. At this point, we introduce the electron microscopic density, such that

$$\mathcal{N}(\theta,\ u,\ t) = \sum_j^{N_e} \delta(\theta - \theta_j(t))\,\delta(u - u_j(t)). \tag{8.97}$$

Using the equations of motion for θ_j and u_j it is then possible to show that this microscopic density satisfies the Liouville equation

$$\frac{d}{dt}\mathcal{N}(\theta, u, t) = 0, \quad \frac{d}{dt} \equiv \left(\frac{\partial}{\partial t} + u\frac{\partial}{\partial \theta} + F\frac{\partial}{\partial u}\right), \tag{8.98}$$

where the force F is

$$F = f(\theta)a + h.\,c. \tag{8.99}$$

This should be coupled with the field operator equation (8.96), now written in the form

$$\frac{d\tilde{a}}{dt} = i\delta\tilde{a} - \frac{i}{4\omega_\nu^2}\int d\theta \int udu\mathcal{N}f(\theta)\tilde{u}_j. \tag{8.100}$$

Equations (8.98)–(8.100) describe the evolution of the electron population confined in the cavity, coupled with the nearly resonant cavity mode. We can study the evolution of this coupled system by introducing the smoothing in the electron distribution, which can be defined as $W(\theta, u, t) = \langle \mathcal{N}(\theta, u, t)\rangle$, and following the procedure followed previously. It is then possible to show that this system is unstable to collective cyclotron emission. This instability will enhance the usual uncorrelated cyclotron emission, and will lead to amplified energy dissipation. The difference with respect to the previous atomic model is that cyclotron electron oscillations were described classically. This is valid as long as the electron temperature (in energy units) is much higher than the photon cyclotron energy $T_e \gg \hbar\omega_c$. At low temperatures however, quantisation of the electron cyclotron motion and the corresponding Landau levels will have to be taken in account. To our knowledge, this has not yet been considered in the literature.

References

[1] Dicke R H 1954 Coherence in spontaneous radiation processes *Phys. Rev.* **93** 99–110

[2] Rehler N E and Eberly J H 1971 Superradiance *Phys. Rev.* A **5** 1735–51

[3] Gross M and Haroche S 1982 Superradiance: an essay on the theory of collective spontaneous emission *Phys. Rep.* **93** 301–96

[4] Garraway B M 2011 The Dicke model in quantum optics: Dicke model revisited *Phil. Trans. R. Soc.* A **369** 1137–55

[5] Kocharovsky Vl V, Zheleznyakov V V, Kocharovskaya E R and Kocharovsky V V 2017 Superradiance: the principles of generation and implementation in lasers *Uspekhi* **60** 345–84

[6] Bonifacio R and De Salvo L 1994 Collective Atomic Recoil Laser (CARL) optical gain without inversion by collective atomic recoiled and self-bunching of two-level atoms *Nucl. Instrum. Methods Phys. Res.* A **341** 360–2

Bonifacio R, De Salvo Souza L, Narducci L M and DÁngelo E J 1994 Exponential gain and self-bunching in a collective atomic recoil laser *Phys. Rev. A* **50** 1716–24

[7] Moore M G and Meystre P 1998 Effects of atomic diffraction on the collective atomic recoil laser *Phys. Rev.* A **58** 3248–58

[8] Klimontovich Y L 1982 *The Kinetic Theory of Non-Ideal Gases and Non-Ideal Plasmas* (Oxford: Pergamon)

[9] O'Neil T M 1980 Cooling of a pure electron plasma by cyclotron radiation *Phys. Plasmas* **23** 725–31

[10] Gabrielse G *et al* 2002 Background-free observation of cold antihydrogen with field-ionization analysis of its states *Phys. Rev. Lett.* **89** 213401

[11] Povilus A P *et al* 2016 Electron plasmas cooled by cyclotron-cavity resonance *Phys. Rev. Lett.* **117** 175001

[12] Gover A *et al* 2019 Superradiant and stimulated-superradiant emission of bunched electron beams *Rev. Mod. Phys.* **91** 035003

[13] Vieira J, Pardal M, Mendonça J T and Fonseca R 2021 Generalized superradiance for producing broadband coherent radiation with transversely modulated arbitrary diluted bunches *Nat. Phys.* **17** 99–104

[14] Zel'dovich Ya B 1971 Generation of waves by a rotating body *JETP Lett.* **14** 180–2
Zel'dovich Ya B 1972 Amplification of cylindrical electromagnetic waves reflected from a rotating body *Sov. Phys. JETP* **35** 1085–87

[15] Penrose R and Floyd R M 1971 Extraction of rotational energy from a black hole *Nat. Phys. Sci.* **229** 177–9

[16] Bekenstein J D and Schiffer M 1998 The many faces of superradiance *Phys. Rev.* D **58** 064014

[17] Brito R, Cardoso V and Pani P 2015 Superradiance, energy extraction, black-hole bombs and implications for astrophysics and particle physics *Lecture Notes in Physics* vol 906 (Berlin: Springer)

[18] Braidotti M C, Faccio D and Wright E M 2020 Penrose superradiance in nonlinear optics *Phys. Rev. Lett.* **125** 193902

[19] Mendonça J T and Serbeto A 2020 Wave-kinetic approach to collective atomic emission *Atoms* **8** 42

[20] Ayllon R, Mendonça J T, Gisbert A T, Piovella N and Robb G R M 2019 Multimode collective scattering of light in free space by a cold atom gas *Phys. Rev.* A **100** 023630

Chapter 9

Light vortices

In this chapter, we review the basic properties of photon orbital angular momentum (OAM), including quantum optical processes, harmonic generation and other nonlinear effects. The existence of light vortices, twisted beams and structured light retained considerable attention in recent years, and were considered in several areas of classical and quantum optics. Twisted laser beams can easily be created, using holograms and phase masks, and their vorticity is due to a finite amount of OAM. This has an important impact on the laser–matter interaction processes.

It is well known that the electromagnetic field of a laser beam carries a finite amount of angular momentum **J**. This quantity can be divided in two distinct parts, the spin **S** associated with the field polarisation, and the orbital angular momentum **L**, or OAM, associated with the field phase structure. Plane waves have spin but no OAM, while twisted beams with an helical phase configuration, have angular momentum, even in the absence of spin (linear polarisation).

Early experiments in the 1930s showed that a transfer of angular momentum between light and matter can occur, when a polarised photon beam interacts with a birefringent plate [1]. But the importance of photon OAM was recognised much later, in the early 1990s, when the term *optical vortices* was coined [2, 3], and it was shown that OAM states of light can be described by Laguerre–Gauss (LG) modes [4]. This opened the way to a variety of different studies on light–matter interaction, in optical media [5] and in plasmas [6]. Even in vacuum, twisted laser beams show remarkable new properties such as light travelling at subluminal speed [7]. Furthermore, superposition of different LG modes leads to the formation of new topological structures, such as light springs [8] and fractional vortex states [9].

doi:10.1088/978-0-7503-2786-2ch9

9.1 Photon OAM

It is known that the total linear momentum of the electromagnetic field, $\mathbf{P}$, is given by the volume integral of the Poynting vector $(\mathbf{E} \times \mathbf{B})$. A similar definition can be used for the total angular momentum $\mathbf{J}$, as [10]

$$\mathbf{P} = \epsilon_0 \int_V (\mathbf{E} \times \mathbf{B})\, d\mathbf{r}, \quad \mathbf{J} = \epsilon_0 \int_V [\mathbf{r} \times (\mathbf{E} \times \mathbf{B})]\, d\mathbf{r}. \tag{9.1}$$

If we assume that the field is described by a superposition of waves, with momentum $\mathbf{k}$ and frequency $\omega_\mathbf{k} = kc$, it is possible to show that the total angular momentum contains two different terms. Using

$$\mathbf{J} = \int_V \mathbf{J}_k \frac{d\mathbf{k}}{(2\pi)^3}, \tag{9.2}$$

the total angular momentum per mode can be written in a form such that $\mathbf{J}_k = \mathbf{S}_k + \mathbf{L}_k$, as

$$\mathbf{J}_k = \epsilon_0 \left[(\mathbf{E}_k{}^* \times \mathbf{A}_k) + \sum_j E_{j,\mathbf{k}} (\mathbf{k} \times \nabla_\mathbf{k}) A_{j,\mathbf{k}} \right], \tag{9.3}$$

where $\mathbf{A}_k$ represents the vector potential, and the sum is made over the field components $j = x, y, z$. The first term in this expression can be easily identified as the photon spin per mode, $\mathbf{S}_k$. The second term is the orbital angular momentum. The corresponding quantum operators can also be established. It is known that the linear momentum operator is given by

$$\mathbf{P} = \sum_\mathbf{k} \mathbf{P}_k, \quad \mathbf{P}_k = \hbar \mathbf{k} (a_\mathbf{k}^\dagger a_\mathbf{k} + a_\mathbf{k} a_\mathbf{k}^\dagger) = \hbar \mathbf{k} \left(a_\mathbf{k}^\dagger a_\mathbf{k} + \frac{1}{2} \right). \tag{9.4}$$

As for the photon spin operator, we have

$$\mathbf{S}_k = -i\hbar a_\mathbf{k}^\dagger a_\mathbf{k} (\mathbf{e}_k{}^* \times \mathbf{e}_k), \tag{9.5}$$

where $\mathbf{e}_k$ is the unit polarisation vector. For propagation along the z-axis, we have two independent polarisation states per mode, given by $\mathbf{e}_\pm = (\mathbf{e}_x \pm i\mathbf{e}_y)/\sqrt{2}$. For a Fock state $|\, n\rangle$ with n identical photons in mode $\mathbf{k}$, the expectation values are

$$\langle \mathbf{P}_k \rangle \equiv \langle n \,|\, \mathbf{P}_k \,|\, n \rangle = \hbar \mathbf{k} \left(n + \frac{1}{2} \right), \quad \langle \mathbf{S}_k \rangle \equiv \langle n \,|\, \mathbf{S}_k \,|\, n \rangle = \pm n\hbar \mathbf{e}_z. \tag{9.6}$$

As for the corresponding photon OAM operator, we obtain (see [11] for details)

$$\mathbf{L}_\mathbf{k} = -\frac{i\hbar}{2} \sum_j a_{j,\mathbf{k}}^\dagger (\mathbf{k} \times \nabla_\mathbf{k}) a_{j,\mathbf{k}} - h.\, c. \tag{9.7}$$

Plane waves are not always the most adequate way to describe the electromagnetic field. This is clear for laser beams where propagation takes place preferentially along a well-defined direction $\mathbf{e}_z$. Ideally, the laser beam has a transverse intensity profile $I(\mathbf{r}_\perp)$, described for instance by a radial Gaussian function, but the intensity profile can be modified by a series of reasons. In general terms, an arbitrary function in the perpendicular plane (x, y) can be represented by a set of orthogonal functions. The most popular representations in laser physics are the Hermite–Gauss (HG) and the Laguerre–Gauss (LG) functions [12]. Here we mainly focus on the Laguerre–Gauss functions, because they have helical phase surfaces, and each mode corresponds to a well-defined quantum OAM state. They can be created quite easily in the laboratory, at least for low intensity beams, starting with a Gaussian beam and using a spiral phase plate (or, alternatively, a holographic phase mask). This increases their practical interest (see figure 9.1 for an illustration). It is needless to say that any HG function can be represented by a superposition of LG functions, and vice-versa. Therefore, a single HG mode is not a pure OAM state, and for that reason are ignored here.

In order to discuss laser beams, or light beams in general, we can use the *paraxial approximation*. We start with the wave equation for the electric field, $\mathbf{E} \equiv \mathbf{E}(\mathbf{r}, t)$, for beams propagating in vacuum, as

$$\left(\nabla^2 + \frac{1}{c^2}\frac{\partial^2}{\partial t^2}\right)\mathbf{E} = 0. \tag{9.8}$$

We then assume propagation along the z-direction, such that

$$\mathbf{E} = \mathcal{E}\mathbf{e}_\mathbf{k} \exp(ikz - i\omega t) + c.\,c., \tag{9.9}$$

where $\mathcal{E}(\mathbf{r})$ is a slowly varying amplitude, $\mathbf{e}_\mathbf{k}$ is the unit polarisation vector, k is the axial momentum and $\omega = kc$ the corresponding frequency. Replacing this in the above equation, the amplitude can be approximately described by

$$\left(\nabla_\perp^2 + 2ik\frac{\partial}{\partial z}\right)\mathcal{E} = 0. \tag{9.10}$$

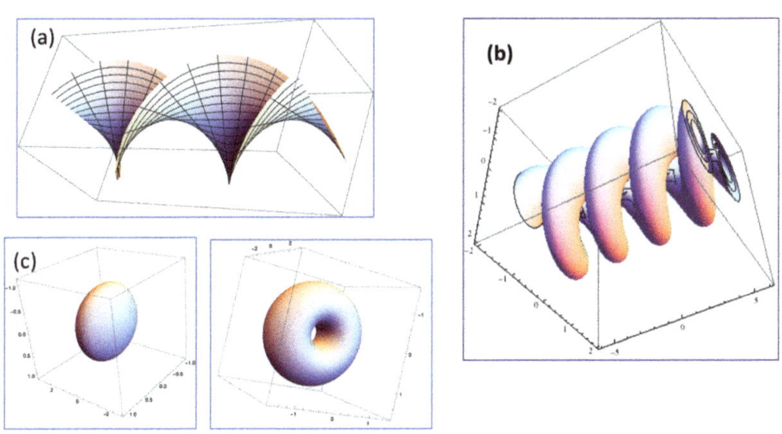

Figure 9.1. Twisted light: (a) surfaces of constant phase, for a LG mode $l = 1$; (b) electric field amplitude for a LG mode $l = 1$, $p = 0$; (c) intensity of a short LG pulse $l = 1$ (right), as compared with a Gaussian pulse $l = 0$ (left).

This is the so-called *paraxial equation*, describing the slow variations of the field. Using cylindrical coordinates $\mathbf{r} \equiv (r, \varphi, z)$, the paraxial wave solutions can be represented by a linear superposition of orthogonal LG modes $F_{p,l}(r, z)$, of the form

$$\mathcal{E}(\mathbf{r}) = \sum_{p,l} u_{p,l} F_{p,l}(r, z) \exp(il\varphi + ikz). \tag{9.11}$$

The integers p and l represent the radial and the azimuthal quantum numbers, and the quantities $u_{p,l}$ are the corresponding amplitudes. The radial number p determines the number of nodes in the radial direction, and the azimuthal number l, also called the *topological charge*, determines the beam vorticity. The LG mode functions are defined as

$$F_{p,l}(r, z) = C_{p,l} X^{|l|/2} L_p^{|l|}(X) \exp\left(-\frac{X}{2}\right), \tag{9.12}$$

where L_p^l are the *associated Laguerre polynomials*. Here, we have used the normalised radial variable $X = (r/w)^2$, where w is the beam waist, allowed to vary slowly along the propagation axis z. The coefficients $C_{p.l}$ are introduced in order to satisfy the orthonormality condition

$$\int_0^\infty r dr \int_0^{2\pi} d\varphi F_{p,l}^* F_{p',l'} e^{i(l'-l)\varphi} = \delta_{pp'} \delta_{ll'}, \tag{9.13}$$

where Kronecker delta symbols were used. The explicit expressions for the normalisation factor, and for the associated Laguerre polynomials, are

$$C_{p,l} = \frac{1}{2\sqrt{\pi}} \left[\frac{(l+p)!}{p!}\right]^{1/2}, \quad L_p^l(X) = \frac{1}{p!} X^{-l} e^X \frac{d^p}{dX^p}[X^{l+p} e^{-X}]. \tag{9.14}$$

Notice that, in the particular case of $(p = 0,\ l = 0)$ we are reduced to the case of a purely Gaussian beam. A remarkable property of the LG modes is that the z-component of the total angular momentum per unit power is determined by [4]

$$J_z \equiv S_z + L_z = \frac{\sigma_z r}{2\omega} \frac{\partial}{\partial r} W_{p.l} + \frac{l}{\omega} W_{p,l}, \tag{9.15}$$

where $W_{p.l} \propto |u_{p,l}|^2$ is the energy density, and $\sigma_z = \pm 1$ represent the spin states associated with (left and right) circular polarisation. For linearly polarised light, $\sigma_z = 0$, the spin term is absent. We should notice that the spin term depends in the radial dependence of the beam intensity, and is absent for a uniform plane wave solution. In the case of a single photon, we have $W_{p,l} = \hbar\omega$, and the expressions for spin and orbital angular momentum become

$$S_z = \sigma_z \hbar, \quad L_z = l\hbar. \tag{9.16}$$

The above LG wave solutions are valid in the paraxial approximation, for beam propagation in free space, but exact wave solutions with similar properties can also be found for beam propagation in optical fibres and cylindrical waveguides.

To illustrate this, we consider an empty cylinder with radius a and perfectly conducting walls. Assuming that the electric field goes to zero at the boundary $r = a$, and using $\mathbf{E} \propto R_l(r)\exp(il\varphi + i\beta z - i\omega t)$, we can reduce the wave equation (9.1) to

$$\left[r\frac{d}{dr}\left(r\frac{d}{dr}\right) + \left(k_\perp^2 r^2 - l^2\right)\right]R_l = 0, \quad k_\perp^2 = \frac{\omega^2}{c^2} - \beta^2. \tag{9.17}$$

The solutions are Bessel functions, $R_l(r) = J_l(k_\perp r)$, which satisfy the boundary conditions $J_l(k_\perp a) = 0$. This means that $k_\perp = \alpha_{p,l}/a$, where $\alpha_{p,l}$ is the pth zero of the Bessel function J_l. The field solutions are therefore of the form

$$\mathcal{E}(\mathbf{r}) = \sum_{p,l} u_{p,l} F_{p,l}(r)\exp(il\varphi + i\beta z), \quad F_{p,l}(r) = c_{p,l} J_l(\alpha_{p,l} r/a). \tag{9.18}$$

The new normalisation factor $c_{p,l}$ is defined as

$$c_{p,l}^{-1} = \sqrt{\pi}\, a J_{l+1}(\alpha_{p,l}), \tag{9.19}$$

in such a way that the orthonormal relations (9.13) are still satisfied by the new radial functions. Similar solutions, given by combinations of Bessel functions, are also valid for non-paraxial beams in unbounded space, as noted by [13] (see also [14]). In this case, the quantity $k_\perp$ appearing in the Bessel field solutions is not quantised. Another possible configuration, where the field is not carried by a beam, is the radiation field emitted by a point source. In this case, the appropriate field representation is a superposition of spherical harmonics, where again quantum photons states of the angular momentum can be specified [15].

An interesting aspect of LG modes is related to the photon (group) velocity. To establish this quantity, we go back to the solutions for a single LG mode in the focal region, and write the total field phase $\Phi(\mathbf{r}, t)$. An appropriate expression for the phase, valid in the beam focal region, needs to take several terms into account, as [16]

$$\Phi(\mathbf{r}, t) = \omega t - \Phi(\mathbf{r}), \quad \Phi(\mathbf{r}) = \frac{\omega r^2}{2cR(z)} + l\Phi + \frac{\omega}{c}z - (2p + |l| + 1)\zeta(z). \tag{9.20}$$

Here, the radius of curvature of the beam front $R(z)$, and the *Gouy phase* $\eta(z)$, both depend on the *Rayleigh length* z_R, as

$$R(z) = \frac{z}{z^2 + z_R^2}, \quad \zeta(z) = \arctan\left(\frac{z}{z_R}\right). \tag{9.21}$$

The Rayleigh length characterises the size of the focal region, and the Gouy phase is the well-known phase anomaly near focus, as explained in [17]. Equation (9.20) allows us to determine the surfaces of constant phase, $\Phi(\mathbf{r}) = \text{const.}$. The phase velocity v_{ph}, characterising the motion of such surfaces, is determined by the condition

$$\delta\Phi(\mathbf{r}, t) = \omega\delta t - \nabla\Phi(\mathbf{r}) \cdot \delta\mathbf{r} = 0. \tag{9.22}$$

This leads to the expression

$$v_{\mathrm{ph}} = \frac{\omega}{|\nabla\Phi(\mathbf{r})|}. \tag{9.23}$$

Similarly, the group velocity v_{gr}, which characterises the motion of wavepackets with a finite spectral width $\delta\omega$, is determined by the condition

$$\delta\left(\frac{\partial}{\partial\omega}\Phi(\mathbf{r})\right) = \left(\delta t - \frac{\partial}{\partial\omega}\nabla\Phi(\mathbf{r}) \cdot \delta\mathbf{r}\right) = 0, \tag{9.24}$$

which leads to

$$v_{\mathrm{gr}} = \left|\nabla\frac{\partial}{\partial\omega}\Phi(\mathbf{r})\right|^{-1}. \tag{9.25}$$

Using the phase function $\Phi(\mathbf{r})$, given by equation (9.20), we can then show that the group velocity of LG modes propagating, in vacuum are smaller than c. This slower than light motion in vacuum was experimentally demonstrated [7, 16, 18], and should not be confused with the phenomenon of *slow light*, which implies light propagation in a nearly resonant medium, where a much stronger reduction of the group velocity can be found [19]. This effect, similar to what occurs in empty waveguides, but observed in unbounded vacuum, is not really surprising, because it is known that any light beam with finite transverse size moves at a speed slower than light, $v_{\mathrm{gr}} < c$. Such a property, valid even for a Gaussian beam, can easily be understood using the position–momentum uncertainty relations applied to photon beams.

9.2 Light springs and fractional vorticity

We have described some of the remarkable properties of light vortices associated with LG modes propagating in vacuum and considered individually. Even more surprising properties can be found for vortices described by a superposition of two or more LG modes. For this purpose, we consider the superposition of an arbitrary number of modes, and then discuss some of the configurations that have been discussed in the literature. We start with a generic vortex field

$$\mathcal{E}(\mathbf{r}, t) = \sum_j u_j(z, t)F_j(\mathbf{r}_\perp, z)\exp\left(i\phi_j(\mathbf{r}, z, t)\right), \tag{9.26}$$

with

$$\phi_j(\mathbf{r}, z, t) = l_j\varphi + (k_j z - \omega_j t), \tag{9.27}$$

and $j \equiv (p_j, l_j)$. One of the simplest but non-trivial examples is that of a light vortex containing two distinct LG modes $j = 1, 2$, with the same frequency ($\omega_1 = \omega_2$), but different vorticities ($l_1 \neq l_2$), and different time-varying amplitudes, $u_1(t)$ and $u_2(t)$. This case corresponds to the so-called *self-torque configuration*, which was proposed for generation of extreme-UV sources with a time-varying

OAM [20]. Assuming that $p_1 = p_2$ and $|l_1|, |l_2| \gg 1$, the total field amplitude of the vortex evolves according to

$$\mathcal{E}(\mathbf{r}, t) \propto u_1(t)\exp(il_1\varphi) + u_2(t)\exp(il_2\varphi). \tag{9.28}$$

The corresponding field intensity is given by

$$|\mathcal{E}(\mathbf{r}, t)|^2 \simeq |E_1|^2\left\{\left[1 + \mathcal{R}(t)^2\right] + 2\mathcal{R}(t)\cos\left[(l_1 - l_2)\varphi\right]\right\}, \tag{9.29}$$

where $\mathcal{R}(t) = u_2(t)/u_1(t)$ is the relative amplitude of the two LG components. The interest of this configuration with two modes is that the effective angular momentum with topological charge $l_{\rm eff}(t)$ seems to vary continuously in time, from l_1 to $l_2 \neq l_1$, due to the temporal variation of the two amplitudes (figure 9.2).

A quite distinct configuration with two modes can be obtained when their amplitudes are identical, but they have different frequencies and helicities. In that case we have $\delta l = l_1 - l_2 \neq 0$, $\delta k = k_1 - k_2 \neq 0$, and $\delta\omega = \omega_1 - \omega_2 \neq 0$. We get the so-called *light springs*, which have been proposed to accelerate helical electron beams [21]. The resulting intensity will vary in space and time as (figure 9.3)

$$|\mathcal{E}(\mathbf{r}, t)|^2 \simeq 2|E_1|^2\{1 + \cos(\delta l\varphi - \delta kz - \delta\omega t)\}. \tag{9.30}$$

Let us now consider a different kind of light vortices, resulting from the superposition of an infinite number of LG modes. As a result of this infinite number, vortices with an arbitrary topological charge, or vorticity, μ, can be generated, as first proposed by Berry [9]. To show this, let us consider the field of an hypothetical light vortex of the form

$$\mathcal{E}(\mathbf{r}, t) = R_\mu(r)\exp\left[i\mu\theta + i\varphi(z, t)\right], \tag{9.31}$$

where μ is any real number and $R_\mu(r)$ a given radial amplitude. Comparing this expression with equation (9.18), and using the orthonormality conditions (9.13), it is

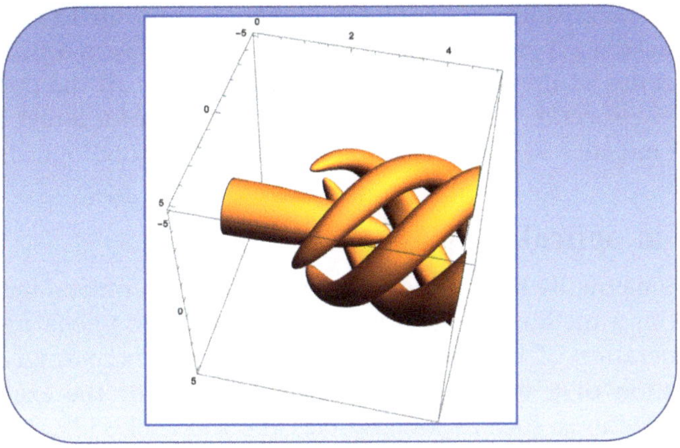

Figure 9.2. Self-torque beams, resulting from the superposition of modes with the same frequency $\omega_j = \omega$, but different amplitudes $u_j(t)$. Transition between a purely Gaussian mode $(l = 0, p = 0)$ to a LG mode $(l = 2, p = 0)$ is shown.

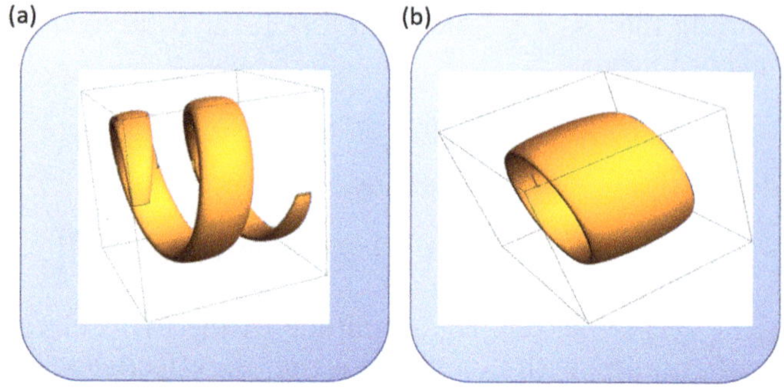

Figure 9.3. Light springs: superposition of three modes and different helicities $l = 13, 14, 15$, (a) with different frequencies ω_l, (b) with the same frequency $\omega_l = \omega$.

possible to show that such a hypothetical vortex can indeed be generated if we use a superposition of LG modes, as in equation (9.18), with coefficients $u_{p,l}$ defined by

$$u_{lp} = c_{lp} \frac{4\mu}{(\mu - l)} e^{il\theta} I_\mu(l, p), \tag{9.32}$$

where the quantity $I_\mu(l, p)$ corresponds to the integral

$$I_\mu(l, p) = \int_0^\infty R_\mu(r) F_{p,l}(r) r dr. \tag{9.33}$$

In particular, this allows us to construct vortices with fractional orbital angular momentum, $\mu = n/m$ where n and m are arbitrary integers. Entanglement of photons with fractional μ was studied experimentally in [22].

An important example of fractional vortices is that of *fermion vortices*, where $\mu = s = \pm 1/2$. They can be associated to fermionic spin states of light, although they result from the superposition of an infinite number of bosonic LG modes of integer topological charge l. A remarkable property of fermion vortices is that they correspond to states of minimum energy $|\mathcal{E}(\mathbf{r}, t)|^2$, among all the possible values of μ [23]. Finally, we should mention the possible existence of a larger variety of light vortices, which includes Möbius strips, and other exotic topological states [24, 25].

9.3 POAM in optical media

Light vortices can equally be considered in vacuum and in optical media. But, when they propagate in a medium, several new interesting phenomena can be observed, such as the generation of high harmonic cascades with increasingly high vorticity [26], the formation of a vortical supercontinuum [27], or the excitation of non-optical vortex structures, such as sound waves and wakefields [29, 30]. Furthermore, propagation in rotating media reveals a possible new version of spin–orbit coupling, not related to atoms but to photons. Finally, excitation of high intensity vortex beams can be achieved in FEL configurations [28]. This will be briefly reviewed here.

We consider propagation in a medium with refractive index $n(\omega)$, where the LG modes maintain the same spatial structure as in vacuum, defined by equation (9.12), but where the frequency ω satisfies a different dispersion relation, of the form

$$k^2 = \frac{\omega^2}{c^2}\epsilon(\omega), \quad \epsilon(\omega) = 1 + \chi(\omega), \tag{9.34}$$

where $\epsilon(\omega) \equiv n^2(\omega)$ is the dielectric function, and $\chi(\omega)$ the linear susceptibility. This quantity depends on the density of matter in the medium, and such a dependence brings the possibility of coupling with phonons, which are density waves. This leads to the *stimulated Brillouin scattering* process, as studied in optical crystals [29] and plasmas [30]. This decay process and similar ones satisfy, not only momentum and energy conservation conditions, which are common to all nonlinear optical processes, but also verify angular momentum conservation.

Configurations involving a pump laser pulse with frequency and momentum $(\omega_0, \mathbf{k}_0)$, scattered photons $(\omega_s, \mathbf{k}_s)$, and sound waves $(\omega_q, \mathbf{q})$, can be described by an interaction Hamiltonian of the form

$$H_{\text{int}} = \hbar g[a(\mathbf{k}_0)a^\dagger(\mathbf{k}_s)b^\dagger(\mathbf{q}) + a^\dagger(\mathbf{k}_0)a(\mathbf{k}_s)b(\mathbf{q})], \tag{9.35}$$

where g is the photon–phonon coupling constant resulting from the perturbations of $\chi(\omega)$, and $a(\mathbf{k}_0)$, $a(\mathbf{k}_s)$, and $b(\mathbf{q})$ are the annihilation operators for the pump scattered photons, and phonons. We then have the three conservation laws, associated with energy, momentum and angular momentum, as

$$\omega_0 = \omega_s + \omega_q, \quad \mathbf{k}_0 = \mathbf{k}_s + \mathbf{q}, \quad L_z \mid l_0\rangle = L_z\big(\mid l_s\rangle + \mid l_q\rangle\big). \tag{9.36}$$

Here, propagation along the z-axis was assumed, with L_z the orbital angular momentum operator component, and $\mid l\rangle$ the photon and phonon states. This nonlinear process is able to excite twisted phonons in the medium. For instance, when sound waves are initially not present and two photon beams are injected from the outside, they can transfer vorticity to the medium, through the excitation oft twisted sound waves.

Purely photon processes can also be considered. For this purpose, we should replace the paraxial equation (9.9) by its nonlinear version

$$\left(\nabla_\perp^2 + 2ik\frac{\partial}{\partial z}\right)\mathcal{E} = -\frac{\omega^2}{c^2}\chi^{(3)}|\mathcal{E}|^2\mathcal{E}, \tag{9.37}$$

where $\chi^{(3)}$ is the third-order nonlinear susceptibility. For convenience, we assume a *Kerr medium*, where the second-order susceptibility is absent, $\chi^{(2)} = 0$. Therefore three-photon coupling is forbidden and only four-photon coupling is allowed. They satisfy the conservation conditions

$$\sum_{j=1}^{4} \omega_j = 0, \quad \sum_{j=1}^{4} \mathbf{k}_j = 0, \tag{9.38}$$

with similar conservation conditions for the angular momentum. For paraxial propagation along the same direction, this additional conservation relation involves vorticity of the four interacting modes, and takes the form

$$\sum_{j=1}^{4} l_j = 0. \tag{9.39}$$

Of particular importance is the possible occurrence of an *optical cascade*, made of a sequence of several four-photon processes with increasing values of vorticity. Such beams can be generated by two pump laser pulses with frequencies ω_0 and ω_1, such that secondary pulses appear due to four-wave mixing, with frequencies $\omega_\nu = \omega_0 + \nu\Delta\omega$, where ν is a positive or negative integer, and $\Delta\omega = \omega_1 - \omega_0$. It was shown that, for purely Gaussian beams, these secondary beams can be used to construct a single-cycle optical pulse in the visible [31]. For twisted beams, we should expect high vorticity numbers, satisfying $l_\nu = l_0 + \nu\Delta l$, with $\Delta l = l_1 - l_0$.

When the difference frequency is equal to the initial frequency, $\omega_0 = \Delta\omega$, this cascaded process leads to *high harmonic generation* (HHG), with $\omega_\nu = \nu\omega_0$ and $l_\nu = \nu l_0$. This excitation of high orbital angular momentum states results from the nonlinear conservation laws, and was demonstrated experimentally by [32]. This could be relevant to the generation of XUV vortices with very high vorticity, $l \gg 1$ [33].

9.4 Quantum optics with OAM

Light vortices have all the ingredients of what we call a *collective photon state*. Other vortices can also be defined, such as the low frequency vortices of photon condensates, and those of superfluid light. Although these different vortex structures usually require the existence of a large number of photons, vorticity can still be present at the single photon or few-photons level, as shown by quantum optics experiments. These experiments use a *parametric down-conversion*, which produces entanglement of different vorticity states. Parametric down-conversion, and the subsequent emission of entangled photons, was first considered theoretically in [34] and observed in [35] (figure 9.4).

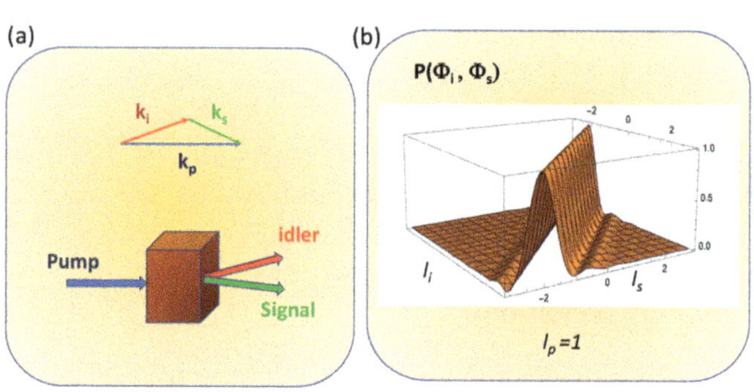

Figure 9.4. Parametric down-conversion of a pump photon beam: (a) principle of the experiment; (b) coincidence probability, as determined by equation (9.45).

The parametric process occurs when an intense laser beam with frequency ω_p, called the *pump*, propagates in a nonlinear crystal, with second-order susceptibility $\chi^{(2)}$, and decays into two photons of lower energy, ω_i and ω_s, usually called the *iddler* and the *signal*. They satisfy the energy and momentum conditions, $\omega_p = \omega_i + \omega_s$ and $\mathbf{k}_p = \mathbf{k}_i + \mathbf{k}_s$. This is called a *spontaneous decay* when no idler beam is injected from the outside, and photons with frequencies ω_i and ω_s are both spontaneously excited in the nonlinear crystal. The decay rate is obviously proportional to the nonlinear susceptibility $\chi^{(2)}$, and to the pump intensity. The two secondary photons form an entangled pair and the corresponding quantum state can be described in terms of Bell states, as seen previously. What is relevant here is the possible entanglement in terms of orbital angular momentum states (figure 9.4).

Ignoring polarisation issues, which depend on the explicit form of the nonlinear susceptibility, the parametric decay can be described simply, as follows. The creation of idler and signal photons are described by the creation operators, $a^\dagger(\mathbf{k}_i)$ and $a^\dagger(\mathbf{k}_s)$, applied to the vacuum state $|\,0\rangle$. The quantities $\mathbf{k}_i$ and $\mathbf{k}_s$ represent the idler and signal wavevectors. In the position representation, we can define the idler and signal states, $|\,\psi_i\rangle$ and $|\,\psi_s\rangle$, as

$$\left|\,\psi_{i,s}\right\rangle = \int d\mathbf{r}_{i,s}\Phi_{i,s}(\mathbf{r}_{i,s})a^\dagger(\mathbf{r}_{i,s})\,|\,0\rangle, \tag{9.40}$$

where the quantities $\Phi_{i,s}(\mathbf{r}_{i,s})$ determine the transverse modes associated with LG functions, and $a^\dagger(\mathbf{r})_{i,s}$ the Fourier transform of the creation operators, as given by

$$a^\dagger(\mathbf{r}) = \int a^\dagger(\mathbf{k})\exp(-i\mathbf{k}_{i,s}\cdot\mathbf{r})d\mathbf{r}. \tag{9.41}$$

Similarly, we define a two-photon state vector as

$$|\,\psi\rangle = \int d\mathbf{r}_i \int d\mathbf{r}_s \Phi_p\left(\frac{\mathbf{r}_i + \mathbf{r}_s}{2}\right)a^\dagger(\mathbf{r}_i)a^\dagger(\mathbf{r}_s)\,|\,0\rangle, \tag{9.42}$$

where $\Phi_p(\mathbf{r})$ describes the transverse distribution of the pump. The joint probability $P(\Phi_i, \Phi_s)$ of finding one iddler photon in a given transverse mode Φ_i and one signal photon in another mode Φ_s will then be determined by

$$P(\Phi_i, \Phi_s) \equiv |\,\langle\psi_s\,|\,\langle\psi_i\,|\,\psi\rangle\,|^2 = \left|\int d\mathbf{r}_i \int d\mathbf{r}_s \Phi_i^*(\mathbf{r}_i)\Phi_s^*(\mathbf{r}_s)\Phi_p\left(\frac{\mathbf{r}_i + \mathbf{r}_s}{2}\right)\right|^2. \tag{9.43}$$

The normalised probability can be obtained, if we divide this expression by the square root of the product $P(\Phi_i)P(\Phi_s)$. Here, the quantities $P(\Phi_j) = \left|\left\langle\psi_j\,\middle|\,\psi\right\rangle\right|^2$, for $j = (i, s)$, are the individual probabilities of the two secondary photons. Let us assume that the transverse functions Φ_p, and Φ_j, are LG functions of the form $F_{lp}\exp(il\varphi)$. Assuming that the secondary beam waists are the same, $w_j = w$, and the pump waist is defined as $w_p = w/W$, we then get [36]

$$P(\Phi_i, \Phi_s) = \mathrm{sinc}^2\left[(l_i + l_s - l_p)\pi\right]\mathcal{R}^2, \tag{9.44}$$

with the normalisation factor

$$\mathcal{R} = \int_0^\infty F_{p_p l_p}(rW^2) F_{p_i l_i}(r) F_{p_s l_s} r^{l_t/2} \exp\left[-r\left(1 + \frac{W^2}{2}\right)\right] dr, \tag{9.45}$$

with $l_t = |l_p| + |l_i| + |l_s|$. It is obvious that this expression of the coincidence probability $P(\Phi_i, \Phi_s)$ implies the conservation of orbital angular momentum, such that $l_p = l_i + l_s$, as expected from classical theory. This shows that the OAM properties of light persist at the single photon level, which is the lowest possible quantum level, and can be observed in entanglement experiments involving only a small number of photons. Notice however that a single twisted photon state results from the superposition of an infinite number of oblique plane-wave photon states. In that sense, it can be seen as a collective quantum state of light.

References

[1] Beth R 1936 Mechanical detection and measurement of the angular momentum of light *Phys. Rev.* **50** 115–25
[2] Coullet P, Gil G and Rocca F 1989 Optical vortices *Opt. Commun.* **73** 403–8
[3] Lugiato L A, Oppo G-L, Tredicce J R, Narducci L M and Pernigo M A 1990 Instability and spatial complexity in a laser *J. Opt. Soc. Am.* B **7** 1019–33
[4] Allen L, Beijersbergen M W, Spreeuw R J C and Woerdman J P 1992 Orbital angular-momentum of light and the transformation of Laguerre-Gaussian laser modes *Phys. Rev.* A **45** 8185–9
[5] Yao A M and Padgett M J 2011 Orbital angular momentum: origins, behavior and applications *Adv. Opt. Photon.* **3** 161–204
[6] Mendonça J T 2012 Twisted waves in a plasma *Plasma Phys. Control. Fusion* **54** 124031
[7] Lyons A, Roger T, Westerberg N, Vezzoli S, Maitland C, Leach J, Padgett M J and Faccio D 2018 How fast is a twisted photon *Optica* **5** 682–6
[8] Pariente G and Quéré F 2015 *Opt. Lett.* **40** 2037
[9] Berry M V 2004 Optical vortices evolving from helicoidal integer and fractional phase steps *J. Opt. A: Pure Appl. Opt.* **6** 259–68
[10] Jackson J D 1975 *Classical Electrodynamics* 2nd edn (New York: Wiley)
[11] Calvo G F, Picón A and Bagan E 2006 Quantum theory of photons with orbital angular momentum *Phys. Rev.* A **73** 013805
[12] Siegmann A E 1986 *Lasers* (Sausalito, CA: University Science Books)
[13] Barnett S M and Allen L 1994 Orbital angular momentum and nonparaxial light beams *Opt. Commun.* **110** 670–8
[14] Chen Q, Qin H and Liu J 2017 Photons, phonons and plasmons with orbital angular momentum in plasmas *Sci. Rep.* **7** 41731
[15] Berestetskii V B, Lifshitz E M and Pitaevskii L P 1982 *Quantum Electrodynamics* (Oxford: Butterworth-Heinemann)
[16] Bouchard F, Harris J, Mand H, Boyd R W and Karimi E 2016 Observation of subluminal twisted light in vacuum *Optica* **3** 351–4
[17] Visser T D and Wolf E 2010 The origin of the Gouy phase anomaly and its generalization to astigmatic wavefields *Opt. Commun.* **283** 3371–5

[18] Giovannini D, Romero J, Potocek V, Ferenczi G, Speirits F, Barnett S M, Faccio D and Padgett M J 2015 Spatially structured photons that travel in free space slower than the speed of light *Science* **347** 857–60

[19] Hau L V, Harris S E, Dutton Z and Behroozi C H 1999 Light speed reduction to 17 metres per second in an ultracold atomic gas *Nature* **397** 594–8

[20] Rego L *et al* 2019 Generation of extreme-ultraviolet beams with time-varying orbital angular momentum *Science* **364** 1253

[21] Vieira J, Mendonça J T and Quéré F 2018 Optical control of the topology of laser-plasma accelerators *Phys. Rev. Lett.* **121** 054801

[22] Oemrawsingh S S R, Ma X, Voigt D, Aiello A, Eliel E R, 't Hooft G W and Woerdman J P 2005 Experimental demonstration of fractional orbital angular momentum entanglement of two photons *Phys. Rev. Lett.* **95** 240501

[23] Mendonça J T, Serbeto A and Vieira J 2018 Plasmon excitations with a semi-integer angular momentum *Sci. Rep.* **8** 7817

[24] Bauer T, Banzer P, Karimi E, Orlov S, Rubano A, Marrucci M, Santamato E, Boyd R W and Leuchs G 2015 Observation of optical polarization Möbius strips *Science* **347** 964–6

[25] Pisanty E, Jiménes G, Vicuña-Hernández V, Picón A, Celi A, Torres J P and Lewenstein M 2019 Knotting fractional-order knots with the polarization state of light *Nat. Photon.* **13** 569–74

[26] Vieira J, Trines R M G, Alves E P, Fonseca R A, Mendonça J T, Bingham R, Norreys P and Silva L O 2016 High orbital angular momentum harmonic generation *Phys. Rev. Lett.* **117** 265001

[27] Mendonça J T 2020 Self-phase modulation and superfluid light in twisted fluids of light *Europhys. Lett.* **129** 64004

[28] Arteaga J A, Serbeto A, Mendonça J T, Tsui K H and Monteiro L F 2017 Orbital angular momentum of a π-pulse emission by dense relativistic cold electron beam *Phys. Plasmas* **24** 123108

[29] Zhi Z, Gao W, Mu C and Li H 2016 Reversible orbital angular momentum photon-phonon conversion *Optica* **3** 212–7

[30] Mendonça J T, Thidé B and Then H 2009 Stimulated Raman and Brillouin backscattering of collimated beams carrying orbital angular momentum *Phys. Rev. Lett.* **102** 185005

[31] Weigand R, Mendonça J T and Crespo H M 2009 Cascaded nondegenerate four-wave-mixing technique for high-power single-cycle pulse synthesis in the visible and ultraviolet ranges *Phys. Rev.* A **79** 063838

[32] Gariepy G, Leach J, Kim K T, Hammond T J, Frumker E, Boyd R W and Corkum P B 2014 Creating high-harmonic beams with controlled orbital angular momentum *Phys. Rev. Lett.* **113** 153901

[33] Hernández-García C, Vieira J, Mendonça J T, Rego L, San Román J, Plaja L, Ribic P R, Gauthier D and Picón A 2017 Generation and applications of extreme-ultraviolet vortices *Photonics* **4** 28

[34] Arnaut H H and Barbosa G A 2000 Orbital and intrinsic angular momentum of single photons and entangled pairs of photons generated by parametric down-conversion *Phys. Rev. Lett.* **85** 286–9

[35] Mair A, Vazari A, Weihs G and Zeilinger A 2001 Entanglement of orbital angular momentum states of photons *Nature* **412** 313–6

[36] Franke-Arnold S, Barnett S M, Padgett M J and Allen L 2002 Two-photon entanglement of orbital angular momentum states *Phys. Rev.* A **65** 033823

Chapter 10

Superfluid light

The concept of Superfluidity is intimately related with that of Bose–Einstein condensation, as noticed immediately after the first experiments on superfluid helium performed in the 30s [1]. But it is also something different, and we can eventually find superfluids which are not condensed fluids. This is particularly true in the case of photons, where superfluid states of light can be achieved without the need of condensation [2]. This is very interesting from the experimental point of view, because you can observe typical condensate phenomena, such as Bogoliubov excitations, in much easier experimental conditions. Furthermore, superfluid light is an interesting approach to study quantum turbulence, and other hot topics of the present time. These questions are analysed here, where we first review the concept of superfluid light, and then consider the case of photon instabilities that lead to the formation of vortex pairs and turbulence.

Superfluidity is a remarkable property of fluids, where the viscosity drops to zero. In such conditions, a fluid can flow through an obstacle without energy dissipation. This was originally discovered in liquid helium, with independent observations by Kapitza [3], and Allen and Misener [4], first reported in 1938. Recent historical accounts are given in [5, 6]. But superfluidity strongly reemerged in more recent years after the discovery of Bose–Einstein condensation in dilute alkali vapours.

Our present knowledge of superfluity is solidly based on the *theory of two fluids*, first proposed by Tisza [7], and refined later by Landau [8]. The Landau theory, based on the concept of *rotons*, became a standard model for superfluidity [9, 10], although it was soon recognised that rotons are not new particles, but simply phonons with a modified dispersion. Rotons can indeed be excited in atomic Bose–Einstein condensates (BEC) with non-local interactions, but they can also occur in non-condensed laser-cooled vapours [11]. In the meanwhile, more elaborate quantum theories of superfluidity have been developed [9, 12]. Our current view of the theory of two fluids assumes that the first fluid is the condensate itself, and second fluid is made of quasi-particles, the phonons, independently of the existence or not of

doi:10.1088/978-0-7503-2786-2ch10

roton dispersion. In the dilute condensed gas, the roton properties are absent most of the time, but rotons can exist in atomic condensates due to enhanced dipole-dipole interactions [13], and in photon condensates due to relativist corrections [14].

The concept of superfluid light emerged recently, through the work of several researchers [15–18]. If superfluidity is well understood in ultra-cold matter, the recent interest on superfluid light is related to the possible study of some of the most relevant aspects of BEC without the need of condensation. Nonlinear optical experiments, much simpler than those of BECs, can be conceived to study several properties of superfluid matter, such as suppression of diffraction and opto-mechanical deformations, drag-force cancellation and persistent currents (see, for a short review [19]). Simple optical experiments on vortex-pair formation and quantum turbulence excitation can also be envisaged, as described in this chapter.

10.1 Fluid equations of light

We first show that propagation of light in a nonlinear medium can be described, under some conditions, by a Gross–Pitaevskii equation. From there, we can derive the fluid equations of light, formally analogous to those of a condensate. As previously stated, this is valid in the absence of condensation. We consider light beam propagation in a medium along the z-axis, with the electric field determined by

$$\mathbf{E}(\mathbf{r}, t) = \mathcal{E}(\mathbf{r}_\perp, z, t)\exp(ikz - \omega t). \tag{10.1}$$

We also assume that light propagates in a nonlinear medium, with Kerr nonlinearity. The envelope amplitude $\mathcal{E}$ can then be described by

$$\left(\nabla_\perp^2 + 2ik\frac{\partial}{\partial z}\right)\mathcal{E} = -\frac{\omega^2}{c^2}[\delta\epsilon + \chi^{(3)}(\omega)|\,\mathcal{E}\,|^2]\mathcal{E}. \tag{10.2}$$

Here, we have used the third order susceptibility $\chi^{(3)}(\omega)$, characterising the *Kerr nonlinearity*, and a dielectric perturbation, $\delta\epsilon \equiv \delta\epsilon(\mathbf{r}_\perp, z)$, which can be used to model some local obstacle, mechanical or optical. We ignore polarisation effects, and assume the linear dispersion relation of the medium, in the usual form

$$k^2c^2 = \omega^2\epsilon(\omega), \qquad \epsilon(\omega) = 1 + \chi^{(1)}(\omega), \tag{10.3}$$

where $\epsilon(\omega)$ is the dielectric function, and $\chi^{(1)}(\omega)$ the linear susceptibility. Let us now use a temporal variable $\tau = z/v_g$, where $v_g = \partial\omega/\partial k$ is the group velocity, and introduce a new amplitude variable such that

$$\Psi(\mathbf{r}_\perp, \tau) = \mathcal{E}(\mathbf{r}_\perp, z = v_g\tau). \tag{10.4}$$

Equation (10.2) can then be transformed into a two-dimensional nonlinear Schrödinger equation, of the form

$$i\hbar\frac{\partial}{\partial\tau}\Psi = \left[-\frac{\hbar^2}{2m_{\mathrm{eff}}}\nabla_\perp^2 + V(\mathbf{r}_\perp, \tau) + g\,|\,\Psi\,|^2\right]\Psi, \tag{10.5}$$

with the two-dimensional Laplacian $\nabla_\perp^2 = \partial^2/\partial x^2 + \partial^2/\partial y^2$. Here, the potential $V(\mathbf{r}_\perp, \tau)$ is due to the optical perturbation, and is defined as

$$V(\mathbf{r}_\perp, \tau) = -\frac{\hbar k}{2\epsilon(\omega)} v_g \, \delta\epsilon(\mathbf{r}_\perp, z = v_g \tau). \tag{10.6}$$

In equation (10.5), we have also introduced an effective photon mass $m_{\rm eff}$, and a nonlinear coupling parameter g, such that

$$m_{\rm eff} = \frac{\hbar k}{v_g}, \qquad g = -\frac{v_g}{c} \frac{\hbar k}{2\epsilon(\omega)} \chi^{(3)}(\omega). \tag{10.7}$$

This envelope equation (10.5) is formally analogous to the Gross–Pitaevskii equation governing the evolution of a two-dimensional Bose–Einstein condensate. But here, in contrast, no photon condensation exists. We should notice that the temporal variable τ actually describes the spatial evolution along the propagation direction z. The Planck constant $\hbar$ is strictly not necessary in equation (10.5), but is useful for dimensional purposes, and stresses the formal similarities with condensates. Notice also that the use of a photon mass $m_{\rm eff}$ is associated here with the envelope approximation, and that the coupling parameter g results from photon–photon interactions associated with the nonlinearity. A focusing nonlinearity, such that $\chi^{(3)}(\omega) > 0$, would correspond to attractive photon–photon collisions, $g < 0$, while a defocusing medium with $\chi^{(3)}(\omega) < 0$ would lead to repulsive interactions, $g > 0$, as it should be expected.

Let us now consider the photon fluid equations, valid in a Kerr medium. We follow the standard procedure, introducing a *Madelung transformation* such that $\Psi = \sqrt{\rho}\exp(i\phi)$, and defining the fluid density ρ, and mean velocity $\mathbf{v}$, as

$$\rho = |\Psi|^2 = |\mathcal{E}|^2, \qquad \mathbf{v} = \frac{\omega}{c\epsilon(\omega)} \nabla_\perp \phi. \tag{10.8}$$

Replacing this into the envelope equation (10.5), we obtain the fluid equations

$$\frac{\partial \rho}{\partial \tau} + \nabla_\perp(\rho \mathbf{v}) = 0, \qquad \frac{\partial \mathbf{v}}{\partial \tau} + \mathbf{v} \cdot \nabla \mathbf{v} = -\frac{g}{m_{\rm eff}} \nabla_\perp \rho + \frac{1}{m_{\rm eff}} \nabla_\perp (V + V_B), \tag{10.9}$$

where V is the potential defined in equation (10.6), and V_B is the Bohm potential

$$V_B = \frac{\hbar^2}{2m_{\rm eff}} \frac{\nabla_\perp^2 \sqrt{\rho}}{\sqrt{\rho}}. \tag{10.10}$$

They describe the photon fluid motion in the perpendicular plane, inside the beam envelope, whereas propagation takes place in the z-direction (parametrised by the 'temporal' variable τ). According to the above definition of the fluid velocity $\mathbf{v}$, for nearly plane-wave modes it is non-zero only if the incident beam is injected at a finite angle $\theta \neq 0$ with respect to the z-axis. In this case, for propagation in the Oxz plane (see figure 10.1), the mean fluid velocity $\mathbf{v} = \mathbf{v}_0$ existing in the absence of any perturbation, will be

$$\mathbf{v}_0 = \frac{\sin\theta}{\sqrt{\epsilon(\omega)}} \mathbf{e}_x. \tag{10.11}$$

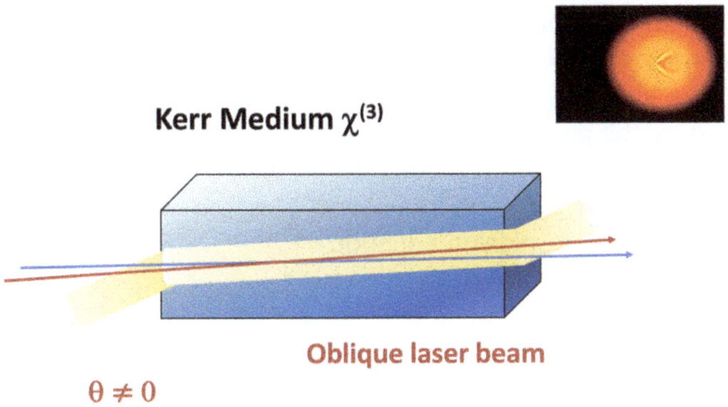

Figure 10.1. Superfluid light geometry: plane wave or Gaussian configuration, with a non-zero angle of incidence $\theta \neq 0$.

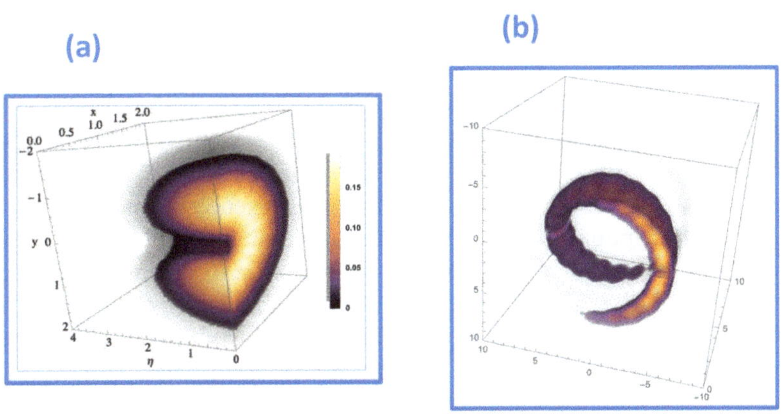

Figure 10.2. Bogoliubov oscillations in twisted light beam with axial propagation, $\theta = 0$: (a) single vortex state, $l = 1$—donut configuration; (b) double vortex states with equal amplitude, $l_1 = 4$, $l_2 = 5$—light springs.

This limitation does not apply to Laguerre–Gauss beams, where a finite fluid velocity always exists, even when $\theta = 0$. This is due to the existence of a transverse photon current, intrinsically associated with the helical phase properties of such beams. For a generic LG mode, with vorticity l and amplitude $\mathcal{E}_0$, the equilibrium photon density and velocity are given by [27].

$$\rho_0 = |\mathcal{E}_0|^2 |F_{pl}(\mathbf{r}\perp)|^2, \quad \mathbf{v}_0 = \frac{l}{r\sqrt{\epsilon(\omega)}}\mathbf{e}_\theta. \tag{10.12}$$

This is illustrated in figure 10.2, where the structure of poloidal excitations is shown for two vortex beams, and for purely axial propagation $\theta = 0$, in clear contrast with the case of a Gaussian or a plane-wave beam.

The quantum fluid equations (10.9) can be used to describe photon fluid perturbations, excited by some mechanical or optical obstacle, and to understand the transition from a fluid to a superfluid flow. We notice that, in the absence of perturbations, the beam geometry imposes an equilibrium value for the flow velocity $\mathbf{v}_0$, as given by equation (10.11) for plane-wave and Gaussian beams, and equation (10.12) for twisted beams. Let us then consider a low frequency perturbation of the light intensity, with frequency $\Omega \ll \omega$, described by the perturbed quantities $\delta\rho = \rho - \rho_0$ and $\delta\mathbf{v} = \mathbf{v} - \mathbf{v}_0$. Replacing this in equations (10.9), ignoring the potential V, and assuming evolution in space and time of the form $\propto \exp(i\mathbf{q} \cdot \mathbf{r}_\perp - i\Omega\tau)$, where $\mathbf{q}$ is the wavevector, we obtain after linearisation, the following dispersion equation

$$(\Omega - \mathbf{v}_0 \cdot \mathbf{q})^2 = C_s^2 q^2 + \frac{\hbar^2 q^4}{4m_{\text{eff}}^2}, \tag{10.13}$$

where $C_s = \sqrt{g\rho_0/m_{\text{eff}}}$ is the Bogoliubov sound speed in the two-dimensional photon gas. This quantity increases with the beam intensity, as shown by its explicit expression

$$C_s^2 = -\chi^{(3)} \frac{|\mathcal{E}_0|^2}{2\epsilon(\omega)}. \tag{10.14}$$

The dispersion relation (10.13) characterises the sound-like oscillations of the perpendicular photon flow. If we take the drift velocity $\mathbf{v}_0$ equal to zero, this is formally identical to the Bogoliubov dispersion of sound perturbations in a Bose–Einstein condensate. Except that we are not in the presence of a condensate, but only dealing with light propagation in a nonlinear medium. Another important difference is that the sound waves in a condensate evolve both in space and time, $\mathbf{r}$ and t, whereas here these sound-like waves evolve in the transverse plane along the z-direction, described by $\mathbf{r}_\perp$ and by the time-like variable $\tau = v_g z$. Despite such differences, superfluidity can nevertheless be clearly identified.

In analogy with the condensates, we can also define a *healing length* ξ, which establishes the limits of validity of a fluid description. Here, this is given by

$$\xi = \frac{1}{k}\left[-\frac{2\epsilon(\omega)}{\chi(3)|\mathcal{E}_0|^2}\right]^{1/2}. \tag{10.15}$$

For small wavenumbers, $q\xi \ll 1$, the above dispersion reduces to that of typical non-dispersive sound waves, $\Omega \simeq C_s|q|$. This shows that the photon beam behaves as a real fluid, with sound waves that capture their collective oscillations. This collective behaviour is mediated by photon–photon interactions resulting from the nonlinear coupling. In contrast, for large wavenumbers, $q\xi \gg 1$, we are reduced to $\Omega \simeq \hbar q^2/2m_{\text{eff}}$. This is what should be expected from a free stream of non-interacting particles, with momentum $\hbar q$ and mass m_{eff}. In this long wavenumber limit, the photon gas really behaves as a collection of free particles. The healing length therefore sets the scale for the collective processes.

Defining the group velocity of sound waves as $\mathbf{V}_g = \partial\Omega/\partial\mathbf{q}$, and using the dispersion relation (10.13), we obtain, for $\mathbf{v}_0 = 0$, the following relation between phase and group velocities

$$v_\Omega V_g = C_s^2 + \frac{\hbar^2}{2m_{\text{eff}}^2} q^2, \tag{10.16}$$

where $v_\Omega = \Omega/q$ is the phase velocity. This expression, which has been confirmed by a number of recent experiments [20–22], clearly identifies two regions: the non-dispersive region of long scale perturbations where collective interactions dominate and $(v_\Omega V_g) \simeq C_s^2$, and the dispersive region of short scale lengths, where quasi-particle free streaming is observed, and $(v_\Omega V_g) \propto q^2$.

Let us now look at the way these photon sound or Bogoliubov waves can be excited. When the potential V describing a given obstacle is included in our analysis, the fluid equations show that a range of Bogoliubov perturbations will be excited, when the flow velocity is larger than the wave phase velocity, or $C_s < v_0$. This is a particular form of *Cherenkov emission*. The energy carried away by these perturbations corresponds to beam dissipation. This is similar to what occurs in a normal fluid flow, but in the case of a photon fluid it can also be identified with *diffraction*, given that these are just light intensity oscillations created around an obstacle.

Surprisingly, this Cherenkov emission of photon sound waves, or light diffraction, and its corresponding dissipation of energy, can only exist at low intensities and disappears for high intensity beams, when the Cherenkov condition cannot be fulfilled, and $C_s \geqslant v_0$. This is the celebrated *Landau criterium*, which is nothing but a particular form of Cherenkov criterium, according to which superfluid flow will occur for high beam intensities satisfying

$$C_s \geqslant v_0, \quad \rho \geqslant \rho_{\text{crit}} \equiv \frac{m_{\text{eff}}}{g} v_0. \tag{10.17}$$

Suppression of Cherenkov cones could then be an undeniable signature of superfluid light. The absence of Bogoliubov waves excited by an obstacle, means that no energy transfer between light and the medium will occur, and therefore beams will propagate with no losses in the presence of an obstacle. Superfluidity could also be described as the absence of *viscosity* of the photon fluid. As an example, we consider a local dielectric perturbation, described by the potential (10.6), with

$$\delta\epsilon(\mathbf{r}_\perp, z) = \delta\epsilon_0 \exp\left(-\frac{r_\perp^2}{2\sigma^2}\right), \tag{10.18}$$

where $\delta\epsilon_0$ is the maximum amplitude of a perturbation located at some position $\mathbf{r}_\perp = 0$, and width σ, much smaller than the beam width. This could be caused by the nonlinear perturbation of an additional laser beam with much shorter beam waist. The result is illustrated in figure 10.3, showing simulations of light beam propagating in a nonlinear medium. Suppression of Cherenkov cone associated with the photon flow through the local obstacle occurs for beam intensity above the threshold I_{crit}, in agreement with the Landau criterion (10.17). In the specific example of this figure, the medium nonlinearity is provided by a 4-level atomic system [23]. Apart from numerical simulations, a number of experiments have confirmed the existence of superfluid light phenomena. The dispersion relation of Bogoliubov waves, equation (10.2) was demonstrated in a number of experiments, as mentioned. And suppression of diffraction and drag-force cancelation above an intensity threshold were also observed [24].

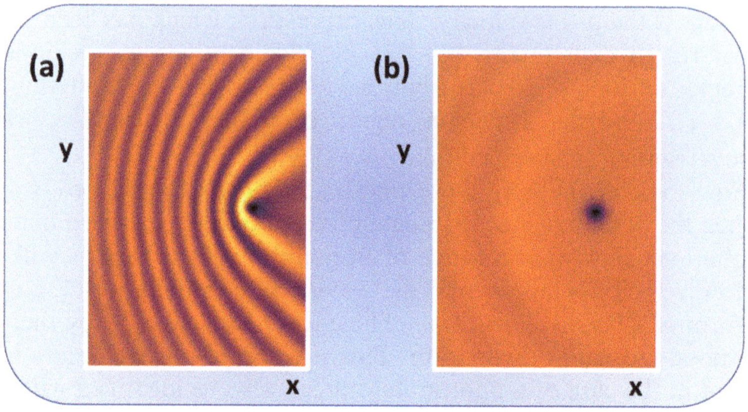

Figure 10.3. Simulations of Superfluid light in nonlinear medium: (a) Cherenkov cone produced by an obstacle, for $I < I_{\text{crit}}$, and (b) their suppression, for $I > I_{\text{crit}}$. (Courtesy Nuno A Silva.)

Let us show how the Cherenkov process can be described. For that purpose, we go back to the fluid equations (10.9). In the fluid frame of reference, the obstacle is moving with some constant velocity $\mathbf{v}_0$, and is described by the potential $V(\mathbf{r}_\perp, \tau) = V(\mathbf{r}_\perp - \mathbf{v}_0\tau)$. We perform a perturbative analysis, such that $\tilde{\rho} = \rho - \rho_0$, is the density ripple created by the obstacle, and we obtain

$$\left(\frac{\partial^2\rho}{\partial\tau^2} - C_s^2\nabla_\perp^2 + \frac{\hbar^2}{4m_{\text{eff}}^2}\nabla_\perp^4\right)\tilde{\rho} = \frac{\rho_0}{m_{\text{eff}}}\nabla_\perp^2 V. \tag{10.19}$$

We assume with generality that the flow velocity is along some direction Ox in the perpendicular plane. It is now useful to make a variable transformation from (x, τ) to (ξ, t), with $\xi = (x - v_0\tau)$ and $t = \tau$. Taking the static condition $\partial/\partial t = 0$, valid in the fluid frame of reference, we obtain

$$\left[(\mathbf{v}_0 \cdot \nabla_\perp)^2 - \left(C_s^2 - \frac{\hbar^2}{4m_{\text{eff}}^2}\nabla_\perp^2\right)\nabla_\perp^2\right]\tilde{\rho} = \frac{\rho_0}{m_{\text{eff}}}\nabla_\perp^2 V(\xi, y), \tag{10.20}$$

where we have $\nabla_\perp^2 = (\partial^2/\partial\xi^2 + \partial^2/\partial y^2)$. Let us now use a Fourier transformation such that

$$\tilde{\rho}(_\perp) = \int \tilde{\rho}(\mathbf{q})\exp(i\mathbf{r}_\perp \cdot \mathbf{q})\frac{d\mathbf{q}}{(2\pi)^2}, \quad V(\xi, y) = \int V(\xi, q_y)\exp(iq_y y)\frac{dq_y}{2\pi}, \tag{10.21}$$

with $\mathbf{r}_\perp \equiv (\xi, y)$. We then get

$$\left\{(\mathbf{q} \cdot \mathbf{v}_0)^2 - \left(C_s^2 q^2 + \frac{\hbar^2}{4m_{\text{eff}}^2}q^4\right)\right\}\tilde{\rho}(\mathbf{q}) = \frac{\rho_0}{m_{\text{eff}}}q^2 V(\mathbf{q}). \tag{10.22}$$

Using $q_\xi = q \sin\theta$, we can see that the density perturbation is maximum at an angle θ, defined by the condition

$$\sin^2\theta \equiv \frac{q_\xi^2}{q^2} = \frac{C_s^2}{v_0^2} \equiv \frac{1}{M^2}, \tag{10.23}$$

where $M = v_0/C_s$ is the Mach number. This corresponds to the Mach (or Cherenkov) cone, observed in the simulations. On this cone, the perturbation attains its extreme value (which is actually a minimum of the total density, for a positive potential perturbation), which can be estimated as

$$\tilde{\rho}_C \sim -4\rho_0 V \frac{m_{\text{eff}}}{\hbar^2 q_{\min}^2}, \tag{10.24}$$

where $q_{\min}^2 \sim 1/L^2$, and L is the size of the system. Let us now assume the one-dimensional problem associated with the case ($y = 0$, $k_y = 0$). Equation (10.22) reduces to

$$\tilde{\rho}(q) = \frac{\rho_0}{m_{\text{eff}}} \frac{q^2 V(q)}{[(v_0^2 - C_s^2)q^2 + (\hbar^2/4m_{\text{eff}}^2)q^4]}. \tag{10.25}$$

Using an inverse Fourier transformation, we obtain the new equation

$$\left\{(v_0^2 - C_s^2)\frac{\partial^2}{\partial\xi^2} + \frac{\hbar^2}{4m_{\text{eff}}^2}\frac{\partial^4}{\partial\xi^4}\right\}\tilde{\rho} = \frac{\rho_0}{m_{\text{eff}}}\frac{\partial^2}{\partial\xi^2}V(\xi, y = 0). \tag{10.26}$$

This can be reduced to a forced oscillator equation, of the form

$$\left(\frac{\partial^2}{\partial\xi^2} + \kappa^2\right)\tilde{\rho} = U(\xi), \tag{10.27}$$

with

$$\kappa^2 = (v_0^2 - C_s^2)\frac{4m_{\text{eff}}^2}{\hbar^2}, \quad U(\xi) = \rho_0 V_0(\xi, y = 0)\frac{4m_{\text{eff}}}{\hbar^2}. \tag{10.28}$$

From here we conclude that the oscillations in the ξ variable can only take place if $v_0^2 > C_s^2$, or $\kappa^2 > 0$, which is nothing but an explicit statement of Landau criterium. Furthermore, these oscillations have a purely diffraction origin, because they vanish if we take the limit $\hbar \to 0$. This means that they are intrinsically associated with the undulatory nature of the fluid particles. An appropriate solution of these equation is given by

$$\tilde{\rho}(\xi, q_y) = \int_\infty^\xi U(\xi', q_y)\sin[\kappa(\xi - \xi')]d\xi'. \tag{10.29}$$

This is valid for arbitrary shapes of the obstacle. The simplest example is that of a point perturbation, as described by $U(\xi, q_y) = U_0(q_y)\delta(\xi)$. In this case the solution is

$$\tilde{\rho}(\xi, q_y) = U_0(q_y)\cos(\kappa\xi). \tag{10.30}$$

Using equation (10.28) and the definition of the Mach number, $M = v_0/C_s$, we can write it in more explicit terms as

$$\tilde{\rho}(\xi, q_y) = U_0(q_y)\cos\left(\frac{2m_{\text{eff}}}{\hbar}C_s\sqrt{M^2 - 1}\ \xi\right). \tag{10.31}$$

Another useful example is that of a Gaussian potential, as defined by

$$V(\xi, y) = V_0 \exp\left(-\frac{\xi^2 + y^2}{2\sigma^2}\right). \tag{10.32}$$

The Cherenkov–Bogoliubov emission given by the solution (10.29) then leads to

$$\tilde{\rho}(\xi, y) = V_0 e^{-y^2/2\sigma^2} \int_{\infty}^{\xi} \exp\left(-\frac{\xi^2}{2\sigma^2}\right) \sin\left[\kappa(\xi - \xi')\right] d\xi'. \tag{10.33}$$

This result is illustrated numerically in figure 10.4, for two different values of the potential width σ, which determines the size of the obstacle. In the limit of a very small size (as compared with $1/\kappa$) this will approach the result of equation (10.31). These analytical solutions are in good qualitative agreement with the simulations of figure 10.3, and with the observations. Apart from these linear waves, theory and simulations also show the formation of nonlinear structures, such as vortices, shock fronts and solitons. In particular, it can be shown, using the nonlinear fluid equations (10.9), that the Mach cone behaves as a one-dimensional dark soliton [17]. But the existence of dark solitons in superfluid light can be discussed in more general terms, using a Korteweg–de Vries (KdV) equation, as shown in [26].

The above theoretical discussion was based on the classical equation for the electric field, equation (10.2), and the apparent quantum properties of the light fluid were only present due to the analogy between this equation and the GP equation of

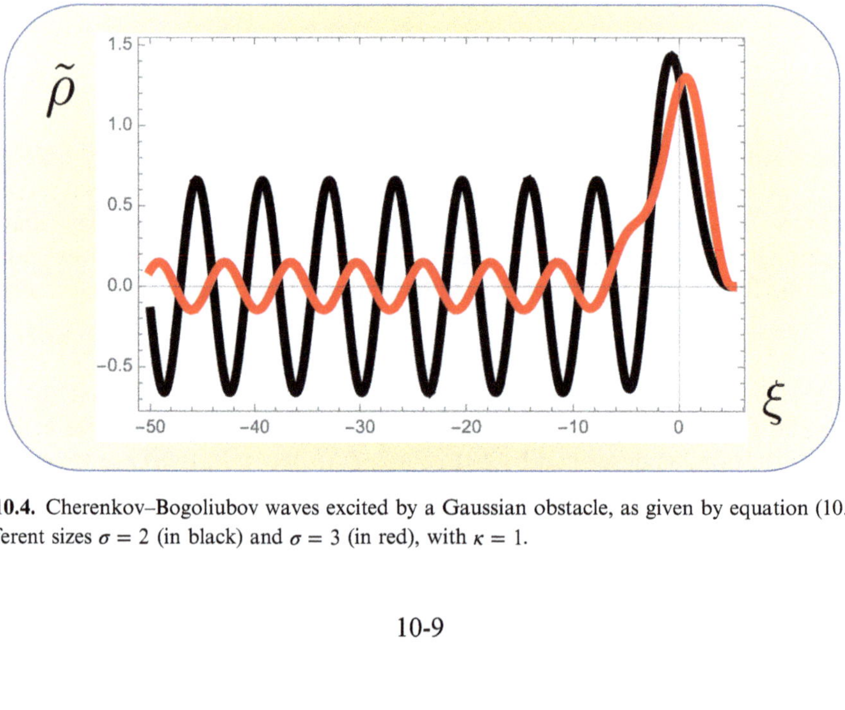

Figure 10.4. Cherenkov–Bogoliubov waves excited by a Gaussian obstacle, as given by equation (10.33), for two different sizes $\sigma = 2$ (in black) and $\sigma = 3$ (in red), with $\kappa = 1$.

Bose–Einstein condensates. As a result, our description was equivalent to the mean-field approximation of a condensate, which neglected quantum fluctuations and only stays valid for a large number of photons. A purely quantum description, valid for an arbitrary number of photons and arbitrarily low beam intensities, can nevertheless be established if we use the quantum version of the GP equation, which can be written in the form [27]

$$i\frac{\partial}{\partial z}\hat{E} = -\frac{1}{2\beta_0}\nabla_\perp^2\hat{E} + \frac{\beta_2}{2}\frac{\partial^2}{\partial t^2}\hat{E} - \frac{\omega}{c}\Delta n(\mathbf{r}_\perp, z) - \frac{\omega}{c}n_2\hat{E}^\dagger\hat{E}\hat{E}, \tag{10.34}$$

where $\hat{E} \equiv \hat{E}(\mathbf{r}_\perp, z, t)$ is the optical field envelope, $\nabla_\perp \equiv (\partial/\partial\mathbf{r}_\perp)$ is the gradient in the plan perpendicular to the propagation direction z, $\Delta n(\mathbf{r}_\perp, z)$ gives the spatial profile of the refractive index of the medium n, and n_2 is the Kerr nonlinear coefficient. For a given position z along propagation, the electric field operators satisfy, the bosonic commutation relations

$$[\hat{E}(\mathbf{r}_\perp, z, t), \quad \hat{E}^\dagger(\mathbf{r}'_\perp, z, t')] = \frac{2}{\epsilon_0}\frac{\hbar\omega}{nc}\delta(\mathbf{r}_\perp - \mathbf{r}'_\perp)\delta(t - t'). \tag{10.35}$$

Such an approach would then be able to confirm the validity, and to establish the limitations, of the mean-field field description.

10.2 Superfluid turbulence

Turbulence in quantum fluids, and possible deviations from classical turbulence is considered nowadays as an important physical problem [28]. It was first approached in liquid helium [29], and more recently in Bose–Einstein condensates of dilute vapours [30]. In this context, experiments using superfluid light could be useful to explore quantum turbulence phenomena, and to understand how quantum turbulence regimes can be achieved.

Here we focus on the processes that can lead to the formation of quantum turbulence. In particular, we show that beam instabilities can be achieved in superfluid light, and that quantum turbulence emerges as a result of the nonlinear instability saturation process. The two-stream instability has been studied in BECs, using both kinetic and fluid dynamics approaches [31, 32]. Generation of a broadband turbulent spectrum, and the excitation of vortex-pairs with opposite vorticity, have been demonstrated by numerical simulations of the superfluid light [33]. The process of vortex-pair formation can also be illustrated by a simple analytical model, which confirms the simulation results.

In order to study the two-stream instability, we assume that the total field $\mathcal{E}$ results from the superposition of two light beams with amplitudes $\mathcal{E}_j$, and wavevectors $\mathbf{k}_j$, for $j = 1, 2$, propagating in nearly opposite directions. For practical reasons, we assume that the two beams have the same frequency, $\omega_j = \omega_0$. Going back to the GP equation (10.5), and using two distinct wavefunctions Ψ_j, we get two coupled equations of the form

$$i\hbar\frac{\partial}{\partial\tau}\Psi_j = \left[-\frac{\hbar^2}{2m_{\text{eff}}}\nabla_\perp^2 + g\mid\Psi_j\mid^2 + g\mid\Psi_1\Psi_2\mid\right]\Psi_j. \tag{10.36}$$

Here we ignore the potential V, which plays no role in the instability process, and assume the same photon mass for both beams. Coupling between the two GP equations comes from the nonlinear coupling, which contains the cross-field term $|\Psi_1\Psi_2|$. We then use Madelung transformations, as before, defined by $\Psi_j = \sqrt{\rho_j}\exp(i\phi_j)$. This will lead to two pairs of fluid equations, as

$$\frac{\partial\rho_j}{\partial\tau} + \nabla_\perp(\rho_j\mathbf{v}_j) = 0,$$
$$\frac{\partial\mathbf{v}_j}{\partial\tau} + \mathbf{v}_i\cdot\nabla\mathbf{v}_i = -\frac{g}{m_{\text{eff}}}\nabla_\perp\left(\rho_j + \sqrt{\rho_1\rho_2}\right) + \nabla_\perp\frac{\hbar^2}{2m_{\text{eff}}^2}\frac{\nabla_\perp^2\sqrt{\rho_j}}{\sqrt{\rho_j}}. \tag{10.37}$$

Comparing this with equations (10.9), we see that beam coupling is due to the product $\rho_1\rho_2$. Let us consider the geometry represented in figure 10.5(a), where the velocities associated with the two beams point to nearly opposite directions. Defining the perturbed densities, $\delta\rho_j = \rho_j - \rho_0/2$, and the perturbed velocities, $\delta\mathbf{v}_1 = \mathbf{v}_1$ and $\delta\mathbf{v}_2 = \mathbf{v}_2 - \mathbf{v}_0$, assuming a space-time evolution of the form $\exp(i\mathbf{q}\cdot\mathbf{r}_\perp - i\Omega\tau)$, and performing a perturbative analysis of the above fluid equations, we are able to derive a dispersion relation of the form

$$1 - \frac{1}{2}C_s^2q^2\left[\frac{1}{\Omega^2 - q^4/4} + \frac{1}{(\Omega - \mathbf{v}_0\cdot\mathbf{q})^2 - H^2q^4/4}\right] = 0, \tag{10.38}$$

where $H = \hbar/2m_{\text{eff}}$ is the quantum parameter. It should be noticed that, for $\mathbf{v}_0 = 0$ this expression reduces to equation (10.13). Solution of this two-stream dispersion equation shows the existence of two branches, determined by

$$\Omega_\pm = \frac{1}{2}qC_s\left[\beta \pm \sqrt{2 + \beta^2 + q^2C_s^2 \pm 2\sqrt{1 + 2\beta^2 + \beta^2q^2C_s^2}}\right], \tag{10.39}$$

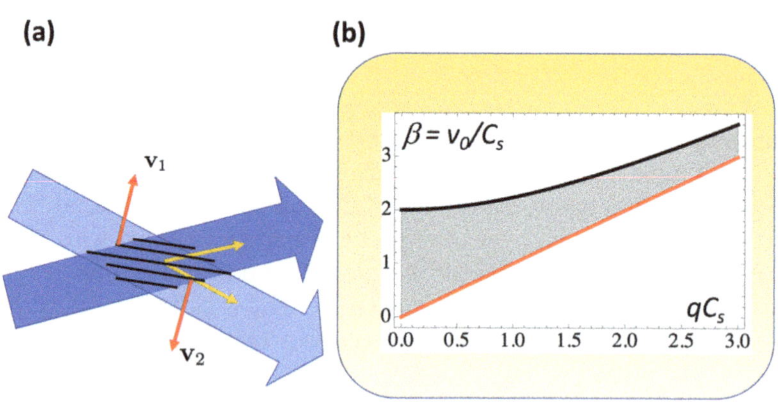

Figure 10.5. Two-stream instability: (a) physical geometry of two interacting laser beams in a Kerr medium; (b) stability diagram.

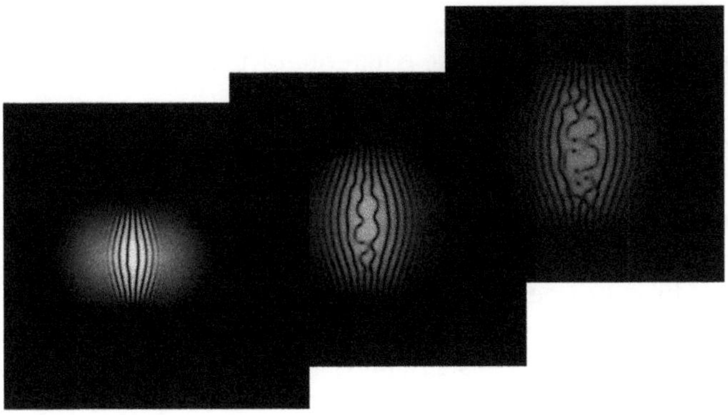

Figure 10.6. Turbulence generated by the two-stream instability of fluids of light, as described by equation (10.5): three different steps of the process of instability growth, saturation and vortex formation. (Courtesy: João D Rodrigues.)

where we have used the stream parameter $\beta = v_0/C_s$. This shows that unstable modes can be excited in the transverse plane, when $\Im(\Omega) > 0$. This is illustrated in figure 10.5(b). Once the sound mode is excite, it grows and saturates by decaying nonlinearly into other modes, thus generating turbulence. Of particular interest is the decay into pairs of vortices with opposite vorticity, as shown in numerical simulations [33] and illustrated in figure 10.6.

What is relevant in quantum turbulence is that its statistical properties differ from those of classical turbulence. In photon fluid turbulence, a detailed analysis of the turbulent spectrum can detect these differences. In classical turbulence, incompressible flow usually exhibits a remarkable energy spectrum, with the typical momentum power law of $q^{-5/3}$. This is the celebrated Kolmogorov spectrum, resulting from successive mode cascades. In contrast, the quantum regime can take place in the high momentum range of the spectrum, shows a much faster decay, as q^{-4}. Simulation experiments performed by [34] seem to confirm these two distinct decay ranges.

10.3 A tale of two fluids

Following the ideas of Landau and Tisza, but using the wave-kinetic model, we now show that superfluid light can be generically described by a mixture of two fluids, where one is the superfluid itself, and the other is the turbulent fluid. The simplest form of a turbulent fluid is to describe it as a gas of Bogoliubov excitations, the bogolon gas. This can exist in photon condensates at finite temperature, as well as in superfluid light. It will be shown that the two co-existing fluids, the perfect superfluid and the bogolon fluid, can be described by similar wave-kinetic equations.

We have seen that superfluid light was described by a nonlinear Schrödinger equation, formally identical to the GP equation for a two-dimensional condensate. Such a description is valid in the mean-field approximation, where quantum fluctuations are negligible. It means that it is valid for beams of light containing a large number of photons. This is precisely the regime where nonlinear optical effects become relevant.

We now use a different but equivalent description, where the nonlinear Schrödinger equation is replaced by a wave-kinetic equation, of the form [25]

$$i\hbar\left(\frac{\partial}{\partial\tau} + \mathbf{v}\cdot\nabla_\perp\right)W = \int\int V(\mathbf{q}, \tau)\,[W_- - W_+]\exp(i\mathbf{q}\cdot\mathbf{r})\frac{d\mathbf{q}}{(2\pi)^3}, \tag{10.40}$$

where $W \equiv W(\mathbf{u}; \mathbf{r}_\perp, \tau)$ is the Wigner function, or quantum quasi-distribution, and $\mathbf{u}$ is the photon group velocity in the transverse plane. This function is related to the wave function $\Psi(\mathbf{r}_\perp, \tau)$, by a Fourier transformation of the field correlation function, as

$$W(\mathbf{u}; \mathbf{r}_\perp, \tau) = \int \Phi^*(\mathbf{r}_\perp - \mathbf{s}/2, t)\Phi(\mathbf{r}_\perp + \mathbf{s}/2, t)\exp(-i\mathbf{q}\cdot\mathbf{s})\,d\mathbf{s}, \tag{10.41}$$

where the auxiliary vector $\mathbf{s}$, and the photon momentum $\mathbf{q} = m_{\text{eff}}\mathbf{u}/\hbar$, are both defined in the transverse plane. In equation (10.40), we use the Fourier transform of the external potential, $V(\mathbf{q}, \tau)$, and define $W_\pm \equiv W(\mathbf{u} \pm \hbar\mathbf{q}/2m_{\text{eff}})$. From the wave-kinetic equation (10.40), we can derive a set of fluid equations describing the evolution of the mean density ρ, and velocity $\mathbf{v}$ of the photon fluid, which coincide to equations (10.9), using the definitions

$$\rho = \int W(\mathbf{v}; \mathbf{r}_\perp, \tau)\,d\mathbf{v}, \quad \mathbf{v} = \frac{1}{\rho}\int \mathbf{u}\,W(\mathbf{u}; \mathbf{r}, t)\,d\mathbf{u}. \tag{10.42}$$

On the other hand, we can describe a turbulent spectrum of bogolon oscillations as the fluid of quasi-particles. This spectrum can also be described by a wave-kinetic equation, similar to that of the photon fluid, as shown below. We start with the quantum fluid equations (10.37), and linearise them with respect to density perturbations $\tilde{\rho} = \rho - \rho_0$. We obtain

$$\left(\frac{\partial^2}{\partial\tau^2} - C_s^2\nabla_\perp^2 + \frac{\hbar^2}{4m_{\text{eff}}^2}\nabla_\perp^4\right)\tilde{\rho} = 0. \tag{10.43}$$

Here we find again the Bogoliubov sound speed, $C_s = \sqrt{g\rho_0/m_{\text{eff}}}$. This is the wave equation for sound waves in the photon fluid. For perturbations evolving as $\exp(i\mathbf{q}\cdot\mathbf{r}_\perp - i\Omega t)$, it leads to the dispersion relation (10.13) in the fluid reference frame, $\mathbf{v}_0 = 0$.

At this point, we should notice the existence of their different time-scales: (i) the fast time-scale of the photon field, which is nothing but the inverse of the photon frequency $t_1 \sim 1/\omega$; (ii) the slow time-scale of the sound waves, which is the inverse of the bogolon frequency $t_2 \sim 1/\Omega$; and (iii) an even slower time-scale, dictated by a possible evolution of the equilibrium density, as defined by $t_3 \sim (1/\rho)\partial\rho/\partial t$. The physical meaning of this third time-scale will become clear below. In order to accommodate this new time-scale, we replace ρ_0 by $\rho_0 + \rho'(\mathbf{r}_\perp, t)$, where ρ' represents that slowly varying part. In such conditions, the dispersion relation (10.13) becomes a local relation, and the sound speed changes as

$$C_s^2(\mathbf{r}_\perp, \tau) = (g/M)[\rho_0 + \rho'(\mathbf{r}_\perp, \tau)]. \tag{10.44}$$

The quantity $\rho'(\mathbf{r}_\perp, t)$ can be associated with long wavelength density perturbations of the second sound. It should be stressed that equation (10.43), can be also reduced to a wave-kinetic equation, formally identical to equation (10.40). For that purpose, we introduce a new Wigner function $\mathcal{W}'$, formally identical to equation (10.41), as

$$\mathcal{W}'(\mathbf{u}; \mathbf{r}_\perp, \tau) = \int \tilde{\rho}^*(\mathbf{r}_\perp - \mathbf{s}'/2, \tau)\tilde{\rho}(\mathbf{r}_\perp + \mathbf{s}'/2, \tau)\exp(-i\mathbf{q}' \cdot \mathbf{s}')\, d\mathbf{s}'. \tag{10.45}$$

This is the Wigner function for the turbulent spectrum. Starting from equation (10.43), it is possible to derive a second wave-kinetic equation, similar to equation (10.40), but now referring to sound waves, or phonons, and not to photons. It takes the form

$$\begin{aligned}\left(\frac{\partial}{\partial t} + \mathbf{v}_q \cdot \nabla_\perp\right)\mathcal{W}' &= -\frac{i}{2\omega}\frac{g}{m_{\text{eff}}}\int \frac{d\mathbf{k}'}{(2\pi)^3} \\ &\times \int \frac{d\omega'}{2\pi}\rho'(\omega')[\mathcal{W}'^{-} - \mathcal{W}'^{+}]\exp(i\mathbf{k}' \cdot \mathbf{r}_\perp - i\omega' t),\end{aligned} \tag{10.46}$$

where $\mathbf{v}_q$ is the group velocity of sound waves. Here, we have used the Fourier components of the spectrum of density fluctuations, as given by

$$\rho'(\mathbf{r}, \tau) = \int \frac{d\mathbf{q}}{(2\pi)^2}\int \frac{d\Omega'}{2\pi}\rho'(\Omega')\exp(-i\Omega' t), \tag{10.47}$$

where the frequencies ω' describe the slow evolution of the background. It should be emphasised that ρ' is distinct from the density perturbations associated with phonons, $\tilde{\rho}$. These two equations, (10.43) and (10.46), provide a consistent kinetic description for the superfluid turbulence in the theory of two fluids. They describe, using two similar wave-kinetic models, the evolution the superfluid (photon) and the turbulent (phonon) fluids. Using these two equations is then possible to show that a new kind of very-long wavelength and very-slow oscillations can be excited in the turbulent superfluid. They satisfy a more general dispersion relation [35], of the form

$$\left(\Omega^2 - \Omega_q^2\right)(\Omega^2 - u_{\text{th}}^2 q^2) = \frac{g}{2m_{\text{eff}}}\frac{C_s^2}{\rho_0}q^2, \tag{10.48}$$

where we have used the quantities

$$\Omega_q^2 = C_s^2 q^2 + \frac{\hbar^2 q^4}{4m_{\text{eff}}^2}, \quad c_{s0}^2 = \frac{g n_0}{m_{\text{eff}}}, \tag{10.49}$$

and

$$u_{\text{th}}^2 = \int (\mathbf{v}_q \cdot \mathbf{u}_{\text{th}})\mathcal{W}(\mathbf{q})\frac{d\mathbf{q}}{(2\pi)^2}. \tag{10.50}$$

The dispersion relation (10.48) describes generalised sound waves, resulting from the coupled oscillations of two fluids, the photon fluid and the turbulent one, where

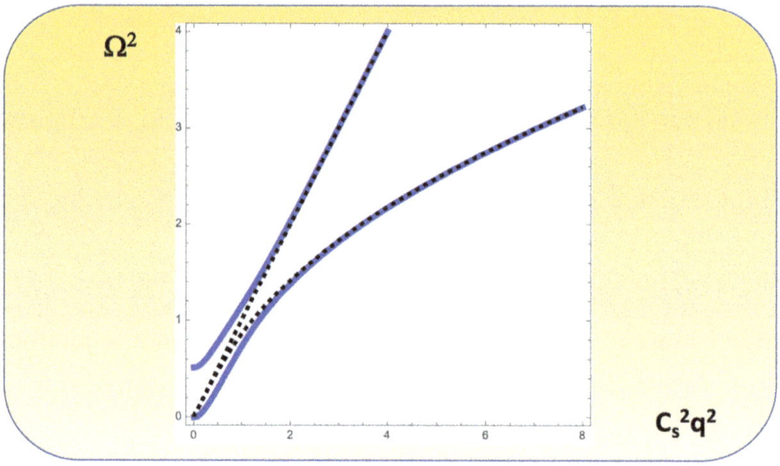

Figure 10.7. Dispersion relations of the generalised sound waves in a superfluid. The upper branch corresponds to normal sound waves, and the lower branch to second sound. The two decoupled solutions are also represented in black-dashed.

the latter is nothing by a phonon gas. Coupling between these two fluids is provided by the term on the r.h.s. Furthermore, on the l.h.s., we recognise the existence of two different modes. One is the usual phonon mode, which exits in pure superfluids, when the turbulent spectrum is not present. It is described by the dispersion relation $(\Omega^2 - \Omega_q^2) = 0$, as previously discussed. The other mode can be called the *second sound*, and corresponds to coupled oscillations of both photons and shorter wavelength phonons. It is described by the new dispersion relation $(\Omega^2 - u_{\rm th}^2 q^2/3) = 0$. We therefore conclude that, in general, the first and the second sound are coupled by the quantity on the l.h.s. of equation (10.48), and can only be considered separately in the limits of very-short wavelength (normal sound) and very-long wavelengths (second sound). This generalised dispersion is illustrated in figure 10.7.

10.4 Superfluid currents

Let us consider the transport properties of a superfluid, taking the contributions of the two different fluids (the condensate and the phonon fluid) into account. We can first define the total mass density ρ, as

$$\rho = M(n_0 + n_x) \equiv Mn, \qquad (10.51)$$

where n_0 is the density of the condensed phase, $n_x = n - n_0$ is the density of the thermal fluid, and n the total density. It is plausible to assume that n_x is proportional to the density of phonons $n_{\rm th}$, because phonon are the constituents of the thermal gas. The above definition of ρ derives from our assumption of two fluids, which were considered above on equal grounds, and described by similar wave-kinetic equations. We can then define, $n_x = \alpha\ n_{\rm th}$, where α is a constant of proportionality still to be determined. Now, if we add the continuity equations for both these two fluids condensate and the phonon gas we obtain a global equation of conservation for the total system, of the form

$$\frac{\partial \rho}{\partial t} + \nabla \cdot (\mathbf{J}_0 + \mathbf{J}_{\mathrm{th}}) = 0, \tag{10.52}$$

where

$$\mathbf{J}_0 = M n_0 \mathbf{v}, \quad \mathbf{J}_{\mathrm{th}} = \alpha n_{\mathrm{th}} \mathbf{u}_{\mathrm{th}}, \tag{10.53}$$

where $\mathbf{v}$ is the mean velocity of the condensate, and $\mathbf{u}_{\mathrm{th}}$ the mean velocity of the thermal fluid. Obviously, we can define a total current, or total mass flow, as $\mathbf{J} = \mathbf{J}_0 + \mathbf{J}_{\mathrm{th}}$. Using the definition of $\mathbf{u}_{\mathrm{th}}$, the thermal mass flow can be described in terms of the phonon quasi-distribution $\mathcal{W}$, as

$$\mathbf{J}_{\mathrm{th}} = M\alpha \int \mathbf{v}_{\mathbf{q}} \mathcal{W}(\mathbf{q}) \frac{d\mathbf{q}}{(2\pi)^2}. \tag{10.54}$$

But, at the same time this thermal current can also be calculated as with the momentum associated with the phonons. We can then write

$$\mathbf{J}_{\mathrm{th}} = M \int \hbar \mathbf{q} \mathcal{W}(\mathbf{q}) \frac{d\mathbf{q}}{(2\pi)^2}, \tag{10.55}$$

where $\hbar\mathbf{q}$ is the momentum of individual phonon quasi-particles. On the other hand, we can use group velocity as defined by equation (10.48), leading to

$$\mathbf{J}_{\mathrm{th}} = \int M_{\mathbf{q}} \mathbf{v}_{\mathbf{q}} \mathcal{W}(\mathbf{q}) \frac{d\mathbf{q}}{(2\pi)^2}. \tag{10.56}$$

The new quantity $M_{\mathbf{q}}$, represents an effective bogolon mass, and is determined by

$$M_{\mathbf{q}} = \frac{\hbar \Omega(\mathbf{q})}{c_s^2 + (\hbar^2 q^2 / 2M)}. \tag{10.57}$$

It obviously depends on the momentum $\mathbf{q}$ of the corresponding bogolon or phonon mode, and weights the contribution of each quasi-particle to the total momentum of the system. But this quantity is somewhat misleading, because the bogolon dispersion shows that these quasi-particle have no finite rest mass, because their energy $\hbar\Omega(\mathbf{q})$ tends to zero when their momentum tends to zero, $|\mathbf{q}| \to 0$. We therefore conclude that the effective bogolon mass is indeed a kinetic mass, associated with a finite momentum, and not a rest mass.

Equating the two expressions established for the thermal mass flow $\mathbf{J}_{\mathrm{th}}$, equations (10.54) and (10.56), we obtain an expression for the quantity α, of the form

$$\alpha \equiv \frac{n_x}{n_0} = \frac{\mathbf{u}}{n_{\mathrm{th}} M u^2} \int M_{\mathbf{q}} \mathbf{v}_{\mathbf{q}} \mathcal{W}(\mathbf{q}) \frac{d\mathbf{q}}{(2\pi)^2}. \tag{10.58}$$

This is the constant of proportionality between the density of thermal fluid, and the bogolon density. It is useful to consider the small wavenumber regime, where collective effects are dominant, and the quantum corrections to the bogolon dispersion containing q^4 can be neglected. In this case, the expressions for the phonon mass and group velocity, become

$$M_{\mathbf{q}} \simeq \frac{\hbar|\mathbf{q}|}{c_s}, \quad \mathbf{v}_{\mathbf{q}} \simeq c_s \frac{\mathbf{q}}{|\mathbf{q}|}. \tag{10.59}$$

As a result, the thermal mass flow (10.54), or thermal current, reduces to

$$\mathbf{J}_{\text{th}} \simeq M\alpha c_s \int \frac{\mathbf{q}}{|\mathbf{q}|} \mathcal{W}(\mathbf{q}) \frac{d\mathbf{q}}{(2\pi)^2}. \tag{10.60}$$

The constant of proportionality α is of the order of $\hbar/Mu$, as shown by equation (10.58). In the presence of dissipation, this current will eventually decay in time, due to the absorption of phonons, and the reduction of the integral over the phonon distribution $\mathcal{W}(\mathbf{q})$. But, in a pure superfluid, with no thermal component, the total current is just given by $\mathbf{J}_0$, and will persist indefinitely.

Another important question is related with the ratio between the thermal and the condensed density, n_x/n_0, or n_{th}/n_0. This can be studied with the above wave-kinetic formalism. In analogy with the Cherenkov emission of radiation by moving electric charges, we can say here that photons can emit or absorb bogolons, thus increasing n_x/n_0, if their (transverse) velocity v, matches the bogolon phase velocity $\Omega_{\mathbf{q}}/|\mathbf{q}|$. The balance between such emission and the reverse absorption process will depend on the details of the photon quasi-distribution, and will eventually lead to Landau damping or growth of bogolons, in a way similar to Landau damping of bogolons in atomic Bose–Einstein condensates.

References

[1] London F 1938 The λ-phenomenon of liquid helium and the Bose-Einstein degeneracy *Nature* **141** 643–4

[2] Chiao R Y and Boyce J 1999 Bogoliubov dispersion relation and the possibility of superfluidity for weakly interacting photons in a two-dimensional photon fluid *Phys. Rev.* A **60** 4114–21

[3] Kapitza P 1938 Viscosity of liquid helium below the λ-point *Nature* **141** 74

[4] Allen J F and Misener A D 1938 Flow of liquid helium II *Nature* **141** 75

[5] Balibar S 2007 The discovery of superfluidity *J. Low Temp. Phys.* **146** 441–70

[6] Griffin A 2009 New light on the intriguing history of superfluidity in liquid ^4He *J. Phys.: Condens. Matter* **21** 164220

[7] Tisza L 1938 Transport phenomena in helium II *Nature* **141** 913

[8] Landau L 1941 The theory of superfluidity of helium II *Phys. Rev.* **60** 356–8

[9] Leggett A J 2006 *Quantum Liquids* (Oxford: Oxford University Press)

[10] Khalatnikov I M 2000 *Introduction to the Theory of Superfluidity* (New York: Westview Press)

[11] Terças H, Mendonça J T and Guerra V 2012 Classical rotons in cold atomic traps *Phys. Rev.* A **86** 053630

[12] Hohenberg P C and Martin P C 1965 Microscopic theory of superfluid helium *Ann. Phys., NY* **34** 291–359

[13] Giovanazzi S and O'Dell D H J 2004 Instabilities and the roton spectrum of a quasi-1D Bose-Einstein condensed gas with dipole-dipole interactions *Eur. Phys. J. D* **31** 439–45
[14] Mendonça J T and Terças H 2017 Bose-Einstein condensation of photons in a plasma *Phys. Rev.* A **95** 063611
[15] Pomeau Y and Rica S 1993 Nonlinear diffraction *C. R. Acad. Sci., Paris* **317** 1287–92
[16] Hakim V 1997 Nonlinear Schrödinger flow past an obstacle in one dimension *Phys. Rev.* E **55** 2835–45
[17] Khamis E G, Gammal A, El G A, Gladush Yu G and Kamchatnov A M 2008 Nonlinear diffraction of light beams propagating in photorefractive media with embedded reflecting wire *Phys. Rev.* A **78** 013829
[18] Leboeuf P and Moulieras S 2010 Superfluid motion of light *Phys. Rev. Lett.* **105** 163904
[19] Carusotto I 2014 Superfluid light in bulk nonlinear media *Proc. R. Soc. A* **470** 20140320
[20] Vocke D, Roger T, Marino F, Wright E M, Carusotto I, Clerici M and Faccio D 2015 Experimental characterization of nonlocal photon fluids *Optica* **2** 484–90
[21] Fontaine Q, Bienaimé T, Pigeon S, Giacobino E, Bramati A and Glorieux Q 2018 Observation of the Bogoliubov dispersion in a fluid of light *Phys. Rev. Lett.* **121** 183604
[22] Fontaine Q, Larré P-E, Lerario G, Bienaimé T, Pigeon S, Faccio D, Carusotto I, Giacobino E, Bramati A and Glorieux Q 2020 Interferences between Bogoliubov excitations in superfluids of light *Phys. Rev. Res.* **2** 043297
[23] Silva N A, Mendonça J T and Guerreiro A 2017 Persistent currents of superfluid light in a four-level coherent atomic medium *J. Opt. Soc. Am.* B **34** 2220
[24] Michel C, Boughdad O, Albert M, Larré P-E and Bellec M 2018 Superfluid motion and drag-force cancellation in a fluid of light *Nat. Commun.* **9** 2108
[25] Mendonça J T and Terças H 2013 *Physics of Ultra-cold Matter* (New York: Springer)
[26] Ali S, Mendonça J T and Kourakis I 2022 *Dark Solitons in Paraxial Superfluid Light* (to be published)
[27] Larré P-E and Carysotto I 2015 Propagation of a quantum fluid of light in a cavityless nonlinear optical medium: general theory and response to quantum quenches *Phys. Rev.* A **92** 043802
[28] Barenghi G F, Skrbek L and Sreenivasan K R 2014 Introduction to quantum turbulence *Proc. Natl Acad. Sci.* **111** 4647–52
[29] Skrbek L, Schmoranzer D, Midlik S and Sreenivasan K P 2021 Phenomenology of quantum turbulence in superfluid light *Proc. Natl Acad. Sci.* **118** e2018406118
[30] Madeira L, Cidrim A, Hemmerling H, Caracanhas M A, dos Santos F E A and Bagnato V S 2020 Quantum turbulence in Bose–Einstein condensates: present status and new challenges ahead *AVS Quantum Sci.* **2** 035901
[31] Terças H, Mendonça J T and Robb G R M 2009 Two-stream instability in quasi-one-dimensional Bose-Einstein condensates *Phys. Rev.* A **79** 065601
[32] Abad M, Recati A, Stringari S and Chevy F 2015 Counter-flow instability of a quantum mixture of two superfluids *Eur. Phys. J. D* **69** 126
[33] Rodrigues J D, Mendonça J T and Terças H 2020 Turbulence excitation in counterstreaming paraxial superfluids of light *Phys. Rev.* A **101** 043810
[34] Silva N A, Ferreira T D and Guerreiro A 2021 Exploring quantum-like turbulence in a two-component paraxial fluid of light *Results Opt.* **2** 100025
[35] Mendonça J T 2013 Second sound in a laser cooled gas *Phys. Lett.* A **377** 1961–5

Part III

Quantum vacuum

The third part of this book concerns the physics of quantum vacuum, as presently explored by ultra-intense lasers. In contrast with the low energy version of quantum electrodynamics (QED), considered in the previous two parts, where quantum vacuum states only concern photons, here we consider the high energy branch of QED, where electron–positron states are included.

These topics are usually absent from quantum optics books, which concentrate on low energy quantum phenomena. But the analogies between low and high energy QED are very strong, and the recent inclusion of both topics in laser–matter interaction studies justify its inclusion here. In particular, QED vacuum can be seen as a special case of quantum optics where the optical medium is vacuum, but where similar nonlinear quantum photon processes also take place. Furthermore, photons are sensitive to static electric and magnetic fields in vacuum, leading to the occurrence of optical birefringence and photon splitting. If we further assume that these quasi-static fields are time-dependent, then time-refraction will also take place.

We mainly concentrate on moderately intense fields, much lower than the Schwinger limit, such that electron–positron pair creation is negligible and only virtual states are considered. But, the problem of emission of particle-pairs and vacuum disruption are becoming more relevant to the ultra-intense laser sources that are now available around the world, and will also be discussed.

In the last chapter, we abandon QED and enter the area of quantum chromodynamics (QCD), considering hypothetical particles called *axions*. We describe the origin of this concept, and its eventual relation with the problem of *dark matter*. We discuss axion interactions with photons and plasmons, and the formation of mixed states that can be called *axion-polaritons*. The possible use of intense laser beams to excite axions, the existence of axion–photon instabilities, and the challenging search for dark matter will conclude this Part III.

IOP Publishing

The Quantum Nature of Light
From photon states to quantum fluids of light
J T Mendonça

Chapter 11

Basic QED concepts

In the quantum formalism that we have considered so far, photons cannot interact with each other in vacuum. However, this is only strictly true at low intensities, because a more complete version of QED shows that electron–positron (e-p) pairs can be excited by intense electromagnetic fields in vacuum. This opens the door for possible nonlinear optical effects in vacuum, which include photon–photon scattering and particle-pair creation.

In the present chapter, we review the basic QED theory, in a more complete version valid for intense fields, where effects associated with virtual and real e-p pairs are taken into account. This could be called *relativistic QED*. We first introduce the Klein–Gordon equation, which is able to describe massive relativistic particles with no spin. It anticipates the Dirac equation and can be used to describe charged particles when spin effects are negligible. Despite the name, the Klein–Gordon equation was first considered by Schrödinger, and nearly after proposed by Klein [1] and Gordon [2] to describe relativistic quantum particles.

We discuss the Dirac equation, first introduced in 1928 [3–5], which is valid for relativistic particles with a semi-integer spin, and discuss its basic properties. We also quantise the Dirac field, and show that it can be decomposed into electron and positron states, and introduce the Volkov states, which are exact solutions of the Dirac equation for charged particles in the presence of an electromagnetic wave [6]. They can be useful to describe the energy exchange between photons and charged particles in vacuum, such as those related to Compton scattering. Volkov solutions have recently been generalised to the case of plasmas, where the photon mass introduces qualitative changes and allows for additional Volkov states associated to electrostatic waves [7, 8].

Another important piece of relativistic QED is the so-called *Euler–Heisenberg effective Lagrangian* [9]. We consider it in the limit of weak fields (for an interesting historical account see [10]). This limit can be used to describe QED effects induced by ultra-intense lasers in vacuum. Starting from this Lagrangian, it is possible describe vacuum as a nonlinear optical medium. Such an approach is valid when the electric field associated with intense light pulses is very large, but stays bellow the *Schwinger field* [11]. This field would be sufficiently intense to produce vacuum disruptions, but is still far from our present experimental capabilities. Possible strategies leading to the Schwinger field, and to the breakdown of the weak-field approximation, have nevertheless been discussed in the literature [12–14]. The vacuum nonlinearities described by the Heisenberg–Euler Lagrangian can be associated with the excitation of virtual electron–positron pairs. Intense photon beams propagating in vacuum are able to create a kind of *virtual plasma state*, made of e-p virtual pairs, where nonlinear optical phenomena take place. In a later chapter, we will discuss a variety of optical processes associated with the nonlinear vacuum. This opens the way to promising studies of QED vacuum, using the ultra-intense laser systems that are being developed around the world.

11.1 Klein–Gordon equation

This equation describes relativistic quantum particles with no spin, as first proposed by Schrödinger, and soon after by Klein and Gordon. It is a straightforward generalisation of the Schrödinger equation, and at the same time, a generalisation of the wave equation for massive fields. In many different problems where spin effects are not relevant, it can be used in alternative to the more difficult, but also more accurate, Dirac equation. Similarities and differences between these two equations will become apparent in this chapter.

In classical relativistic mechanics, the energy of a particle is determined by the relation $E^2 = \mathbf{p}^2c^2 + m^2c^4$, where $\mathbf{p}$ is the particle momentum. If, in this expression, we replace the energy and momentum, E and $\mathbf{p}$, by their quantum operators, $(i\hbar\partial_t)$ and $(-i\hbar\nabla)$, and apply the resulting operator to the particle wavefunction ψ, we obtain a relativistic wave equation of the form

$$\frac{\hbar^2}{c^2}\frac{\partial^2}{\partial t^2}\psi = [\hbar^2\nabla^2 - m^2c^2]\psi. \tag{11.1}$$

In contrast with the Schrödinger equation, this so-called Klein–Gordon, or KG equation, contains a double time derivative, which has important physical consequences. The KG equation satisfies a continuity equation, and can be used as a wave equation for relativistic quantum particles under certain conditions. First of all, it contains no information about the spin, and can strictly only be used for spin-zero particles, such as the π-meson or the Higgs boson. We can nevertheless use it as a simplified model for electrons and positrons, to describe processes where spin effects are irrelevant. Second, this equation satisfies negative energy solutions, which means that the resulting density function is not positive definite, and its meaning can only be clarified by field quantisation.

Let us first write the equation in covariant form. For that purpose, we introduce the 4-position $x^\mu = (ct, \mathbf{r})$, where $x^0 = ct$ is the time variable, and $x^i = (x, y, x)$ for $i = 1, 2, 3$ are the Cartesian components of the vector position $\mathbf{r}$. We also use the 4-momentum, $p^\mu = (E/c, \mathbf{p})$, where the $\mu = 0$ component gives the particle energy $p^0 = m\gamma c$, where $\gamma = 1/\sqrt{1 - (v^2/c^2)}$ is the relativistic gamma factor. The corresponding covariant components are obtained using the Minkowski metric tensor $g_{\mu\nu} = \mathrm{diag}\,(+1, -1, -1, -1)$, such that $x_\mu = g_{\mu\nu}x^\nu = (ct, -\mathbf{r})$, and $p_\mu = g_{\mu\nu}p^\nu = (E/c, -\mathbf{p})$. We notice that the modulus of the 4-momentum is an invariant, and its value is determined by the particle mass, $p_\mu p^\mu = m^2c^2$. We can then rewrite equation (11.1) as

$$\left[\partial_\mu\partial^\mu + k_C^2\right]\psi = 0, \tag{11.2}$$

where $k_C = mc/\hbar$ is the Compton wavenumber. This is obviously Lorentz invariant. In order to understand the meaning of this equation, we multiply it by ψ^*. After subtracting its complex conjugate and multiplying by a factor, $-i\hbar/2m$, we get the continuity equation

$$\frac{\partial\rho}{\partial t} + \nabla \cdot \mathbf{J} = 0. \tag{11.3}$$

The density ρ, and the current $\mathbf{J}$, are defined by

$$\rho = \frac{i\hbar}{2mc^2}(\psi^*\partial_t\psi - \psi\partial_t\psi^*), \quad \mathbf{J} = \frac{\hbar}{2im}(\psi^*\nabla\psi - \psi\nabla\psi^*), \tag{11.4}$$

with $\partial_t = \partial/\partial t$. Notice that $\mathbf{J}$ is defined in the same way as with the Schrödinger equation. However, the density ρ is not positive definite, and cannot be identified with a probability density. Furthermore, the KG equation satisfies plane-wave solutions, with positive and negative energies, of the form

$$\psi(\mathbf{r}, t) = \psi_\pm \exp\,(i\mathbf{p}\cdot\mathbf{r}/\hbar - iE_\pm t/\hbar), \quad E_\pm = \pm\sqrt{\mathbf{p}^2c^2 + m^2c^4}, \tag{11.5}$$

where $\psi_\pm$ are constants. This means that the energy spectrum is not bounded from below.

Our aim in this chapter is to understand the possible coupling between charged particles and photons. This means that we have to discuss the implications of the presence of an electromagnetic field. If we describe the field using scalar and vector potentials, V and $\mathbf{A}$, the energy of the particle E, in a classical description, becomes $(E - eV) = \sqrt{(\mathbf{p} + e\mathbf{A})^2c^2 + m^2c^4}$. For electrons, the charge will be $e = -|e|$, and for positrons, $e = |e|$. If we now apply the corresponding operator to the particle wavefunction ψ, we get

$$\frac{1}{c^2}(i\hbar\partial_t - eV)^2\psi = [(i\hbar\nabla - e\mathbf{A})^2 + m^2c^2]\psi, \tag{11.6}$$

which generalises equation (11.1). It is sometimes useful to introduce dimensionless scalar and vector potentials, $\mathcal{U}$ and $\mathbf{a}$, as defined by

$$\mathcal{U} = \frac{eV}{mc^2}, \quad \mathbf{a} = \frac{e\mathbf{A}}{mc}. \tag{11.7}$$

We can write the KG equation in covariant form, as

$$[(i\partial - ak_C)^2 - k_C^2]\psi = 0, \tag{11.8}$$

where the four-potential $a = (\mathcal{U}, \mathbf{a})$, and the four-derivative $\partial = (\partial_t/c, \nabla)$, were used. An alternative, and more common, version of the KG equation can also be obtain, using natural units, $\hbar = 1$ and $c = 1$. Equation (11.6) is then reduced to

$$[(i\partial - ea)^2 - m^2]\psi = 0. \tag{11.9}$$

The interest of this relativistic wave equation, in any of its equivalent forms, will be exemplified later.

11.2 Dirac equation

In contrast with the Klein–Gordon equation, we now consider a single time-derivative equation, formally identical to the Schrödinger equation, but containing a different Hamiltonian operator. We assume an equation of the form

$$i\hbar\partial_t\psi = H\psi, \quad H = \vec{\alpha} \cdot \mathbf{p}\, c + \beta mc^2, \tag{11.10}$$

where $\vec{\alpha}$ and β are 4×4 vector and scalar matrices, and ψ is a vector wavefunction, called a *spinor* (or more appropriately, a *bi-spinor*). It can easily be seen that, if we want the Hamiltonian operator H to be Hermitian, we need to assume that $\vec{\alpha}$ and β are also Hermitian. This guarantees that, in contrast with the Klein–Gordon case, the new probability density associated with ψ will be a positive definite quantity. If, on the other hand, we apply the operator H twice, we get an equation that can be directly compared with Klein–Gordon's, as

$$-\hbar^2\partial_t^2\psi = c^2\left(-i\hbar\alpha^j\partial_j + \beta mc\right)^2\psi, \tag{11.11}$$

where α^j, with $j = 1, 2, 3$, are the components of $\vec{\alpha}$. This will be identical to equation (11.1) if we assume that

$$\alpha^j\alpha^{j'} + \alpha^{j'}\alpha^j = 2\delta^{jj'}I, \quad \alpha^j\beta + \beta\alpha^j = 0, \quad (\alpha^j)^2 = \beta^2 = I, \tag{11.12}$$

where I is the identity matrix. From this, we conclude that the eigenvalues of α^j and β can only be equal to ± 1. Multiplying equation (11.10) by $\psi^\dagger$ and subtracting the hermitic conjugate, we obtain a continuity equation, formally identical to (11.3), but where the new definitions of probability density ρ, and probability current $\mathbf{J}$, are now given by

$$\rho = \psi^\dagger\psi, \quad \mathbf{J} = c\, \psi^\dagger\vec{\alpha}\psi. \tag{11.13}$$

We can also introduce a four-vector current, $J^\mu = (J^0, J^i)$, and write the continuity equation in a covariant form, as

$$\partial_\mu J^\mu \equiv \frac{1}{c}\partial_t J^0 + \partial_i J^i = 0. \tag{11.14}$$

A particularly useful choice of the matrices $\vec{\alpha}$ and β is given by

$$\alpha^j = \begin{bmatrix} 0 & \sigma^j \\ \sigma^j & 0 \end{bmatrix}, \quad \beta = \begin{bmatrix} I & 0 \\ 0 & -I \end{bmatrix}, \tag{11.15}$$

where σ^j are the Pauli matrices, and I is the 2×2 identity matrix. This provides the so-called *standard representation* of the Dirac equation. Generalisation of equation (11.10) in order to include electromagnetic potentials is straightforward, using the new Hamiltonian operator

$$H = \vec{\alpha} \cdot (\mathbf{p} - e\mathbf{A})c + \beta mc^2 + eV. \tag{11.16}$$

It is also possible to write the Dirac equation in a covariant form. For this purpose, we multiply it by β and define new 4-matrices γ^μ such that

$$\gamma^0 = \beta, \quad \gamma^j = \beta\alpha^j = \begin{bmatrix} 0 & \sigma^j \\ -\sigma^j & 0 \end{bmatrix}. \tag{11.17}$$

We then get

$$\left[\gamma^\mu\left(i\hbar\partial_\mu - eA_\mu\right) - mc\right]\psi = 0. \tag{11.18}$$

It is now useful to introduce the kinetic momentum $\vec{\pi} = \mathbf{p} - e\mathbf{A}$. Assuming that ψ can be decomposed into two different vectors φ and χ, such that

$$\psi = \begin{bmatrix} \varphi \\ \chi \end{bmatrix}, \tag{11.19}$$

we obtain the following coupled equations

$$\begin{aligned} i\hbar\partial_t \varphi &= c(\vec{\sigma} \cdot \vec{\pi})\chi + (mc^2 + eV)\varphi, \\ i\hbar\partial_t \chi &= c(\vec{\sigma} \cdot \vec{\pi})\varphi - (mc^2 - eV)\chi. \end{aligned} \tag{11.20}$$

In the weak-field regime, the dominant term on the r.h.s. of these equations is mc^2. We can then assume solutions of the form

$$\varphi = \bar{\varphi} \exp(-i\omega_C t), \quad \chi = \bar{\chi} \exp(-i\omega_C t), \tag{11.21}$$

where $\omega_C = ck_C$ is the Compton frequency, and $\bar{\varphi}$ and $\bar{\chi}$ are slowly varying amplitudes. Replacing this in equations (11.20), we obtain

$$\begin{aligned} i\hbar\partial_t \bar{\varphi} &= c(\vec{\sigma} \cdot \vec{\pi})\bar{\chi} + eV\bar{\varphi}, \\ i\hbar\partial_t \bar{\chi} &= c(\vec{\sigma} \cdot \vec{\pi})\bar{\varphi} - (2mc^2 - eV)\bar{\chi}. \end{aligned} \tag{11.22}$$

Assuming that the scalar potentials are small, $|eV| \ll mc^2$, we get a typical amplitude for these two vector functions, valid for weakly relativistic particles

$$\bar{\chi} \simeq \frac{\vec{\sigma} \cdot \vec{\pi}}{2mc} \bar{\varphi} \ll \bar{\varphi}. \tag{11.23}$$

This means that the vector function $\bar{\varphi}$ becomes dominant in the weakly relativistic limit. Using this limit in the first of equations (11.22), we then get

$$i\hbar\partial_t \bar{\varphi} = c\left[\frac{1}{2m}(\vec{\sigma} \cdot \vec{\pi})^2 + eV\right]\bar{\varphi}. \tag{11.24}$$

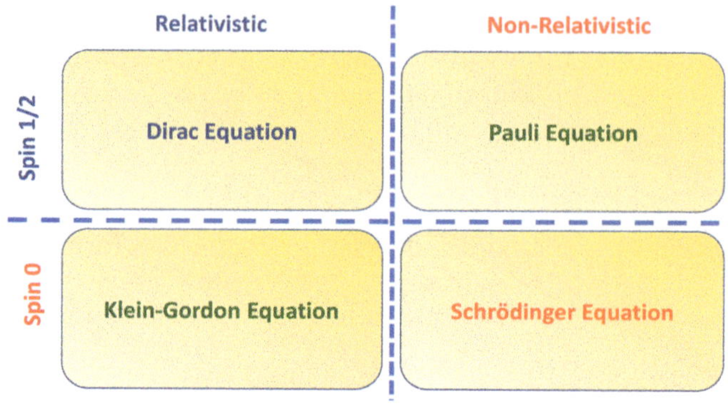

Figure 11.1. Table of the different wave equations of quantum mechanics.

Noting that

$$(\vec{\sigma} \cdot \vec{\pi})^2 = \vec{\pi}^2 + i\vec{\sigma} \cdot (\vec{\pi} \times \vec{\pi}) = \vec{\pi}^2 - \frac{e\hbar}{c}\mathbf{B}, \tag{11.25}$$

where $\mathbf{B} = \nabla \times \mathbf{A}$, we finally obtain the Pauli equation

$$i\hbar\partial_t \bar{\varphi} = \left[\frac{1}{2m}(\mathbf{p} - e\mathbf{A})^2 - \frac{e\hbar}{2mc}\vec{\sigma} \cdot \mathbf{B} + eV\right]\bar{\varphi}. \tag{11.26}$$

It is known that the spinor function $\bar{\varphi}$ in this equation describes the evolution of a charged particle (electron or positron) with spin 1/2. The two extra-dimensions of the Dirac wavefunction ψ are associated with the positive and negative energy solutions. Neglecting the Pauli term containing $\vec{\sigma}$, we would be reduced to the Schrödinger equation. This shows the close connection between the Dirac equation and the well-known wave equations of non-relativistic quantum mechanics. See figure 11.1 for an illustration.

11.3 Volkov states

In order to understand the connections between the particle states (electron and positron), and the quantum states of light, it is useful to discuss a particular class of solutions of the Dirac equation. These are exact solutions, found by Volkov in 1936 [6], which are valid for particles in the field of an electromagnetic wave. For simplicity, we discuss them using the Klein–Gordon equation, but a comment will be added concerning its extension to the Dirac equation.

We assume an electromagnetic wave in vacuum, described by the normalised vector potential $\mathbf{a}$, with $\mathcal{U} = 0$. The four-potential will just be $a = (0, \mathbf{a})$, and its square $a^2 = -\mathbf{a}^2$. In the Lorentz gauge, we also have the condition $(\nabla \cdot \mathbf{a}) = 0$. Equation (11.8) will then be transformed into

$$[(\nabla^2 - \partial_t^2) - 2ik_C(\mathbf{a} \cdot \nabla) - (1 + a^2)k_C^2]\psi = 0. \tag{11.27}$$

We now assume that the vector potential **a** describes an electromagnetic wave propagating along an arbitrary z-direction, with wavenumber k and frequency $\omega = kc$. We can then write

$$\mathbf{a}(\mathbf{r}, t) = \mathbf{a}_0 f(\tau), \quad \tau = t - \frac{kz}{\omega}, \tag{11.28}$$

where $\mathbf{a}_0$ is the wave amplitude and $f(\tau)$ a function to be specified. For a simple monochromatic wave, we simply have $f(\tau) = \cos(\omega\tau)$. In order to solve equation (11.27) and obtain the particle wavefunction ψ in the presence of this potential, we assume a solution of the form

$$\psi(\mathbf{r}, t) = \Psi(\tau)\exp[-i(p \cdot x)], \tag{11.29}$$

where $(p \cdot x) = p_0 c\, t - \mathbf{p} \cdot \mathbf{r}$, with particle energy $E = p_0 c$ and momentum $\mathbf{p}$ to be specified. These quantities are not arbitrary, because they should comply with the condition

$$p^2 \equiv (p_0^2 - \mathbf{p}^2) = m^2c^2 = \hbar^2 k_C^2. \tag{11.30}$$

Replacing this assumed solution in equation (11.27), we derive an equation for the function $\Psi(\tau)$, of the form

$$2i(p \cdot k)\Psi' + F(\tau)\Psi 2k_C(a \cdot p) - \mathbf{a}^2 k_C^2 \psi = 0, \tag{11.31}$$

where $\Psi' = d\Psi/d\tau$, and

$$F(\tau) = \hbar\omega[(a \cdot p)k_C + \mathbf{a}^2 k_C^2]. \tag{11.32}$$

At this point it should be noticed that, due to the transverse nature of electromagnetic wave, we have $(\mathbf{a} \cdot \mathbf{e}_z) = 0$. We also notice that $(a \cdot p) = -(\mathbf{a} \cdot \mathbf{p})$. Integration of this equation, and the use of equation (11.29), then lead to the final solution

$$\psi(x) = u \exp[iS(\tau)], \tag{11.33}$$

where u is a constant, and the phase $S(\tau)$ is determined by

$$S(\tau) = -(p \cdot x) + \int^{\tau} \left[\frac{e(a \cdot p)}{(p \cdot k)} k_C + \frac{\mathbf{a}^2 k_C^2}{2(p \cdot k)} \right] d\tau. \tag{11.34}$$

If, instead of the Klein–Gordon equation we had used the Dirac equation, we would get a similar solution, but where the function ψ is replaced by a bi-spinor, and a new exponential factor associated with spin would appear.

11.4 Quantisation of the Dirac field

For constant electromagnetic potentials, V and A, and therefore no electromagnetic waves, we have positive and negative energy solutions of the Dirac equation, of the form

$$\begin{aligned} \psi^{(+)}(\mathbf{r}, t) &= u(\mathbf{p})\exp(i\mathbf{p} \cdot \mathbf{r}/\hbar - i\omega_{\mathbf{p}} t), \\ \psi^{(-)}(\mathbf{r}, t) &= v(-\mathbf{p})\exp(-i\mathbf{p} \cdot \mathbf{r}/\hbar + i\omega_{\mathbf{p}} t), \end{aligned} \tag{11.35}$$

where, for a given value of the particle momentum $\mathbf{p}$, we have wavefunction amplitudes $u(\mathbf{p})$ and $v(-\mathbf{p})$ corresponding to positive and negative energy solutions, as we can define a positive quantity $E = \hbar\omega_{\mathbf{p}}$, defined by the relation

$$E = eV + \sqrt{(\mathbf{p} - e\mathbf{A})^2 c^2 + m^2 c^4}. \tag{11.36}$$

Quantisation of the Dirac field then leads to the following field operator

$$\hat{\psi}(x) = \sum_{s,\mathbf{p}} \left[\hat{b}_{s\mathbf{p}} u_{s,\mathbf{p}} e^{i\mathbf{p}\cdot\mathbf{r}/\hbar - i\omega_{\mathbf{p}} t} + \hat{d}^{\dagger}_{s,\,-\mathbf{p}} v_{s,-\mathbf{p}} e^{-i\mathbf{p}\cdot\mathbf{r}/\hbar + i\omega_{\mathbf{p}} t} \right], \tag{11.37}$$

where the positive and negative energy states $\pm\hbar\omega_{\mathbf{p}}$ are defined in accordance with equation (11.36), and s specifies the two possible up ($s = \uparrow$) and down ($s = \downarrow$) spin states. The operators $\hat{b}^{\dagger}_{s\mathbf{p}}$ and $\hat{b}_{s\mathbf{p}}$ are the creation and annihilation of electron states, while the operators $\hat{d}^{\dagger}_{s\mathbf{p}}$ and $\hat{d}_{s\mathbf{p}}$ are creation and annihilation of positron states. The formal analogy with the quantised photon field is striking, if we notice that the antiparticle of a photon is also a photon with the same frequency and momentum. Therefore, for photons, the difference between particles and anti-particles vanishes, and $\hat{b}_{s\mathbf{p}}$ and $\hat{d}_{s\mathbf{p}}$ are replaced by a single operator. On the other hand, spin s plays the same role for the Dirac field as polarisation plays for the photon field.

Rearranging the terms inside the sum in equation (11.37), we can write it in another form, as

$$\hat{\psi}(x) = \sum_{s,\mathbf{p}} \left[\hat{b}_{s\mathbf{p}} u_{s\mathbf{p}} e^{-i\omega_{\mathbf{p}} t} + \hat{d}^{\dagger}_{s\mathbf{p}} v_{s\mathbf{p}} e^{+i\omega_{\mathbf{p}} t} \right] e^{i\mathbf{p}\cdot\mathbf{r}/\hbar}. \tag{11.38}$$

The creation and annihilation pairs of operators, $(b^{\dagger}, d^{\dagger})$ and (b, d), satisfy the anti-commutation relations

$$\left\{ \hat{b}_{s\mathbf{p}}, \hat{b}_{s'\mathbf{p}'} \right\} = (2\pi)^3 \delta^3(\mathbf{p} - \mathbf{p}')\delta_{ss'}, \quad \left\{ \hat{d}_{s\mathbf{p}}, \hat{d}_{s'\mathbf{p}'} \right\} = (2\pi)^3 \delta^3(\mathbf{p} - \mathbf{p}')\delta_{ss'}, \tag{11.39}$$

where $\{\hat{a}, \hat{b}\} = \hat{a}\hat{b} + \hat{b}\hat{a}$. Again, this is formally similar (but contrasts with) the case of photons, or bosons in general, where these anti-commutation relations are replaced by commutation relations of the form $[\hat{a}, \hat{b}] = \hat{a}\hat{b} - \hat{b}\hat{a}$, as seen before. In the presence of static and uniform potentials, the spinors $u_{\uparrow\mathbf{p}}$ and $u_{\downarrow\mathbf{p}}$, and their corresponding antiparticle spinors, are determined by

$$u_{\uparrow\mathbf{p}} = \begin{bmatrix} 1 \\ 0 \\ \dfrac{\pi}{E+m} \\ 0 \end{bmatrix}, \quad u_{\downarrow\mathbf{p}} = \begin{bmatrix} 0 \\ 1 \\ 0 \\ \dfrac{-\pi}{E+m} \end{bmatrix}, \quad v_{\uparrow\mathbf{p}} = \begin{bmatrix} \dfrac{-\pi}{E+m} \\ 0 \\ 1 \\ 0 \end{bmatrix}, \quad v_{\downarrow\mathbf{p}} = \begin{bmatrix} 0 \\ \dfrac{\pi}{E+m} \\ 0 \\ 1 \end{bmatrix}, \tag{11.40}$$

where $E = eV + \sqrt{\boldsymbol{\pi}^2 + m^2}$, with $\boldsymbol{\pi} = (\mathbf{p} - e\mathbf{A})$, and the charge is $e = \mp|e|$ for positive or negative energy states.

11.5 Euler–Heisenberg Lagrangian

The Euler–Heisenberg Lagrangian is an approximate Lagrangian density of vacuum, valid for an electromagnetic field configuration that stays constant over distances larger that the electron Compton wavelength, $\lambda_C = h/mc \simeq 2.4 \times 10^{-12}$ m. Because this scale length is much sorter than usual laser wavelengths, it becomes a very useful approximation for laser experiments. This Lagrangian density is equal to the usual classical electromagnetic one, $\mathcal{L}_0 = -\epsilon_0 \mathcal{F}$, plus a nonlinear quantum correction, $\delta\mathcal{L}$, of the form

$$\mathcal{L} = \mathcal{L}_0 + \delta\mathcal{L}(\mathcal{F}, \mathcal{G}), \tag{11.41}$$

which depend on two quantities, $\mathcal{F}$ and $\mathcal{G}$. These are Lorentz invariants, defined by

$$\mathcal{F} = \frac{1}{4} F_{\mu\nu} F^{\mu\nu} = \frac{1}{2}(c^2 B^2 - E^2), \quad \mathcal{G} = \frac{1}{4} F_{\mu\nu} \tilde{F}^{\mu\nu} = c(\mathbf{E} \cdot \mathbf{B}). \tag{11.42}$$

Here, $F^{\mu\nu}$ is the electromagnetic field tensor, $\tilde{F}^{\mu\nu} = (1/2)\epsilon^{\mu\nu\alpha\beta} F_{\alpha\beta}$ its dual, and $\epsilon^{\mu\nu\alpha\beta}$ is the antisymmetric unit tensor, which changes sign under the interchange of any pair of indices. The nonlinear quantum correction $\delta\mathcal{L} \equiv \delta\mathcal{L}(\mathcal{F}, \mathcal{G})$ is determined by the expression [9, 15]

$$\delta\mathcal{L} = -\frac{1}{8\pi^2} \int_0^\infty e^{-m_e^2 s} \left[\frac{e^2 abs^2}{\tan(eas)\tanh(ebs)} + \frac{e^2 s^2}{3}(a^2 - b^2) - 1 \right] \frac{ds}{s^3}. \tag{11.43}$$

This is the celebrated *Euler–Heisenberg Lagrangian*, which results from the Dirac action [5, 9, 11]. The quantities a and b are defined as

$$a = \left(\sqrt{\mathcal{F}^2 + \mathcal{G}^2} - \mathcal{F} \right)^{1/2}, \quad b = \left(\sqrt{\mathcal{F}^2 + \mathcal{G}^2} + \mathcal{F} \right)^{1/2}. \tag{11.44}$$

This Lagrangian is valid for fields that are constant over a distance larger than the Compton wavelength, λ_C, and over a time-scale longer than the Compton time $\tau_C = \lambda_C / c$, where we have

$$\lambda_C = \frac{h}{m_e c} \simeq 3.8 \times 10^{-13} m, \quad \tau_C = \frac{\lambda_C}{c} s \simeq 1.3 \times 10^{-21} s. \tag{11.45}$$

Given the extremely small values of these Compton quantities, the Euler–Heisenberg Lagrangian stays valid for most laser and XUV experiments performed nowadays. It can also be seen that the quantities appearing in equation (11.43) are such that

$$a^2 - b^2 = -2\mathcal{F}, \quad ab = \sqrt{\mathcal{G}^2}. \tag{11.46}$$

It is important to note that the nonlinear correction $\delta\mathcal{L}$ is complex, and that its imaginary part reveals the possible occurrence of particle-pair creation due to the presence of fields in vacuum. These fields are represented by the two quantities (a, b), or by the invariants $(\mathcal{F}, \mathcal{G})$. The persistence amplitude probability of the vacuum state is determined by

$$\langle 0 \mid 0 \rangle_{(a,b)} = \exp\left\{ i \int d^4x \, \delta\mathcal{L}(a, b) \right\}. \tag{11.47}$$

Therefore, the probability for vacuum decay can be written as

$$P(a, b) = 1 - \left| \langle 0 \mid 0 \rangle_{(a,b)} \right|^2 = 1 - \exp(-w), \tag{11.48}$$

where

$$w = 2\, \Im[\delta \mathcal{L}(a, b)], \tag{11.49}$$

is the vacuum decay rate. Of course, for $w \ll 1$, we have $P(a, b) \simeq w$. By definition, the imaginary part of the nonlinear Lagrangian satisfies the relation

$$\Im[\delta \mathcal{L}(a, b)] = \frac{1}{2i}[\delta \mathcal{L}(a, b) - \delta \mathcal{L}^*(a, b)]. \tag{11.50}$$

Looking at the expression of this Lagrangian in equation (11.43), we notice that its integrand is real, and that its imaginary part can only result from the contribution of the poles. These poles are associated with the tangent in the denominator. Let us then focus on the quantity

$$\frac{eas}{\tan(eas)} = \frac{eas}{\sin(eas)} \cos(eas). \tag{11.51}$$

The poles are located at the points $(eas) = n\pi$, where n is an integer. Using the Cauchy theorem, and summing over all the poles, it is then possible to obtain

$$\Im[\delta \mathcal{L}(a, b)] = \frac{(ea)^2}{8\pi^3} \sum_{n=1}^{\infty} \frac{1}{n^2} \frac{(bn\pi/a)}{\tanh(bn\pi/a)} \exp\left(-n\pi \frac{m_e^2}{ea}\right). \tag{11.52}$$

This quantity is proportional to a^2, and therefore only non-zero for a non-zero field, such that $a \neq 0$. In order to illustrate the importance of this result, let us consider the special case where the magnetic field is absent $\mathbf{B} = 0$ and we are reduced to a static electric field, $a = |\mathcal{E}|$ and $b = 0$. Noting that $(z/\tanh z) \to 1$, when $z \equiv (bn\pi/a) \to 0$, we can write the probability w for vacuum decay, equation (11.49), as

$$w = \frac{(e\mathcal{E})^2}{8\pi^3} \sum_{n=1}^{\infty} \frac{1}{n^2} \exp\left(-n\pi \frac{m_e^2}{e|\mathcal{E}|}\right). \tag{11.53}$$

This is the celebrated result for particle-pair creation in a constant electric field $\mathcal{E}$, usually known as the *Schwinger formula*. In the exponent of this expression, we recognise the so-called *critical field*, defined as $\mathcal{E}_c = m_e^2/e$, or in explicit units, as

$$\mathcal{E}_c = \frac{m_e^2 c^3}{e\hbar} = 1.3 \times 10^{18}\ V/m. \tag{11.54}$$

Notice that this is a non-perturbative result, because we cannot expand equation (11.53) in power series of the electric field $\mathcal{E}$.

References

[1] Klein O 1926 Quantentheorie und fünfdimensionale Relativitätstheorie *Z Phys.* **37** 895–906

[2] Gordon W 1926 Der comptoneffekt nach der Schrödingerschen theorie *Z Phys.* **40** 117–33

[3] Dirac P A M 1928 Quantum theory of the electron *Proc. R. Soc. A* **117** 610–24

[4] Feshbach H and Villars F 1958 Elementary relativistic wave mechanics of spin 0 and spin 1/2 *Rev. Mod. Phys.* **30** 24–45

[5] Itzykson C and Zuber J B 1980 *Quantum Field Theory* (Montreal: McGraw-Hill)

[6] Wolkow D M 1935 Über eine Klasse von Lösungen der Diracschen Gleichung *Z Phys.* **94** 250–60

[7] Mendonça J T and Serbeto A 2011 Volkov solutions for relativistic quantum plasmas *Phys. Rev.* E **83** 026406

[8] Raicher E and Eliezer S 2013 Analytical solutions of the Dirac and Klein-Gordon equations in plasma induced by high-intensity laser *Phys. Rev.* A **88** 022113

[9] Heisenberg W and Euler H 1936 Folgerungen aus der Diracschen Theorie des Positrons *Z Phys.* **98** 714–32

[10] Dunne G V 2012 The Heisenberg-Euler effective action: 75 years on *Int. J. Mod. Phys.* A **27** 1260004

[11] Schwinger J 1951 On gauge invariance and vacuum polarization *Phys. Rev.* **82** 664–79

[12] Quéré F and Vincenti H 2021 Reflectong petawatt laser off relativistic plasma mirrors: a realistic path to the Schwinger limit *High Power Laser Sci. Eng.* **9** E6

[13] Luo W, Liu W-Y, Yuan T, Chen M, Yu J-Y, Li F-Y, Del Sorbo D, Ridgers C P and Sheng Z-M 2018 QED cascade saturation in extreme high fields *Sci. Rep.* **8** 8400

[14] Efimenko E S, Bashinov A V, Bastrakov S I, Gonoskov A A, Muraviev A A, Meyerov I B, Kim A V and Sergeev A M 2018 Extreme plasma states in laser-governed vacuum breakdown *Sci. Rep.* **8** 2329

[15] Jentschura U D, Gies H, Valluri S R, Lamm D R and Weniger E J 2002 QED effective action revisited *Can. J. Phys.* **80** 267–84

Chapter 12

Particle pair creation

In this chapter we discuss the creation of matter, more particularly of electron–positron (e-p) pairs, from intense electric fields in vacuum. The processes can be useful to study quantum vacuum effects with ultra-intense laser pulses. We first consider the historical Klein paradox, which is associated with the interaction of a charged particle with a scalar potential step. We also use canonical quantisation of the Dirac field to explore the temporal equivalent to the Klein model. This is based on step-wise vector potential discontinuity in time, in contrast with the traditional Klein model based on a scalar potential space discontinuity. We then extend the model by introducing a time-scale for the potential variation. This allows to study the transition from a singular electric field spike, with infinitesimal duration, to the other extreme case of a constant field with infinite duration, typical of the Schwinger configuration. Our results are intrinsically non-perturbative. Explicit expressions for pair creation as a function of the potential time-scales are given. Finally, we discuss the processes of e-p pair emission from photons, and not from static or quasi-static fields.

12.1 Klein paradox

The Klein paradox is one of the oldest problems associated with the Dirac equation, and can be explained by the creation of electron–positron (e-p) pairs at a potential boundary [1–3]. The original model is based on a sharp discontinuity of the scalar potential, which creates a local electric field responsible for particle pair creation. This model has been explored in many different ways along the years, not only in quantum vacuum [4–6] (figure 12.1). In order to illustrate the Klein paradox, we consider a particle of mass m and charge e, with momentum p and energy $E = \sqrt{p^2 + m^2}$, moving in the empty region $z < 0$, where no fields are present, and interacting with a potential step located at $z = 0$, as described by the $V(z) = V_0 H(z)$ where (z) is the Heaviside function. The corresponding electrostatic potential is $\phi_0 = V_0/e$. We assume an incident electron state assumed in a well-defined spin state ($s = \uparrow$), as described by

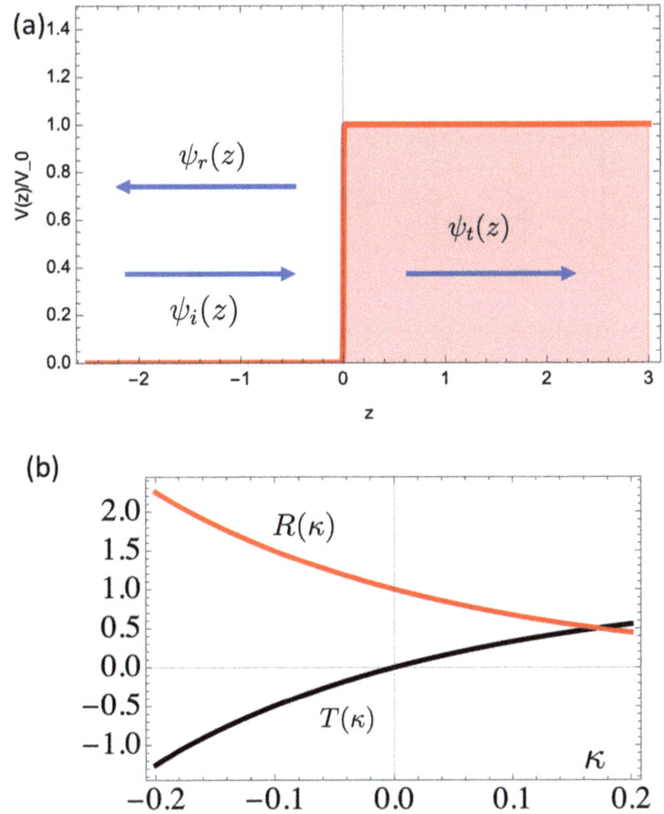

Figure 12.1. The Klein paradox: (a) incident particle with momentum p and energy $E = \sqrt{p^2 + m^2}$, interacts with a step potential V_0 at $z = 0$. (b) Reflection and transmission coefficients $R(\kappa)$ and $T(\kappa)$ show that, for large potentials, the transmission coefficient can be negative.

$$\psi_{\uparrow \mathbf{p}} \equiv \psi_i(z) = \sqrt{\alpha}\begin{bmatrix} 1 \\ 0 \\ \dfrac{p}{E+m} \\ 0 \end{bmatrix}\exp(ipz), \tag{12.1}$$

where $\alpha = (E + m)/p$ is the normalisation factor. With generality, we can assume that the reflected particle is described by a superposition of up and down spin states, as described for $z < 0$ by the spinor

$$\psi_r(z) = r_\uparrow\sqrt{\alpha}\begin{bmatrix} 1 \\ 0 \\ \dfrac{-p}{E+m} \\ 0 \end{bmatrix}e^{-ipz} + r_\downarrow\sqrt{\alpha}\begin{bmatrix} 0 \\ 1 \\ \dfrac{p}{E+m} \\ 0 \end{bmatrix}\exp(-ipz). \tag{12.2}$$

These are states corresponding to positive energy solutions. Similarly, the transmitted positive energy states are described, in the region $z > 0$ where the potential V is nonzero, by the spinors

$$\psi_t(z) = t_\uparrow\sqrt{\alpha'}\begin{bmatrix} 1 \\ 0 \\ \dfrac{p'}{E - V_0 + m} \\ 0 \end{bmatrix} e^{ip'z} + t_\downarrow\sqrt{\alpha'}\begin{bmatrix} 0 \\ 1 \\ \dfrac{-p'}{E - V_0 + m} \\ 0 \end{bmatrix} \exp(-ip'z), \tag{12.3}$$

where the new normalisation factor is $\alpha' = (E - V_0 + m)/|p'|$. In this region, the momentum associated with the incident energy E, is given by $p' = [(E - V_0)^2 - m^2]^{1/2}$. These solutions need to satisfy a continuity condition at $z = 0$, such that

$$\psi_i(0) + \psi_r(0) = \psi_t(0). \tag{12.4}$$

From here, we derive the reflection and transmission coefficients appearing in equations (12.2) and (12.3). First of all, we can immediately see that $r_\downarrow = 0$ and $t_\downarrow = 0$, which means that no spin coupling is observed. We therefore drop reference to spin in the following. From the continuity condition we also conclude that

$$\sqrt{\alpha}(1 + r) = \sqrt{\alpha'}\, t, \quad \sqrt{\alpha}(1 - r) = \sqrt{\alpha'}\frac{\alpha}{\alpha'}t. \tag{12.5}$$

This can also be written as

$$r = \frac{1 - \kappa}{1 + \kappa}, \quad t = \frac{2\sqrt{\kappa}}{1 + \kappa}, \tag{12.6}$$

where we have defined the quantity

$$\kappa = \frac{p'}{p}\frac{E + m}{E - V_0 + m}. \tag{12.7}$$

We see that, for incident particles at low energy, such that $|E - V_0| < m$, the momentum p' is imaginary. This means that the spinor wave is evanescent in the potential region $z > 0$, and transmission through the potential step is prevented, as one would expect from classical theory. However, for large potentials, such that $V_0 \geqslant E + m$, particle transmission becomes possible, even if the energy of the incident particle is smaller than the energy of the potential barrier, $E < V_0$. This paradoxical result is sometimes called *Klein tunnelling*. Using the above formulas, we can also write the transmission and reflection coefficients T and R, as

$$R(\kappa) = |r|^2 = \left(\frac{1 - \kappa}{1 + \kappa}\right)^2, \quad T(\kappa) = |t^2| = \frac{4\kappa}{(1 + \kappa)^2}, \tag{12.8}$$

from where we conclude that $R(\kappa) + T(\kappa) = 1$. We should note that, for a high potential barrier, $V_0 \geqslant E + m$, the quantity κ is negative, which means that the transmission coefficient in this case is also negative, $T(\kappa) < 0$. The reflection coefficient will be larger than one, $R(\kappa) > 1$, which is a signature of particle pair creation. As can easily be seen, the probability for pair creation is determined by the transmission coefficient, as $P_{\text{pair}} = |T(\kappa)|$. However, the case of an infinite potential step is unphysical, and a potential barrier with a finite width Δz should be considered instead [3].

12.2 Temporal Klein model

We now turn to the temporal problem, where the spatial step of a scalar potential is replaced by the temporal step of a vector potential [7]. In the previous model, an electric field located at the boundary $z = 0$ and with infinite duration was created by a scalar potential discontinuity described by $V(z) = V_0 H(z)$, and $\mathbf{A} = 0$ was assumed. Now, we consider an electric field with infinitesimal duration, located at $t = 0$, and uniformly distributed in space, as described by $\mathbf{A}(t) = \mathbf{A}_0 H(t)$, and $V = 0$. This is very similar to time-refraction, already discussed before. In the case of quantum optics, a temporal discontinuity of the refractive index would lead to the emission of photon-pairs. Here, a temporal discontinuity of the vector potential, will lead to the emission of e-p pairs (figure 12.2).

In a constant potential, the positive and negative energy solutions $\psi^{\pm}(x)$, of the Dirac equation (11.18) are

$$\psi^{+}(x) = u(\mathbf{p})\exp(ip_{\mu}x^{\mu}), \quad \psi^{-}(x) = v(-\mathbf{p})\exp(-ip_{\mu}x^{\mu}), \tag{12.9}$$

where $p^{\mu} = (E, \mathbf{p})$ is the 4-momentum. For simplicity, we use natural units, $\hbar = 1$ and $c = 1$. We know that the energy E and momentum $\mathbf{p}$ satisfy a relation of the form

$$E = \sqrt{[\bar{\sigma} \cdot (\mathbf{p} - e\mathbf{A})]^2 + m^2}. \tag{12.10}$$

We assume that $\mathbf{A} = A\mathbf{e}_3$, in the z-direction. In this configuration, the transverse momentum $p_{\perp}$ can be integrated in the mass and therefore ignored. Quantisation of the Dirac field leads to

$$\psi(x) = \sum_{s,\mathbf{p}} \left[b_{s\mathbf{p}} u_{s\mathbf{p}} e^{-iE_{\mathbf{p}}t} + d_{s\mathbf{p}}^{\dagger} v_{s\mathbf{p}} e^{+iE_{\mathbf{p}}t} \right] e^{i\mathbf{p}\cdot\mathbf{r}/\hbar}, \tag{12.11}$$

as shown in the previous chapter. The corresponding energy states $E_{\mathbf{p}}$ are defined in accordance with equation (12.9), and the index s defines the up ($s = \uparrow$) and down ($s = \downarrow$) spin states. The explicit expression for the spinors in this field operator, are given by equations (11.40), with π replaced by $k = (p_z - eA)$, and $E = \sqrt{k^2 + m^2}$.

The Dirac equation remains valid for all times and, in the presence of a potential discontinuity at $t = 0$, this implies the existence of continuity relations such that

$$\psi(x)|_{t=0} \equiv \psi(\mathbf{r}, t = 0^{-}) = \tilde{\psi}(\mathbf{r}, t = 0^{+}). \tag{12.12}$$

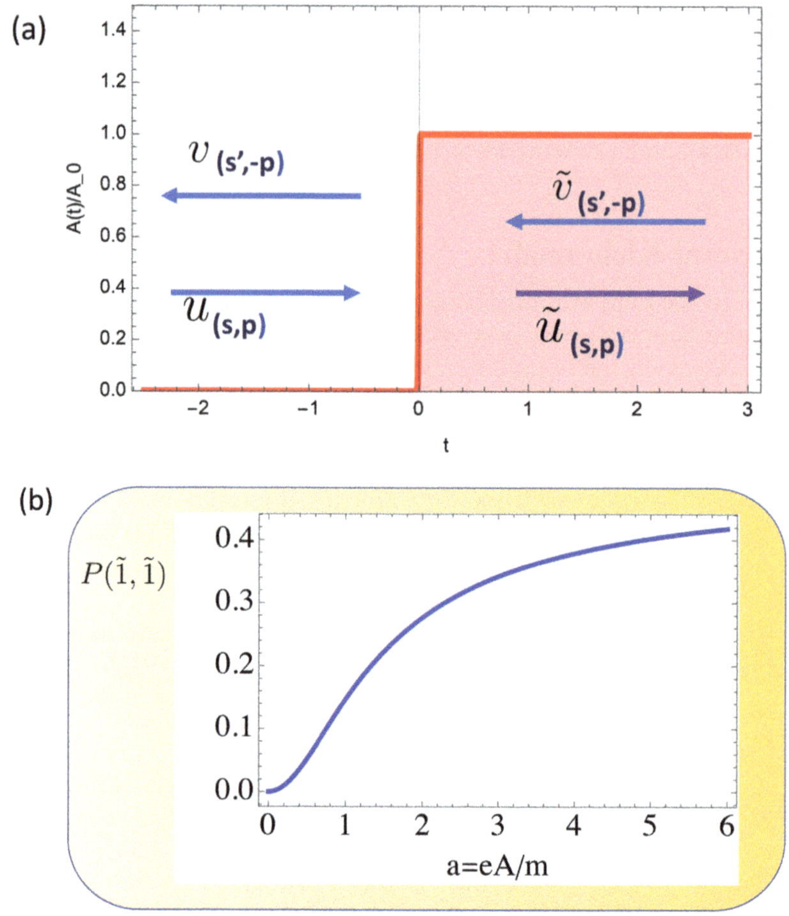

Figure 12.2. Temporal Klein model: (a) initial and final quantum states, for a vector potential discontinuity at time $t = 0$; (b) probability for particle pair creation $P(\tilde{1}, \tilde{1})$ as a function of the potential amplitude $a = eA/m$, as given by equation (12.22).

Using equation (12.11), this can be written more explicitly as

$$\sum_{s,\mathbf{p}} \left[b_{s\mathbf{p}} u_{s\mathbf{p}} + d^{\dagger}_{s\mathbf{p}} v_{s\mathbf{p}} \right] e^{i\mathbf{p}\cdot\mathbf{r}/\hbar} = \sum_{s,\mathbf{p}} \left[\tilde{b}_{s\mathbf{p}} \tilde{u}_{s\mathbf{p}} + \tilde{d}^{\dagger}_{s\mathbf{p}} \tilde{v}_{s\mathbf{p}} \right] e^{i\mathbf{p}\cdot\mathbf{r}/\hbar}, \tag{12.13}$$

where the new quantities are valid for $t \geqslant 0$. The momentum $\mathbf{p}$ remains unchanged, but the energy changes from E to $\tilde{E}$. Following the procedures of non-perturbative field theory [7, 8], we can establish a *Bogoliubov transformation*, between the old ($t = 0^-$) and the new ($t = 0^+$) pairs of operators $(b, d^{\dagger})$ and $(\tilde{b}, \tilde{d}^{\dagger})$. Simplifying the notation, this takes the form

$$\tilde{b} = Ab + B\tilde{d}^{\dagger}, \quad \tilde{d}^{\dagger} = -Bb + Ad^{\dagger}, \tag{12.14}$$

where the coefficients of the transformation satisfy the fermionic relation $|A|^2 + |B|^2 = 1$. Similarly, we get, for the spinors

$$\tilde{u} = Au + Bv, \quad \tilde{v} = -Bu + Av. \tag{12.15}$$

Writing the spinors (u, v) and $(\tilde{u}, \tilde{v})$ in normalised form, we then get

$$A = \sqrt{\alpha\tilde{\alpha}}\left[1 + \frac{k\tilde{k}}{(E + m)(\tilde{E}+m)}\right], \quad B = \sqrt{\alpha\tilde{\alpha}}\left[\frac{\tilde{k}}{(\tilde{E}+m)} - \frac{k}{(E + m)}\right], \tag{12.16}$$

with the normalisation factors α and $\tilde{\alpha}$ given by

$$\sqrt{\alpha} = \frac{E + m}{\sqrt{k^2 + (E + m)^2}}, \quad \sqrt{\tilde{\alpha}} = \frac{\tilde{E}+m}{\sqrt{\tilde{k}^2 + (\tilde{E}+m)^2}}. \tag{12.17}$$

The corresponding expressions for the energy are $E = \sqrt{k^2 + m^2}$, and $\tilde{E} = \sqrt{\tilde{k}^2 + m^2}$, where $k = \sqrt{p_\perp^2 + p_z^2}$ and $\tilde{k} = \sqrt{p_\perp^2 + (p_z - eA)^2}$. Let us now consider the transformation of vacuum states. We can define the unperturbed vacuum as a superposition of e-p pair states $|\, 0, 0\rangle$. This particle pair corresponds to an electron with spin and momentum $(s, \mathbf{p})$, and positron with spin and momentum $(s' \neq s, -\mathbf{p})$. The occurrence of a sudden potential transition, at $t = 0$, transforms the initial pure vacuum into a condensate vacuum made of e-p pairs, such that

$$|\, 0, 0\rangle = \frac{e^{i\varphi}}{\sqrt{1 + (B/A)^2}}\left(|\, \tilde{0}, \tilde{0}\rangle + \frac{B}{A}\, |\, \tilde{1}, \tilde{1}\rangle\right), \tag{12.18}$$

where φ is an arbitrary phase. The probability for e-p pair creation at $t = 0$, is then be given by

$$P(\tilde{1}, \tilde{1}) = |\, \langle 0{,}0\, |\, \tilde{1}, \tilde{1}\rangle\, |^2 = B^2. \tag{12.19}$$

Using equations (12.16) and (12.17), we then get

$$P(\tilde{1}, \tilde{1}) = \alpha\tilde{\alpha}\left(\frac{\tilde{k}}{\tilde{E}+m} - \frac{k}{E + m}\right)^2. \tag{12.20}$$

For particles created with zero momentum, we have $(k \simeq 0, E \simeq m)$ and

$$\tilde{k} = am, \quad \tilde{E} = m\sqrt{1 + a^2}, \tag{12.21}$$

where $a = eA/m$ is the normalised potential. The probability for e-p pair creation becomes

$$P_p(1) = \frac{a^2}{a^2 + \left(1 + \sqrt{1 + a^2}\right)^2}, \tag{12.22}$$

where $a = eA/m$ represents the potential amplitude. This result is illustrated in figure 12.2(b). For each spin state s, this probability is of the order of 1/2, for moderately high potential amplitudes, $a \geqslant 2$, which means that it is nearly equal to 1 for the two available spin states. But, as noted before, an absolutely sharp potential step is unrealistic and, for more practical situations with a finite rise time, the probability for electron–positron pair creation for each quantum state is strongly reduced, as shown next.

12.3 Time-varying fields

Let us now consider a more general situation where the potentials, and therefore the fields, can vary with time. The case of oscillating fields was first considered by [9] and, in more recent years, time-varying field configurations were studied by several authors [10–13]. Space-time variation has also been approached [14].

An arbitrary temporal variation of the vector potential $\mathbf{A}(t)$, can be described by an infinite number of infinitesimal temporal Klein steps. The field operators valid before and after each of these elementary steps are related by discrete Bogoliubov transformation, as defined in equation (12.14). Summing over all these steps, we arrive at a temporal Bogoliubov transformation between the initial operators $b(0)$, $d^\dagger(0)$ and those valid at an arbitrary time t, of the form

$$b(t) = \mathcal{A}(t)b(0) + \mathcal{B}(t)d^\dagger(0), \quad d^\dagger(t) = \mathcal{A}(t)d^\dagger(0) - \mathcal{B}(t)b(0), \tag{12.23}$$

where the coefficients of the transformation are given by the expressions

$$\mathcal{A}(t) = \cos[r(t)], \quad \mathcal{B}(t) = \sin[r(t)]. \tag{12.24}$$

The quantity $r(t)$, which defines the rate of the transformation and can be called the squeezing parameter, is defined as

$$r(t) = \int_{t_0}^{t} R(t')\exp\left[2i\varphi_p(t')\right]dt', \tag{12.25}$$

where the amplitude $R(t)$ is determined by

$$R(t) = \frac{1}{2E}\sqrt{2E(E+m)}\left[\frac{1}{E+m} - \frac{k^2}{E(E+m)^2}\right]\partial_t k, \tag{12.26}$$

and the phase function is given by the time integral $\varphi(t) = \int^t E(t')dt'$. The energy depends on the potential, according to $E = \sqrt{k^2 + m^2}$, with $k = [p_z - eA(t)]$, as we have seen. Notice that the new transformation still obeys the fermionic condition $|A(t)|^2 + |B(t)|^2 = 1$, valid at all times. The probability for pair particle creation can then be written as

$$P(t) = |\,\mathcal{B}(t)|^2 = \sin^2[r(t)|]. \tag{12.27}$$

For small values of $|r_p(t)| \ll 1$, this probability becomes

$$P(t) = \left| \int_0^t R(t')\exp\left[2i\varphi(t')\right]dt' \right|^2. \tag{12.28}$$

This is valid for arbitrary-time-varying potentials, and can also be stated as

$$P(t) = \left| \int_0^t dt' \int d\omega\ R(\omega)\exp[-i\omega t' + 2i\varphi(t')] \right|^2, \tag{12.29}$$

where $R(\omega)$ is the Fourier transform of the function $R(t)$. Noting that, for particles at rest, the phase function behaves as $\varphi(t) \sim mt$, we can see from this expression that particle pair creation will be mainly due to the Fourier components $\omega \simeq 2m$. For instance, in a standing wave oscillating at frequency ω_0, this would give $\omega_0 \simeq m$. It means that two counter-propagating photons with frequency (energy) equal to the rest mass (rest energy) of the particles will be able to create pairs. Furthermore, the function $R(t)$ varies nonlinearly with the applied potential. This also shows that multiphoton pair creation can occur for $n\omega \simeq 2m$, where integer $n \gg 1$. Let us now focus on a particular example, which is a straightforward generalisation of the temporal Klein model, and is described by

$$\mathbf{A}(t) = \mathbf{A}_0 \frac{1}{2}\left[1 + \tanh\left(\frac{t}{\tau}\right)\right]. \tag{12.30}$$

In this model, an electric field with duration τ is applied to vacuum and, if the potential is uniform, no magnetic field is present

$$\mathcal{E}(t) = -\frac{\partial}{\partial t}\mathbf{A}(t), \qquad \mathbf{B} = \nabla \times \mathbf{A} = 0. \tag{12.31}$$

This example is able to connect two important and apparently unrelated cases. One is the case of infinitesimal duration, $\tau \to 0$, which is nothing but the temporal Klein configuration already discussed in the previous section. The other is the case of infinite duration, $\tau \to \infty$, which corresponds to a (nearly) constant electric field, and coincides with the Schwinger–Sauter configuration, already considered in the previous chapter. For $t \geqslant \tau$, the probability (12.28) can be transformed into the following expression

$$P(a) \simeq F(a)^2 \exp\left[-\pi\frac{\mathcal{E}_c}{\mathcal{E}}\right], \tag{12.32}$$

where $\mathcal{E}_c = m^2/e$ is the Schwinger field, $a = eA_0/m = 2e\tau\mathcal{E}/m$ is the reduced potential amplitude, and $\mathcal{E}$ is the maximum electric field as given by equation (12.31), with the auxiliary function

$$F(a) = \frac{1}{2}\int_{-a}^{a} \frac{d\alpha}{\sqrt{1+\alpha^2}\left[\alpha^2 + \left(1 + \sqrt{1+\alpha^2}\right)^2\right]^{1/2}}, \tag{12.33}$$

and the variable $\alpha = a\ t/\tau$ was used. This approximate result shows that the temporal Klein model with a finite temporal width contains the exponential Schwinger factor, multiplied by a form function that depends on the field duration τ. The function $F(a)$ is formally similar to that obtained in the temporal Klein model

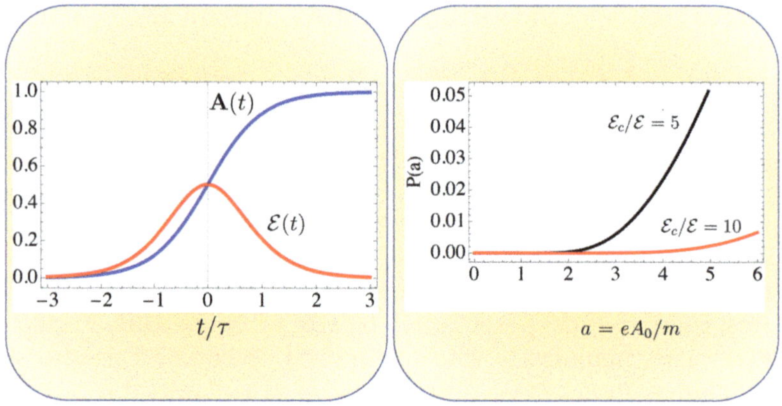

Figure 12.3. Temporal Klein model with a finite duration, τ: (a) potential and field configurations; (b) probability $P(a)$ for $t \geqslant \tau$, given by equation (12.32), for $\mathcal{E}_c/\mathcal{E} = 10$ (in red), and $\mathcal{E}_c/\mathcal{E} = 5$ (in blue).

of the previous section. But, for $\mathcal{E} \ll \mathcal{E}_c$, the Schwinger exponential factor is always dominant and considerably reduces the probability for pair creation. See figure 12.3 for an illustration.

12.4 Nonlinear trident process

Let us now discuss single photon processes leading to pair production. In 1936, Breit and Wheeler were able to demonstrate theoretically that two gamma-ray photons could produce an e-p pair, due to the nonlinear properties of vacuum, if their energy was larger than the rest energy of the electron, $\hbar\omega_\gamma \geqslant m_e c^2$ [15]. Obviously, the photons of a typical laser are not sufficiently energetic to participate in such process. However, given the large number of low energy photons existing in a laser beam, we can imagine a multiphoton version of the *Breit–Wheeler process*, fulfilling the energy gap between the rest energy of the created particles and the energy of two individual photons. This was considered in the early days of laser developments, by Reiss [16] and by Nikishov and Ritus [17].

With the advent of high-intensity laser beams, it became clear that γ ray photons can be emitted by energetic electrons created by laser–matter interaction processes. Bremsstrahlung emission, or nonlinear Compton scattering of laser photons by electron energy tails, then lead to the formation of gamma-ray photons, which are added to the original low energy photons. The modern nonlinear version of the Breit–Wheeler (BW) process includes two steps, one where a gamma-ray photon is emitted by an energetic electron, and the other where the gamma-ray interacts with several low energy photons. These two steps can be represented as:

$$e_{-} + \omega \rightarrow e^{-} + \gamma, \quad \gamma + N\omega \rightarrow e^{-}e^{+}. \tag{12.34}$$

The direct BW process involving just two high energy photons has not been observed so far. But its modern, multiphoton and nonlinear, version was observed in 1997 at SLAC [18], when an electron beam with energy 46.6GeV, produced by a linear

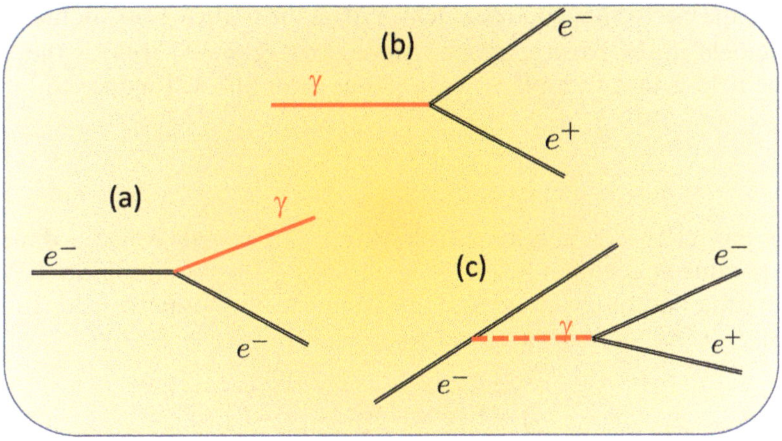

Figure 12.4. Basic mechanisms of e-p pair creation by intense lasers beams: (a) photon spontaneous emission; (b) photon decay into e-p pairs; (c) trident process.

accelerator, was made to collide with a tera-watt laser pulse in the visible, at a wavelength of 527 nm. This sequence of events, where first one high energy photon is emitted and then mixes with several low energy photons to produce pairs, can be seen as a kind of *trident process*, by with an electron in the presence of the intense laser field decays into another electron and an e-p pair, as illustrated in figure 12.4.

Energy conservation requires that the total energy of the interacting photons will be larger than the rest energy of the created particle pair. Noting that the rest mass is modified by the background laser field, energy conservation will depend on the laser amplitude, as

$$\hbar\omega_\gamma + N\omega_0 \geqslant 2mc^2\sqrt{1 + a^2}. \tag{12.35}$$

Here, ω_γ and ω_0 are the frequencies of high and low energy photons, N is the number of low energy photons, and $a = e\mathcal{E}_0/\omega_0 mc$ is the dimensionless laser field amplitude. In the original Breit–Wheeler process we would have $\omega_\gamma = \omega_0$, $N = 1$, and renormalisation of the electron mass would be ignored, $a = 0$. It is obvious that the minimum number of photons N satisfying energy conservation grows with the laser amplitude. For a process involving N low energy photons, the probability for pair creation would be proportional to a^{2N}. In the SLAC experiment, the minimum value would be $N = 6$. However, the observed scaling corresponds to a^{10}, which indicates that a non-perturbative behaviour is observed, even at low intensities, $a \simeq 0.3$ [19].

The interaction of charged particles and fields can be characterised by a relativistic invariant, which is defined in dimensionless form by

$$\chi_e = \frac{E_{\rm pr}}{E_s} = \frac{1}{mcE_S}\sqrt{-\left(F^{\mu\nu}p_\nu\right)^2}, \tag{12.36}$$

where $F^{\mu\nu}$ is the electromagnetic field tensor, p_ν is the particle four-momentum, $E_{\rm pr}$ is the electric field in the particle proper frame, and $E_S = m^2c^3/(e\hbar)$ is the Schwinger field. If the field is that of a plane wave in vacuum, this will reduce to

$$\chi_e = \frac{E}{E_S}\left(\gamma - \frac{p_\parallel}{mc}\right), \tag{12.37}$$

where γ is the relativistic gamma factor. Similarly, we can define a dimensionless quantum parameter for photons. This is important for the nonlinear BW process, where a gamma photon propagates in the presence of an intense field. In that case, we replace p_ν by $\hbar k_\nu \equiv \hbar(\omega_\gamma, \mathbf{k}_\gamma)$, and we get

$$\chi_\gamma = \frac{\hbar\omega_\gamma}{mc^2E_S}\sqrt{-(\mathbf{E} + \mathbf{v}_\gamma \cdot \mathbf{B})^2 - (\mathbf{v}_\gamma \cdot \mathbf{E})^2/c^2}, \tag{12.38}$$

where $\mathbf{v}_\gamma = \mathbf{k}_\gamma/\omega_\gamma$. In the case of a plane wave, this would reduce to

$$\chi_\gamma = \frac{E}{E_S}(\omega_\gamma - k_\parallel c)\frac{\hbar}{mc^2}. \tag{12.39}$$

We can see that for parallel propagation, we have $\chi_\gamma = 0$, while for antiparallel propagation we have a maximum

$$\chi_{\rm max} \simeq \frac{2E}{E_S}\frac{\hbar\omega_\gamma}{mc^2}. \tag{12.40}$$

The probability for pair production by the BW process can be given in terms of the parameter χ_γ, by

$$W_{\rm BW} = \frac{2\alpha^2 c}{3r_e}\frac{mc^2}{\hbar\omega_\gamma}b(\chi_\gamma), \tag{12.41}$$

where α is the fine-structure constant, $r_e = e^2/(4\pi\epsilon_0 mc^2)$ is the *classical electron radius*, and $b(\chi_\gamma)$ is a complicated function of the quantum parameter χ_γ (see [20, 21] for details). In the quasi-classical limit of $\chi_\gamma \ll 1$ this function is approximately given by

$$b(\chi_\gamma) \simeq 0.34\chi_\gamma \exp\left(\frac{8}{3\chi_\gamma}\right), \tag{12.42}$$

and, in the quantum regime of $\chi_\gamma \gg 1$, by

$$b(\chi_\gamma) \simeq 0.57\chi_\gamma^{2/3}. \tag{12.43}$$

In both cases, the pair production rate is solely determined by the value of the quantum parameter χ_γ.

References

[1] Klein O 1929 Die Reflexion von Elektronen an einem Potentialsprung nach der relativistischen Dynamik von Dirac *Z Phys.* **53** 157–65

[2] Holstein B R 1998 Klein's paradox *Am. J. Phys* **66** 507–12

[3] Dombey N and Calogeracos A 1999 Seventy years of the Klein paradox *Phys. Rep.* **315** 41–58

[4] Umul Y Z 2019 A survey on Klein paradox *Optik* **181** 258–63

[5] Katsnelson M I, Novoselov K S and Geim A K 2006 Chiral tunnelling and the Klein paradox in graphene *Nat. Phys.* **2** 620

[6] Gerritsma R, Lanyon B P, Kirchmair G, Zähringer F, Hempel C, Casanova J, Garcia-Ripoll J J, Solano E, Blatt R and Roos C F 2011 Quantum simulation of the Klein paradox with trapped ions *Phys. Rev. Lett.* **106** 060503

[7] Mendonça J T 2021 Temporal Klein model for particle-pair creation *Symmetry* **13** 1361

[8] Umesawa H 1993 *Advanced Field Theory* (New York: AIP)

[9] Brezin E and Itzykson C 1970 Pair production in vacuum by an alternating field *Phys. Rev.* D **2** 1191–9

[10] Schmidt S, Blaschke D, Röpke G, Smolyansky S A, Prozorkevich A V and Toneev V D 1998 A quantum kinetic equation for particle production in the Schwinger mechanism *Int. J. Mod. Phys.* E **7** 709–22

[11] Tanji N 2009 Dynamical view of pair creation in uniform electric and magnetic fields *Ann. Phys., NY* **324** 1691–736

[12] Bell A R and Kirk J G 2008 Possibility of prolific pair production with high-power lasers *Phys. Rev. Lett.* **101** 200403

[13] Deffner S 2016 Shortcuts to adiabaticity: suppression of pair production in driven Dirac dynamics *New J. Phys.* **18** 012001

[14] Aleksandrov I A and Kohlfürst C 2020 Pair production in temporally and spatially oscillating fields *Phys. Rev.* D **101** 096009

[15] Breit G and Wheeler J A 1934 Collision of two light quanta *Phys. Rev.* **46** 1087–91

[16] Reiss H R 1962 Absorption of light by light *J. Math. Phys.* **3** 59–67

[17] Nikishov A I and Ritus V I 1964 Quantum processes in the field of a plane electromagnetic wave and in a constant field *Sov. Phys. JETP* **19** 529–41

[18] Burke D *et al* 1997 Positron production in multiphoton light-by-light scattering *Phys. Rev. Lett.* **79** 1626–9

[19] Hu H, Müller C and Keitel C H 2010 Complete QED theory of multiphoton trident pair production in strong laser fields *Phys. Rev. Lett.* **105** 080401

[20] Erber T 1966 High-energy electromagnetic conversion processes in intense magnetic fields *Rev. Mod. Phys.* **38** 626–59

[21] Mercuri-Baron A, Grech M, Niel F, Grassi A, Lobet M, Di Piazza A and Riconda C 2021 Impact of the laser spatio-temporal shape on Breit-Wheeler pair production *New J. Phys.* **23** 085006

Chapter 13

Nonlinear vacuum

The study of vacuum effects by intense laser pulses became very popular in recent years [1–4]. Not only are laser systems with increasing power being developed in many laboratories around the world, but also intense and ultra-short electric field spikes can be created by an accumulation of high-harmonics, using the existing laser facilities [5, 6]. In parallel, new experimental configurations have been proposed. For instance, it was shown recently that such high-harmonic pulses can excite super-radiant emission of photons in quantum vacuum [7].

An interesting aspect of quantum electrodynamics (QED) vacuum is the emergence of two distinct vacuum properties. They are both described by the same theoretical approach, based on the Dirac equation and the resulting Lagrangian structure. These two properties are: first, the possible emission of particle-antiparticle pairs by electromagnetic fields. These fields can be static, like in the Schwinger–Sauter process, or dynamical, as those associated with laser beams. This means that, although indirectly, photon beams can emit electron–positron (e-p) pairs. But, a single photon or a plane electromagnetic wave cannot emit these particle pairs. Particle emission processes were described in the previous chapter.

Second, the quantum vacuum is a birefringent and nonlinear optical medium. The property of birefringence is manifested in the presence of a static electric or magnetic field and can, in principle, be demonstrated by polarimetric techniques [8]. Furthermore, photon coupling with static fields, can be observed as photon scattering. This is the case of *Delbrück scattering* [9], which is associated with scattering of gamma rays by the Coulomb field of heavy nuclei, and first observed in 1933. A related process is photon splitting in a static electric field, first observed in 2002 [10]. This is the decay of a photon into two different photons, mediated by the static field. Photon splitting in a plane wave, such as an intense laser beam, can also be conceived [11], but not has yet been observed. Photon scattering and photon splitting processes are illustrated in figure 13.1.

doi:10.1088/978-0-7503-2786-2ch13

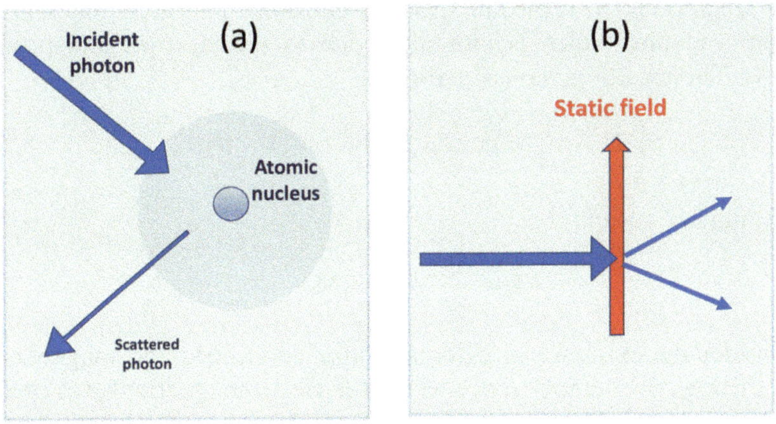

Figure 13.1. Quantum vacuum: (a) photon scattering (Delbrück scattering) in the electric field of heavy nuclei; (b) photon splitting in a static field.

Many other nonlinear optical processes can be considered in vacuum, from four-wave mixing to harmonic generation. Photon acceleration and time-refraction in a non-stationary cavity vacuum are also possible. However, these effects are extremely small, and are still awaiting for experimental proof. Due to the advent of ultra-intense laser systems, at the multi-PetaWatt level [12], this became a very active area of research in recent years, where spectacular new results can be expected [13]. This will be the object of the present chapter.

13.1 Vacuum birefringence

We start with the Heisenberg–Euler Lagrangian, as discussed in chapter 11. This approximate Lagrangian form is valid for field amplitudes much smaller than the Schwinger field E_c, and for times much shorter than the Compton time-scale, $1/\omega_C$. These limitations are not very stringent, because this approximation stays valid for the most laser fields produced nowadays in the laboratory, and for frequencies up to the x-ray domain. Furthermore, this Lagrangian only depends on the relativistic invariants $\mathcal{F}$ and $\mathcal{G}$, as we have seen. The new Maxwell's equations in vacuum, associated with that Lagrangian take the usual form if we define the displacement and induction fields, **D** and **B** as

$$\mathbf{D} = \epsilon_0 \mathbf{E} + \mathbf{P}, \quad \mathbf{B} = \mu_0(\mathbf{H} + \mathbf{M}). \tag{13.1}$$

Vacuum effects are described by the vacuum polarisation **P** and a magnetisation **M** fields, as if vacuum was some sort of material medium. Their explicit expression is

$$\mathbf{P} = 2\zeta(4\mathcal{F}\mathbf{E} + 7c\mathcal{G}\mathbf{B}), \quad \mathbf{M} = -2c^2\zeta(4\mathcal{F}\mathbf{B} + 7\mathcal{G}\mathbf{E}/c). \tag{13.2}$$

Here, we have used the nonlinear parameter

$$\zeta = 2\alpha^2 \frac{\epsilon_0^2 \hbar^3}{45 m_e^4 c^5}, \tag{13.3}$$

and $\alpha = e^2/2\epsilon_0 hc \simeq 1/137$ is the fine structure constant. This gives physical substance to quantum vacuum, which is ultimately due to the electron and positron field fluctuations. The resulting wave equation is

$$\left(\nabla^2 - \frac{1}{c^2}\partial_t^2\right)\mathbf{E} = \mu_0[\partial_t \mathbf{J} + c^2\nabla(\nabla \cdot \mathbf{P})], \tag{13.4}$$

with the nonlinear current

$$\mathbf{J} = \frac{\partial \mathbf{P}}{\partial t} + \nabla \times \mathbf{M}. \tag{13.5}$$

Let us consider the existence of externally applied electric and magnetic fields, $\mathbf{E}_e$ and $\mathbf{B}_e$ in the region of empty space where the electromagnetic waves propagate. If the fields associated with these waves are $\tilde{\mathbf{E}}$ and $\tilde{\mathbf{B}}$, and evolve in space and time as $\exp(i\mathbf{k} \cdot \mathbf{r} - i\omega t)$, where $\mathbf{k}$ is the wavevector and $\omega \simeq kc$ the frequency, we obtain a dispersion relation of the form [14]

$$\omega_\pm = kc\left(1 - \frac{1}{2}\lambda_\pm \mid \mathbf{Q} \mid^2\right). \tag{13.6}$$

Here, the vector $\mathbf{Q}$ represents the effect of the external fields and is determined by

$$\mathbf{Q} = \mathbf{n} \times \mathbf{E} + c \times (\mathbf{k} \times \mathbf{B}/k), \tag{13.7}$$

where $\mathbf{n} = \mathbf{k}/k$ is the unit vector along the direction of propagation. This describes the so-called *vacuum birefringence* [15]. The coefficients $\lambda_\pm$ corresponding to the two modes of propagation are given by $\lambda_+ = 14\zeta$ and $\lambda_- = 8\zeta$. These two modes correspond to two orthogonal polarisation states. The dispersion relation (13.6) shows that the phase and group velocities in vacuum are slightly lower than c, and are independent of the frequency. However, they depend on the direction of propagation with respect to the static fields, through the value of $|\mathbf{Q}|$. This makes the usual comparison of QED vacuum with a virtual electron–positron plasma somewhat inaccurate, because in a plasma the phase velocity would be larger than c. Notice that, if the fields in equation (13.7) were those of the wave mode itself, the unit vector $\mathbf{n}$ would be perpendicular to both $\mathbf{E}$ and $\mathbf{B}$, and the quantity $\mathbf{Q}$ would be identically zero. This shows that a plane wave cannot be nonlinearly perturbed by its own field, and a photon represented by this wave in vacuum cannot split into two photons. The same is not exactly valid for a focused beam, which necessarily contains oblique photons.

Of particular importance is the magnetised vacuum, where $\mathbf{E}_e = 0$ and $\mathbf{B}_e \equiv \mathbf{B}_0 \neq 0$ is a static magnetic field. In this case, equations (13.2) take the form

$$\mathbf{P} = -2c^2\zeta[2B_0^2\tilde{\mathbf{E}} - 7(\tilde{\mathbf{E}}\cdot\mathbf{B}_0)\mathbf{B}_0], \quad \mathbf{M} = 4c^4\zeta[B_0^2\mathbf{B} + 2(\mathbf{B} \cdot \mathbf{B}_0)\mathbf{B}_0]. \tag{13.8}$$

Noting that, for photon propagation along the direction of $\mathbf{k}$, we have $(\mathbf{k} \cdot \tilde{\mathbf{E}}) = 0$, and $\mathbf{B} = (\mathbf{k} \times \tilde{\mathbf{E}})/\omega$, we can easily get

$$k^2 = \frac{\omega^2}{c^2}[1 + \chi_v(\omega, \mathbf{k})], \tag{13.9}$$

where the susceptibility of the magnetised vacuum is given by

$$\chi_v(\omega, \mathbf{k}) = \mu_0 c^4 \zeta \left\{ 14(\mathbf{e} \cdot \mathbf{B}_0)^2 + 8\left[\mathbf{e} \cdot \left(\mathbf{B}_0 \times \frac{\mathbf{k}}{k}\right)\right]^2 \right\}, \tag{13.10}$$

with the unit vector $\mathbf{e} = \tilde{E}/|\tilde{\mathbf{E}}|$. We can also define the refractive index as $N(\omega, \mathbf{k}) = \sqrt{1 + \chi_v(\omega, \mathbf{k})} \simeq 1 + \chi_v(\omega, \mathbf{k})/2$, which depends on the photon polarisation state. For polarisation parallel and perpendicular to the static magnetic field, we obtain

$$N_{\|} = 1 + 7\mu_0 c^4 \zeta B_0^2 \sin^2 \alpha, \quad N_{\perp} = 1 + 4\mu_0 c^4 \zeta B_0^2 \sin^2 \alpha, \tag{13.11}$$

where α is the angle between the static field $\mathbf{B}_0$ and the wavevector $\mathbf{k}$. Notice that, for propagation along the static field we are reduced to $N_{\|} = N_{\perp} = 1$, which means that Faraday rotation is strictly forbidden. But, for a perpendicular field, such as that of figure 13.2, this can eventually be measured. For instance, taking a magnetic field as large as $B_0 = 10$ Tesla, this gives a magnetic birefringence of $\Delta N = N_{\|} - N_{\perp} = 3\zeta B_0^2 \simeq 4 \times 10^{-22}$ would be expected. For a propagation distance L, this corresponds to a phase shift of $\Delta\varphi = k\Delta NL$. If the wave is initially polarised linearly, so that the parallel and perpendicular components have equal amplitudes, this phase difference will generate an elliptic polarisation at the end, proportional to the propagation distance L and to the amplitude of $\mathbf{B}_0$. This is the well-known *Cotton–Mouton effect*. Given the smallness of the nonlinear vacuum parameter ζ, this effect is extremely small, but the phase difference can be amplified using an optical cavity. The amplification factor is proportional to the quality factor of the cavity, which can be very large. Until now, optical cavity experiments have been unable to detect vacuum birefringence in the laboratory [8], but some indirect evidence has already been claimed from polarimetric observation of neutron stars [16].

We now consider the case of a rotating magnetic field $\mathbf{B}_e(t) = B_0\, \mathbf{b}(t)$ with constant amplitude B_0, but with a rotating direction. In this case, the unit vector $\mathbf{b}(t)$ rotates with an angular frequency ω_0 in the plane perpendicular to the wavevector $\mathbf{k}$.

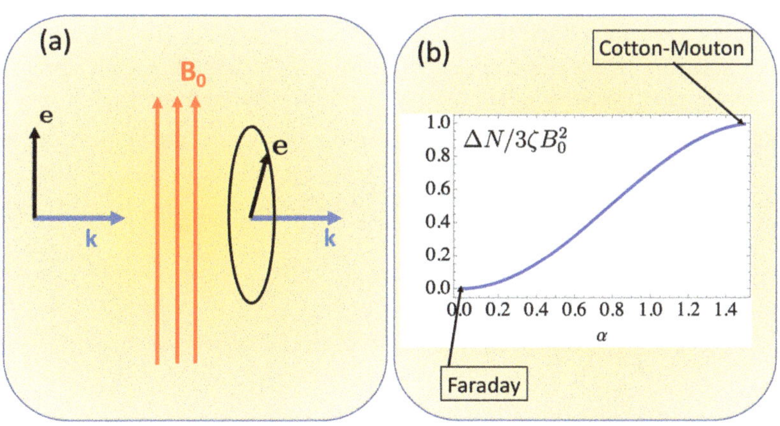

Figure 13.2. Birefringence in a magnetised vacuum: (a) Cotton–Mouton configuration; (b) normalised birefringence $\Delta N/3\zeta B_0^2$, as a function of the angle between the wavevector $\mathbf{k}$ and the static magnetic field $\mathbf{B}_0$.

In other words, it rotates in the plane of wave polarisation. Let us examine the influence of the first order terms of **P** and **M**, proportional to the rotating field amplitude, and use

$$\mathcal{F} = -c^2 B\ |B_0|\ \sin(\omega_0 t), \quad \mathcal{G}(\omega) = cE\ |B_0|\ \cos(\omega_0 t). \tag{13.12}$$

Equations (13.2) show that the magnetic field rotation will couple waves with frequency $\omega_{jn} = j\omega + n\omega_0$, where $j = 1, 2$ and $n = 0, \pm 1, \pm 2, \ldots$. We then solve the wave equation for a field of the form

$$\mathbf{E}(\mathbf{r}, t) = \sum_{j,n} \mathbf{E}(\omega_{jn})\exp(i\mathbf{k}_{jn} \cdot \mathbf{r} - i\omega_{jn} t), \tag{13.13}$$

where $\mathbf{E}(\omega_{jn})$ are slowly varying amplitudes. In this nonlinear analysis we use the linear dispersion for each mode, $k_{jn} = \omega_{jn}/c$. Assuming that the modes are polarised in the same direction $\mathbf{e} \equiv \mathbf{e}_{jn} = \mathbf{E}_{jn}/|E_{jn}|$, and using equation (13.4), we can derive evolution equations for the amplitudes, involving the different Fourier components of **P** and **M**. Assuming that the dominant mode has frequency ω, vacuum effects lead to the excitation of secondary fields with frequencies 2ω and $(\omega + \omega_0)$, described by the coupled equations

$$(\partial_t + \mathbf{v}_{11} \cdot \nabla)E(\omega + \omega_0) = -i(\omega + \omega_0)\frac{2^3 c^2}{\epsilon_0}\zeta B(2\omega)B_0 E^*(\omega)\exp(i\phi), \tag{13.14}$$

$$(\partial_t + \mathbf{v}_{20} \cdot \nabla)E(2\omega) = -i\omega\frac{2^4 c^2}{\epsilon_0}\zeta B(\omega + \omega)B_0^* E(\omega)\exp(-i\phi), \tag{13.15}$$

where the phase mismatch ϕ is determined by

$$\phi = \Delta\mathbf{k} \cdot \mathbf{r} = (\mathbf{k}_{20} - \mathbf{k}_{11} - \mathbf{k}) \cdot \mathbf{r}. \tag{13.16}$$

This phase factor results from the lack of momentum carried by the rotating (but spatially uniform) magnetic field, thus leading to a lack of total momentum in the nonlinear coupling. In principle, if we appropriately choose the angle between the different modes, we can reduce this phase mismatch to zero. It is the case of waves ω and 2ω propagating along the same z-direction, with wave $(\omega + \omega_0)$ propagating at an angle β, such that $\cos\beta = \omega/(\omega + \omega_0)$. It can easily be seen that, for a negligibly small value of ω_0/ω, the phase ϕ tends to zero as $\beta \simeq \sqrt{2\omega_0/\omega}$. For practical reasons, it is useful to consider propagation of the three interacting modes in the z-direction. The coupled mode equations (13.14)–(13.15) become

$$\begin{aligned} \frac{d}{d\tau}E(\omega + \omega_0) &= -iw'\ E(2\omega)\exp(i\phi), \\ \frac{d}{d\tau}E(2\omega) &= -iw''\ E(\omega + \omega_0)\exp(-i\phi), \end{aligned} \tag{13.17}$$

where $\tau = z - ct$, and the coupling coefficients are

$$w' = (\omega + \omega_0)\frac{2^4 c}{\epsilon_0}\zeta B_0 E^*(\omega), \quad w'' = \omega\frac{2^4 c}{\epsilon_0}\zeta B_0^* E(\omega), \tag{13.18}$$

with the phase mismatch

$$\phi = -\frac{\omega_0}{c}z = -2\pi\frac{\omega_0}{\omega}\frac{z}{\lambda}. \tag{13.19}$$

Here, λ is the wavelength of the fundamental wave mode. This phase mismatch can be neglected for relatively small propagating distances, such that $z \ll \lambda(\omega_0/\omega)$. In this case, the coupled mode equations lead to

$$\left(\frac{d^2}{d\tau^2} + \Omega^2\right) E(\omega + \omega_0) = 0, \tag{13.20}$$

where the nonlinear frequency Ω, determined by the vacuum properties, is given by

$$\Omega^2 = w' \, w'' = \frac{2^8 c^2}{\epsilon_0^2}\zeta^2 |B_0|^2 \, |E(\omega)|^2. \tag{13.21}$$

From here we obtain the following solutions,

$$E(\omega + \omega_0, \tau) = -E(2\omega, 0)\frac{\Omega}{w'}\sin(\Omega\tau), \qquad E(2\omega, \tau) = E(2\omega, 0)\cos(\Omega\tau). \tag{13.22}$$

This is compatible with the initial conditions $E(\omega + \omega_0) = 0$ for $\tau = 0$, which means initial absence of a sideband signal. Given the smallness of the QED vacuum parameter ζ, it is useful to consider the case $\Omega\tau \ll 1$, which is valid even for very large interaction distances. We then get the final sideband amplitude

$$|E(\omega + \omega_0, \tau)| = |E(2\omega, 0)|\frac{2^4 c}{\epsilon_0}\zeta|B_0||E(\omega)|\omega\tau. \tag{13.23}$$

In this derivation we have assumed that the amplitude of the fundamental photon mode $|E(\omega)|$ remained constant during the interaction process. This parametric approximation is justified for a very weak nonlinearity, which is the case here. This solution for the sideband at frequency $(\omega + \omega_0)$ implies the presence of a seed signal at the second harmonic frequency, $E(2\omega, 0) \neq 0$. Such a solution could be useful for possible test experiments using ultra-intense lasers. Alternatively, we could consider the excitation of harmonics, but using now a seed field with the sideband frequency, $E(\omega + \omega, 0) \neq 0$. The present analysis could even be extended to a more general case where sidebands $(\omega + n\omega_0)$ are coupled to the harmonic fields $(n + 1)\omega$, for $n > 1$.

13.2 Photon acceleration

Instead of static or rotating external fields, we can consider the case of external fields $\mathbf{E}_0$ and $\mathbf{B}_0$, that are associated with a strong laser pulse with frequency ω_0, propagating along some given direction $\mathbf{n}_0 = \mathbf{k}_0/k_0$. In this case, any probe photon propagating along some other direction $\mathbf{n}$, would be affected by the vacuum nonlinearity associated $\mathbf{Q}$. The photon dispersion relation would then be locally perturbed, in the region occupied by the laser beam, leading to possible reflection and refraction effects. These effects are indeed extremely small, but can be amplified in some configurations, such as those leading to photon acceleration. This effect is well-known in plasmas and optics, and is described in detail in the book [18]. In the present case, pertinent to pure vacuum, the probe photon is allowed to interact for a very long time (or over a very long distance) with the rear or the front of the intense pulse [17]. We can write the dispersion relation for the probe pulse as

$$\omega = kc\left[1 - \frac{1}{2}\epsilon_0\lambda_\pm f(\mathbf{n}, \mathbf{n}_0)|\,\mathbf{E}_0(\mathbf{r}, t)|^2\right]. \tag{13.24}$$

The value of $\lambda_\pm$ is determined by the photon polarisation state, and the geometric factor $f(\mathbf{n}, \mathbf{n}_0)$ is defined by the expression

$$f(\mathbf{n}, \mathbf{n}_0) = 2 - (\mathbf{n}\cdot\mathbf{e}_0)^2 - [\mathbf{n}\cdot(\mathbf{n}_0\cdot\mathbf{e}_0)]^2 - 2(\mathbf{n}\cdot\mathbf{n}_0), \tag{13.25}$$

and $\mathbf{e}_0 = \mathbf{E}_0/E_0$ is the unit polarisation vector of the intense pulse. In equation (13.24), the intense field amplitude vary in space and time, according to the assumed pulse shape. The simplest way to describe the test photon behaviour in is to use the ray equations which can be written in canonical form as

$$\frac{d\mathbf{r}}{dt} = \frac{\partial\omega}{\partial\mathbf{k}}, \quad \frac{d\mathbf{k}}{dt} = -\frac{\partial\omega}{\partial\mathbf{r}}, \tag{13.26}$$

where the frequency ω, is given by equation (13.24). Given its dependence on time t, we can immediately conclude that the photon frequency will vary along propagation. Let us assume a specific geometry. The intense pulse propagates along the z-axis, and the probe photon propagates at an angle θ, such that $(\mathbf{n}\cdot\mathbf{n}_0) = \cos\theta$. Neglecting effects associated with perpendicular propagation, we assume $\mathbf{k}_\perp = k\sin\theta = \text{const.}$, and describe the parallel photon momentum as $k_z = p$. The ray equations are then reduced to

$$\frac{d\eta}{dt} = \frac{\partial}{\partial p}h(\eta, p), \quad \frac{dp}{dt} = -\frac{\partial}{\partial\eta}h(\eta, p), \tag{13.27}$$

where $\eta = z - ut$, and $u \simeq c$ is the group velocity of the intense pulse, and $H(\eta, p)$ is the new Hamiltonian

$$H(\eta, p) = kc\left[1 - \frac{1}{2}\lambda^* f(\theta)I_0 g(\eta)\right] - up, \tag{13.28}$$

where I_0 is the laser pulse intensity and $g(\eta)$ the pulse shape function. Here, we have introduced the quantities $\lambda^* = 4\hbar\lambda_\pm/c$ and $f(\theta) = 2(1 - \cos\theta) - \sin^2\theta$. The angle θ is not constant and evolves along propagation, because the parallel photon momentum p changes, while the perpendicular momentum $\mathbf{k}_\perp$ stays constant. Notice that the new Hamiltonian $H(\eta, p)$ is another constant of motion. From there we conclude that, given an initial frequency $\omega_i \equiv \omega(t = 0)$ of the probe photon, and an initial injection angle $\theta_i \equiv \theta(t = 0)$, the above ray equations will determine a photon frequency shift given by

$$\frac{\omega(t)}{\omega_i} = \frac{(1 - \delta N_i)}{[1 - \delta N(t)]}\frac{(1 - \cos\theta_i - \delta N_i)}{[1 - \cos\theta(t) - \delta N(t)]}, \tag{13.29}$$

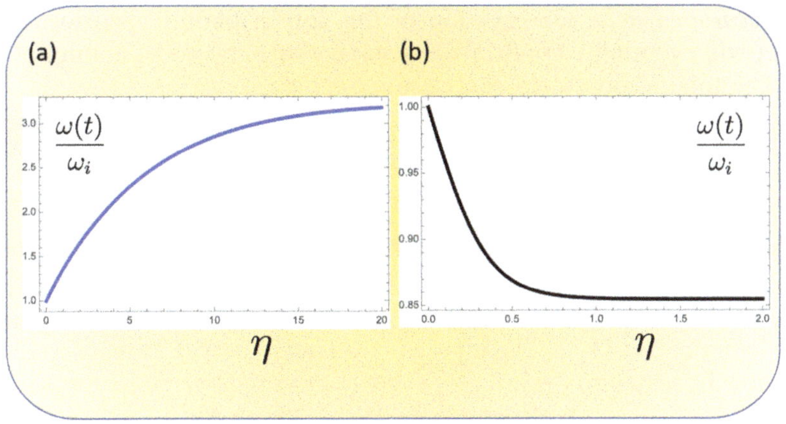

Figure 13.3. Probe photon interacting with an intense pulse in vacuum: (a) photon acceleration at the rear of the intense pulse, $\omega'(t) > 0$; (b) photon deceleration at the front, $\omega'(t) < 0$.

where $\delta N(t) = \lambda^* f(\theta) I(\eta)$ represents the nonlinear refractive index perturbation associated with the intense beam. We can see that, when in the course of the interaction of the probe photon with the intense beam, the propagation angle is such that denominator in this expression tends to zero, which corresponds to $\cos\theta(t) \to 1 - \delta N(t)$, a significant frequency shifts will occur in vacuum. Integration of the above ray equations confirms that this can indeed occur, if the probe photon interacts long enough with the rear or the front of the intense pulse. The rear interaction with lead to a positive frequency shift (acceleration) and the front interaction to a negative shift (deceleration). This is illustrated in figure 13.3. However, the ray equations also show that large propagation distances will be needed to create observable shifts.

13.3 Photon–photon scattering

Let us now examine the case of four-wave mixing in vacuum, associated with the possible occurrence of photon–photon scattering in vacuum. This can be also described by the nonlinear polarisation and magnetisation fields, defined in equation (13.2).The third-order nonlinearity in **P** and **M** allows the coupling of four different fields, which can be either static fields or wave fields. Let us consider the case of purely wave fields, with no static components. The wave equation (13.4) will then be used describe the interaction of four photon states, associated with plane waves with electric fields $\mathbf{E}_j(\mathbf{r}, t)$, evolving in space and time according to different frequencies ω_j and wavevectors $\mathbf{k}_j$, with $j = 1, 2, 3, 4$. This nonlinear interaction will satisfy the usual energy and momentum conservation relations, which are

$$\mathbf{k}_1 + \mathbf{k}_2 = \mathbf{k}_3 + \mathbf{k}_4, \qquad \omega_1 + \omega_2 = \omega_3 + \omega_4. \tag{13.30}$$

These so-called phase matching conditions can be easily satisfied in vacuum, due to the complete absence of dispersion. In the presence of three intense beams associated with the first three fields, $j = (1, 2, 3)$, we can then predict the appearance of a fourth field at $(\omega_4, \mathbf{k}_4)$, created by the nonlinear vacuum properties. This is usually called the *scattered field*. Observation of this fourth photon field in a laboratory experiment will eventually confirm the accuracy of our understanding of quantum vacuum. The total electric field can be represented as

$$\mathbf{E}(\mathbf{r},\, t) = \frac{1}{2}\sum_{j=1}^{4}\left[\mathbf{E}_j(\mathbf{r},\, t)\exp(i\mathbf{k}_j\cdot\mathbf{r} - i\omega_j t) + c.\, c.\right], \tag{13.31}$$

where $\mathbf{E}_j(\mathbf{r}, t)$ are slowly varying amplitude envelopes. Only the terms contained in $\mathbf{P}$ and $\mathbf{M}$ that are associated with products of the form $\mathbf{E}_1\mathbf{E}_2\mathbf{E}_3^*$ and $\mathbf{E}_1^*\mathbf{E}_2^*\mathbf{E}_3$ will contribute to the fourth field mode, $\mathbf{E}_4$. Retaining these terms, assuming that the amplitude of the three intense beams $j = (1, 2, 3)$ is constant, and using the wave equation (13.4), we get a solution for the scattered field, valid far away from the interaction region, as

$$E_4(\mathbf{r},\, t) = \frac{\zeta k_4^2}{\pi r} e^{ik_4 r - \omega_4 t}\int_V (E_1E_2E_3^*)_{(t=t_R)} e^{ik_4(\mathbf{k}_4 - \mathbf{r}/r)\cdot\mathbf{r}'} d\mathbf{r}'. \tag{13.32}$$

where $t_R = t - |\mathbf{r} - \mathbf{r}'|/c$. This will be the scattered field. The total number of scattered photons can estimated using the relation $N_4 = \epsilon_0 |E_4|^2/4\hbar\omega_4$, and integrating it over the entire solid angle. Its value can be estimated by the following practical formula [19]

$$N_4 = \alpha_g\left(\frac{1\mu m}{\lambda_4}\right)^3\left(\frac{L}{1\ \mu m}\right)\Pi_{j=1}^3\left(\frac{P_j}{1\ PW}\right), \tag{13.33}$$

where α_g is a geometric factor and P_j the power of the three intense beams. We can see from here that $N_4 \propto \omega^3 P^3$, where the power of the three injected beams are nearly order $P_j \simeq P$. This can go above the threshold value for detection in a single shot experiment, $N_4 = 1$, if the injected power is above 1 PetaWatt and the light frequency corresponds to the near UV.

A possible geometry of this 4-wave interaction is represented in figure 13.4. The particular interest of this three-dimensional geometry is that the scattered photon

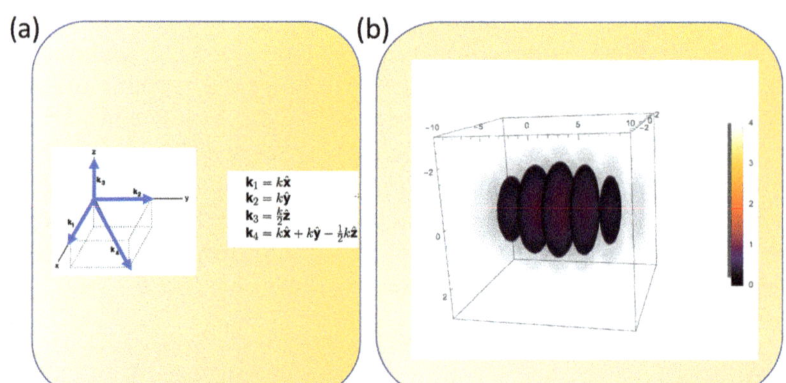

Figure 13.4. Photon–photon scattering due to 4-wave interactions in vacuum: (a) three-dimensional configuration; (b) two-dimensional configuration: total intensity E^2 of two counter-propagating pulses with the same frequency, near focus.

propagates in a direction different from that of the injected beams, thus accessible to direct detection. However, alternative two-dimensional geometries have also been proposed, and the simplest example is that of a standing wave (or two counter-propagating beams), as represented in figure 13.4(b). Notice that, when we replace one of the photon modes by a static field, then three-wave mixing configurations will be allowed, because one static field will work as a kind of provider for three-photon mixing processes. This accounts for the photon splitting effect, mentioned before.

Another interesting wave-mixing process is that of harmonic generation. This is another popular topic in the laser QED literature. For instance, generation of a second harmonic can occur by three-wave mixing in the presence of a static magnetic field [20]. In this case, we return to the wave equation (13.4) and consider the following solutions

$$\mathbf{E}(\mathbf{r}, t) = \sum_{n=1,2} \mathbf{E}(n\omega)\exp(i\mathbf{k}_n \cdot \mathbf{r} - in\omega t), \tag{13.34}$$

where $\mathbf{E}(n\omega)$ are slowly varying amplitudes for a wave with frequency ω and its second harmonic (2ω). Exact phase matching (or exact momentum conservation) will imply that the different waves propagate along the same direction. We also assume that both waves have the same polarisation state $\mathbf{e}$, in order to maximise the nonlinear coupling. Performing a perturbative analysis, we obtain an equation for the second harmonic, as

$$\frac{2i}{c^2}(\partial_t + \mathbf{v} \cdot \nabla)E(2\omega) = -\mu_0 2\omega\left[\mathbf{P}(2\omega) - \frac{1}{c}\left(\frac{\mathbf{k}}{k} \times \mathbf{M}(2\omega)\right)\right] \cdot \mathbf{e} \tag{13.35}$$

with $\mathbf{v} = c\mathbf{k}/k$. Here, we use the Fourier components of the first order vacuum polarisation fields $\mathbf{P}$ and $\mathbf{M}$ at the second harmonic frequency, as given by

$$\mathbf{P}(2\omega) = 2\zeta[4\mathcal{F}^{(1)}(\omega)\mathbf{E}(\omega) + 7c\mathcal{G}^{(1)}(\omega)\mathbf{B}(\omega)] \tag{13.36}$$

$$\mathbf{M}(2\omega) = -2c^2\zeta[4\mathcal{F}(\omega)\mathbf{B}(\omega) + 7\mathcal{G}(\omega)\mathbf{E}(\omega)/c] \tag{13.37}$$

with the quantities

$$\mathcal{F}(\omega) = -c^2\mathbf{B}(\omega) \cdot \mathbf{B}_0, \quad \mathcal{G}(\omega) = c\mathbf{E}(\omega) \cdot \mathbf{B}_0. \tag{13.38}$$

Considering wave propagation along the z-direction, we obtain the evolution equation

$$\frac{d}{d\tau}E(2\omega) = -i\frac{\omega c}{\epsilon_0}\zeta E^2(\omega)(\hat{e} \cdot \mathbf{B}_0). \tag{13.39}$$

Assuming that initially we have a strong field at the fundamental frequency and a negligible harmonic signal, $E(\omega) \gg E(2\omega)$, and noting that the nonlinear coupling is extremely small, we obtain

$$|E(2\omega)| \simeq \frac{\omega c}{\epsilon_0}\zeta|E(\omega)|^2 B_0\, \tau \cos\theta, \tag{13.40}$$

where θ is the angle between the wave field $\mathbf{E}(\omega)$ and the static magnetic field $\mathbf{B}_0$. This effect could be pertinent to ultra-intense laser experiments in the multi-PetaWatt regime.

13.4 Vacuum undulator

We now examine the case of a magnetised vacuum where the static magnetic field is modulated in some direction, thus creating a kind of photon undulator. If an intense photon beam propagates in such vacuum, a different kind of photon scattering will eventually occur. The total electric and magnetic fields associated with this geometry can be written as

$$\mathbf{E} = \mathbf{E}_0 + \mathbf{E}_s, \qquad \mathbf{B} = \mathbf{B}_e + \mathbf{B}_0 + \mathbf{B}_s, \tag{13.41}$$

where scattered radiation is described by the fields $\mathbf{E}_s$ and $\mathbf{B}_s$, and The intense beam is described by the field

$$\mathbf{E}_0(\mathbf{r}, t) = \mathbf{E}_0(\mathbf{r}_\perp, \eta)\exp(i\mathbf{k}_0 \cdot \mathbf{r} - i\omega_0 t) + c.\ c. \tag{13.42}$$

with a similar expression for $\mathbf{B}_0(\mathbf{r}, t)$. We assume propagation along the z-direction and use the variables variable $\eta = (z - ct)$, with v_0. Similarly, we define $\mathbf{P} = \mathbf{P}_0 + \mathbf{P}_s$ and $\mathbf{M} = \mathbf{M}_0 + \mathbf{M}_s$, where the dominant terms $\mathbf{P}_0$ and $\mathbf{M}_0$ are quadratic with respect to the intense field $\mathbf{E}_0$, while the quantities $\tilde{\mathbf{P}}$ and $\tilde{\mathbf{M}}_r$ are quadratic with respect $\mathbf{B}_e$, and proportional to the scattered field $\mathbf{E}_s$. We use $\mathbf{B}_e(\mathbf{r}) = B_e(z)\mathbf{e}_z$, and the wave equation becomes

$$\left(\nabla^2 - \frac{1}{c^2}\partial_t^2\right)\mathbf{E}_s = \mu_0\, \partial_t \mathbf{J}(\mathbf{r}_\perp, \eta), \tag{13.43}$$

where the current responsible for the emission of scattered radiation is determined by

$$\vec{J}_0 = -2\zeta c^2 \mid E_0(\mathbf{r}_\perp, \eta)|^2\, \mathbf{b}(\partial_z B_e), \tag{13.44}$$

where we have defined the auxiliary vector

$$\mathbf{b} = \frac{1}{|B_0||B_e|}[2(\mathbf{B}_0 \cdot \mathbf{B}_e) - 7(\mathbf{e}_0 \cdot \mathbf{B}_e)\mathbf{B}_0], \tag{13.45}$$

with $\mathbf{e}_0 = \mathbf{E}_0/|E_0|$. We solve the above wave equation for a simple wiggler field $B_e(z) = B_e \sin(k_w z)$, where k_w is the wiggler wavenumber. Using a double Fourier transformation of the scattered field, in $\mathbf{r}_\perp$ and time, we can derive an expression for field scattered in the z-direction as [21]

$$E_\omega(z) = -a_\omega(\mathbf{k}_\perp)B_e\frac{\pi k_w}{2k}e^{ikz}\int_{-\infty}^{\infty} \mid E_0(\eta)|^2\, e^{-i(k \pm k_w)\eta}\delta(\omega - \omega_\pm)d\eta, \tag{13.46}$$

where we have used the amplitude

$$a_\omega(\mathbf{k}_\perp) = 2\zeta\mu_0 c(\mathbf{b} \cdot \mathbf{e}_\omega)\int f(\mathbf{r}_\perp)e^{-i\mathbf{k}_\perp \cdot \mathbf{r}_\perp}d\mathbf{r}_\perp, \tag{13.47}$$

where $f(\mathbf{r})$ is a given transverse profile function, for instance a Gaussian. The frequencies appearing in equation (13.46) are given by

$$\omega_\pm = (k \pm k_w)\beta c, \quad \beta = v_0/c, \tag{13.48}$$

where $v_0 \simeq c$ is the group velocity of the intense beam. Even if propagating in vacuum is considered, we always have $\beta \sim 1$ but smaller than one, because the group velocity of a photon beam always stays slightly below c, due to the finite beam waist, and we only have $\beta = 1$ for a plane wave. It should also be noticed that the wavenumber k depends on the frequency, through the relation $k^2 = \omega^2/c^2 - k_\perp^2$. This allows us to replace the above expression by

$$\omega_\pm = \frac{k_w \beta}{(1 - \beta \cos\theta)}, \tag{13.49}$$

where $\theta = \tan^{-1}(k_\perp/k)$ is the angle between the direction of propagation with respect to the static field. This is very interesting, because it shows an unexpected relation with the frequency emitted by a beam of electrons in a wiggler field. Such an analogy becomes more visible when we consider propagation at small angles, $\theta \simeq 0$, where this expression can be written in the form

$$\omega \simeq 2\gamma_0^2 k_w c, \quad \gamma_0 \simeq \frac{1}{\sqrt{1 - \beta^2 \cos^2\theta}}. \tag{13.50}$$

This shows that the natural wiggler frequency $\omega_w = k_w c$ is multiplied by a factor $2\gamma_0 \gg 1$, thus leading to high frequency emission of radiation in vacuum. Here, γ_0 can be seen as an effective gamma factor of the intense laser beam. It replaces the gamma factor of the relativistic electron beam in a magnetic wiggler (figure 13.5). As for the amplitude of the scattered radiation, it can be estimated as

$$|E_\omega| \simeq a_\omega \frac{\pi k_w}{2k} B_e E_0^2. \tag{13.51}$$

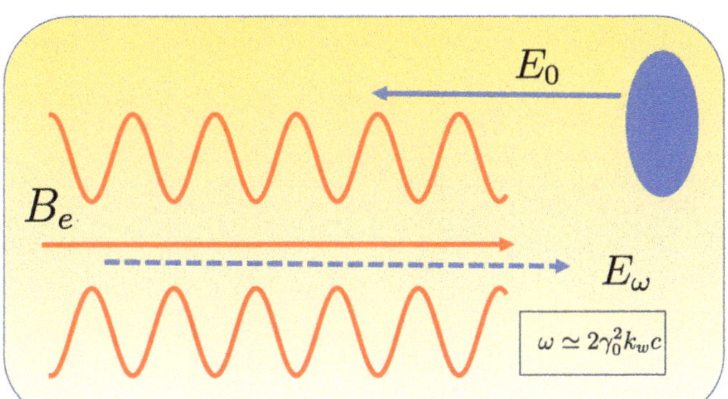

Figure 13.5. Photon undulator geometry: an intense photon beam with frequency ω_0 propagates along the axis of a magnetic wiggler, in vacuum, creating a scattered field at frequency ω, given by equation (13.50).

Another interesting version of this magnetic wiggler leads to the possible emission of electromagnetic vortices from quantum vacuum [22]. This is related to the transverse structure of the magnetic field, which was ignored above. As before, we assume propagation of an intense beam along the magnetic axis Oz, with $\mathbf{k}_0 = k_0\mathbf{e}_z$ and laser frequency $\omega_0 = k_0 c$. But now, we assume a more general wiggler field, with a generic radial structure, as described in cylindrical coordinates $\mathbf{r} = (r, \theta, z)$, by the expression

$$\mathbf{B}_w(\mathbf{r}) = B_w \int \mathbf{b}(\mathbf{r}_\perp, k_w)\exp(ik_w z)\frac{dk_w}{2\pi}, \tag{13.52}$$

where B_w is a constant field amplitude, k_w is the wiggler wavenumber, and the quantity $\mathbf{b}(\mathbf{r}_\perp, k_w)$ is a function of the transverse coordinates. This vector function, can be represented by orthogonal Laguerre–Gauss (LG) functions $F_{lp}(r)$, and we can generally state that

$$\mathbf{B}_w(\mathbf{r}) = \sum_{l,p} \int \mathbf{e}_{lp}(r, k_w)F_{lp}(r)\exp(ik_w z + il\theta)\frac{dk_w}{2\pi}, \tag{13.53}$$

where $F_{lp}(r)$ are normalised and orthogonal, as we have seen in chapter 9. A typical example is the Stanford wiggler [23], with just two values of $k_z = \pm k_w$, and two azimuthal components $l = 0$ and $l = 2$. For simplicity, we consider a single LG mode, such that

$$\mathbf{B}_w(\mathbf{r}) = B_z\mathbf{e}_z F_{lp}(r)\exp(ik_w z + il\theta). \tag{13.54}$$

When an intense Gaussian laser pulse propagates in vacuum, in the presence of this field, the nonlinear quantum properties of vacuum will induce a scattered field, represented by the electric and magnetic fields $(\mathbf{E}_s, \mathbf{B}_s)$. Solving the wave equation for the scattered field then leads to the following final expression, for a given wiggler k_w mode component,

$$E_{lp}(\omega) = -(cB_w)k_w R(\omega_0)\int_{-\infty}^{\infty} f(\eta)e^{i(k_w - k)\eta}, \tag{13.55}$$

where $R(\omega)$ is a dimensionless factor, determining the nonlinear vacuum response to an intense photon beam with intensity a^2 and frequency ω, defined by

$$R(\omega) = \frac{\alpha}{45}\left(\frac{\hbar\omega}{mc^2}\right)^2 a^2, \tag{13.56}$$

where α is the fine structure constant, and $a = eE_0/mc\omega$ is the normalised beam amplitude. In the particular case of a Gaussian laser pulse envelope with duration σ/c, we obtain

$$E_{lp}(\omega) = -\sqrt{2\pi}(cB_w)(k_w\sigma)R(\omega_0)F_{lp}(r)\exp\left(-\frac{k^2\sigma^2}{2}\right). \tag{13.57}$$

The scattered field is proportional to the intensity of the laser pulse, a_0^2, and to the magnetic wiggler field B_w. It also increases with the product $k_w\sigma$, and is maximum

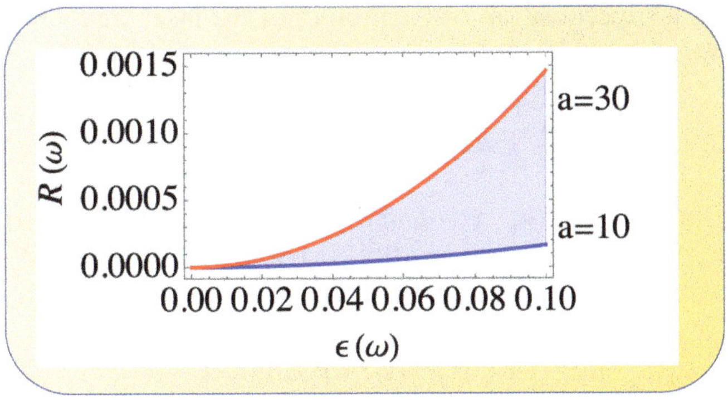

Figure 13.6. Vacuum efficiency factor $R(\omega)$, defined by equation (13.56), for $a = 10$ (blue) and $a = 30$ (red), as a function of the normalised photon energy $\epsilon(\omega) = \hbar\omega/mc^2$.

for $k^2\sigma^2 \ll 1$. The space-time description of the scattered field, corresponds to a twisted field that can be described as

$$\mathbf{E}_{ls}(\mathbf{r}, t) = -\sqrt{2\pi}(k_w\sigma)(cB_w)R(\omega_0)F_{lp}(r)\exp(ikz + il\theta - i\omega t), \tag{13.58}$$

where the frequency is determined by equation (13.48). The number of photons associated with this LG field (l, p), is then equal to $N = \epsilon_0|E_s|^2/4\hbar\omega$. The emission efficiency of the different LG modes is mediated by the dimensionless quantity $R(\omega)$, as defined by above. This is an important quantity, because it determines the emission efficiency of the quantum vacuum process. The dependence of $R(\omega)$ with respect to the energy of the incident laser photons is illustrated in figure 13.6.

Polarisation of the emitted radiation will be maximum along the directions of the electric and magnetic fields of the incident laser pulse, for $(\mathbf{e}_\omega^* \cdot \mathbf{e}_0) \simeq 1$ or $(\mathbf{e}_\omega^* \cdot \mathbf{b}_0) \simeq 1$. For practical estimates, we can assume that $B_w \sim 1$Tesla, and $a_0 = 10$, which leads to

$$\frac{|E_s|}{cB_w} \simeq 4 \times 10^{-2}(k_w\sigma)\left(\frac{\hbar\omega_0}{mc^2}\right)^2. \tag{13.59}$$

This estimate remains valid for any wiggler field structure (l, p), including the wiggler with no helicity, and shows that emission of twisted photons in quantum vacuum seems possible. The potential advantage of using helical wiggler configurations in QED experiments is that they could reduce the noise level, given the absence of twisted light sources of noise in these experiments.

13.5 Superradiant vacuum

Recent advances in laser technology have shown that a sequence of very short electric field spikes could be produced by intense laser-target interactions. This is due to an accumulation of a large number of harmonics of the incident field, in a way similar to that used in laser mode-locking. However, instead of a superposition of a large number of laser cavity modes, with nearby frequencies, as described in section

6.5, here we have a superposition of harmonics of the laser frequency, $\omega_n = n\omega_1$, for $n = 1, 2, \ldots, N_h$, as

$$\mathbf{E}_h(\mathbf{r}, t) = \frac{1}{2}\sum_n \mathbf{e}_n \mathcal{E}_n(\mathbf{r})\exp(in\varphi) + c.\,c., \tag{13.60}$$

with the phase function $\varphi = (\mathbf{k}_1 \cdot \mathbf{r} - i\omega_1 t)$. Assuming that all the harmonics have the same amplitude, polarisation and spatial distribution, such that $\mathcal{E}_n(\mathbf{r}) = \mathcal{E}_1(\mathbf{r})$ and $\mathbf{e}_n = \mathbf{e}_1$, we obtain a sequence of pulses of the form

$$\mathbf{E}_h(\mathbf{r}, t) = \frac{N_h}{2}\mathbf{e}_1\mathcal{E}_1(\mathbf{r})\exp(iN_h\varphi_1/2)Sinc\left(\frac{N_h}{2}\varphi\right), \tag{13.61}$$

where we have used the sine-cardinal function $Sinc(x) = (\sin x/x)$. This defines a series of $N \neq N_h$ consecutive electric field spikes, with amplitude N_h and duration $\delta t \sim 1/(N_h\omega_1)$, occurring at $\varphi = 2\nu\pi$, for ν integer. The interval between consecutive spikes is determined by the fundamental field frequency, and the number of spikes can be large for a sufficiently large duration of the high-harmonic pulse $\Delta t \gg 1/\omega_1$. When $N_h \gg 1$, we can simply describe the high-harmonic field as

$$\mathbf{E}_h(\mathbf{r}, t) = \pi\mathbf{e}_1\mathcal{E}_1(\mathbf{r})\exp(iN_h\varphi_1/2)\sum_\nu \delta(\varphi - 2\pi\nu). \tag{13.62}$$

Let us then assume that this field collides with an intense laser pulse with field $\mathbf{E}_0(\mathbf{r}, t)$, oscillating at the frequency ω_0, as illustrated in figure 13.7. The resulting scattered field, emitted at a frequency ω, is determined by [7]

$$E_s(\omega) = -4i(k\sigma_z)\frac{\omega_0}{\omega_1}\mathcal{E}_1 R(\omega_0)\sum_\nu \exp(i\nu\theta_{\text{mix}}), \tag{13.63}$$

where $\sigma_z \simeq c\Delta t$ is the width of the incident pulse, the vacuum factor $R(\omega_0)$ is defined by equation (13.56), and the mixing angle θ_{mix} is given by

$$\theta_{\text{mix}} = \left(2\omega_0 + \frac{N_h}{2}\omega_1 - \omega\right)\frac{\lambda_1}{c}, \tag{13.64}$$

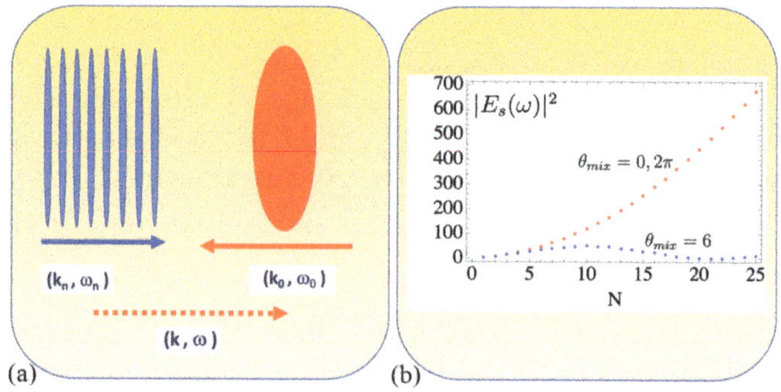

Figure 13.7. Superradiance in vacuum: (a) collision of an intense laser field $\mathbf{E}_0$, with a high-harmonic pulse with frequencies $\omega_n = n\omega_1$; (b) dependence of the scattered energy $|E_s(\omega)|^2$ (in arbitrary units) on the number of electric field spikes N, at resonance $\theta_{\text{mix}} = 0, 2\pi$ (in red), and off-resonance, $\theta_{\text{mix}} = 6$.

where λ_1 is the wavelength of the first harmonic field. We can see that the scattered field is proportional to the intensity a_0^2 of the incident field, and to the amplitude $\mathcal{E}_1$ of the harmonic field, as expected, with the vacuum conversion factor R. What is new here is the existence of an interference between terms associated with the different spikes ν. Interference becomes constructive when the mixing angle becomes $\theta_{\mathrm{mix}} \simeq 2n\pi$, with n integer. This occurs for frequencies such that $\omega = 2\omega_0 + (N_h/2 - n)\omega_1$. In this case, the sum in equation (13.63) is equal to N, and the scattered energy attains its maximum value. In this case all the spikes emit in phase and the overlapped scattered signal behaves as $|E_s(\omega)|^2 \propto N^2$, which is typical of superradiance. This is also illustrated in figure 13.7.

Superradiant scattering can therefore amplify the nonlinear scattered process by a factor of $N^2 \gg 1$. A closer analysis of the phase matching condition shows that superradiance indeed occurs for $\omega = 2\omega_0$ and $k = -2k_0$, which corresponds to second harmonic backscattering. It is also interesting to see that a large number of four-wave processes due to vacuum nonlinearity can lead, by field superposition, to an effective three-wave process where a second harmonic backscattered signal is produced.

References

[1] DiPiazza A, Muller C, Hatsagortsyan K Z and Keitel C H 2012 Extremely high-intensity laser interactions with fundamental quantum systems *Rev. Mod. Phys.* **84** 1177

[2] Marklund M and Shukla P K 2006 Nonlinear collective effects in photon-photon and photon-plasma interactions *Rev. Mod. Phys.* **78** 591

[3] Ehlotzky F, Krajewska K and Zaminski J Z 2009 Fundamental processes in quantum electrodynamics in laser fields of relativistic power *Rep. Prog. Phys.* **72** 046401

[4] Gu Y-J, Jirka M, Klimo O and Weber S 2019 Gamma photons and electron-positron pairs from ultra-intense laser-matter interaction: a comparative study of proposed configurations *Matter Radiat. Extremes* **4** 064403

[5] Paul P M, Toma E S, Berger P, Mullot G, Augé F, Balcou P, Muller H G and Agostini P 2001 Observation of a train of attosecond pulses from high harmonic generation *Science* **292** 1689–92

[6] Vincenti H and Quéré F 2012 Attosecond lighthouses: how to use spatiotemporally coupled light fields to generate isolated attosecond pulses *Phys. Rev. Lett.* **108** 113904

[7] Mendonça J T 2021 Superradiance in quantum vacuum *Quantum Rep.* **3** 42–52

[8] Ejlli A, Della Valle F, Gastaldi U, Messineo G, Pengo R, Ruoso G and Zavattini G 2020 The PVLAS experiment: a 25 year effort to measure vacuum magnetic birefringence *Phys. Rep.* **871** 1–74

[9] Papatzacos P and Mork K 1975 Delbrück scattering *Phys. Rep.* **21** 81

[10] Akhmadaliev Sh Zh *et al* 2002 Experimental investigation of high-energy photon splitting in atomic fields *Phys. Rev. Lett.* **89** 061802

[11] Affleck I and Kruglyak L 1987 Photon splitting in a plane-wave field *Phys. Rev. Lett.* **59** 1065

[12] Danson C N *et al* 2019 Petawatt and exawatt class lasers worldwide *High Power Laser Sci. Eng.* **7** 54

[13] Zhang P, Bulanov S S, Seipt D, Arefiev A V and Thomas A G R 2020 Relativistic plasma physics in supercritical fields *Phys. Plasmas* **27** 050601

[14] Bialynicka-Birula Z and Bialynicka-Birula I 1970 Nonlinear effects in quantum electrodynamics. Photon propagation and photon splitting in an external field *Phys. Rev.* D **2** 2341

[15] Klein J J and Nigam B P 1964 Birefringence of vacuum *Phys. Rev.* **135** B1279

[16] Mignani R P *et al* 2016 Evidence for vacuum birefringence from the first optical-polarimetry measurement of the isolated neutron star RXJ1 856.5-3754 *Mon. Not. R. Astron. Soc.* **465** 492

[17] Mendonça J T, Marklund M, Shukla P K and Brodin G 2006 Photon acceleration in vacuum *Phys. Lett.* A **359** 700

[18] Mendonça J T 2001 *Theory of Photon Acceleration* (Bristol: Institute of Physics Publishing)

[19] Lundström E, Brodin G, Lundin J, Marklund M, Bingham R, Collier J, Mendonca J T and Norreys P 2006 Using high-power lasers for detection of elastic photon-photon scattering *Phys. Rev. Lett.* **96** 083602

[20] Kaplan A E and Ding Y J 2000 Field-gradient-induced second-harmonic generation in magnetized vacuum *Phys. Rev.* A **62** 043805

[21] Mendonça J T 2009 New effects in quantum vacuum: photon undulator and transition radiation *J. Phys. A: Math. Theor.* **42** 375403

[22] Mendonça J T 2017 Emission of twisted photons from quantum vacuum *Europhys. Lett.* **120** 61001

[23] Kincaid B M 1977 A short-period helical wiggler as an improved source of synchrotron radiation *J. Appl. Phys.* **48** 2684–91

IOP Publishing

The Quantum Nature of Light
From photon states to quantum fluids of light
J T Mendonça

Chapter 14

The axions

In this chapter we discuss the possible existence of axions, and their coupling to photons. Although speculative, this problem has attracted a growing attention of researchers in different fields, and is interesting for a series of reasons. First, the concept of axion was introduced in theoretical physics in order to solve the so-called *strong charge-parity (CP) problem* of quantum chromodynamics (QCD) [1]. Second, because of the smallness of its mass and weak interaction with matter, the axion is a good candidate for *dark matter* [2]. The proof of its existence could therefore solve simultaneously two of the major problems of contemporary physics, the QCD problem and the cosmological problem. Finally, the recent development of ultra-intense lasers could eventually help to discover the axion or other axion-like particles.

The axion concept resulted from a theoretical proposal made by Peccei and Quinn in 1977, to solve a contradiction found in QCD, which concerns the non-existence of dipole moment for the neutron [3]. The problem is due to a QCD Lagrangian term, which changes sign under a CP transformation, thus violating the CP invariance of the theory. Such a violation would lead to a finite electric dipole moment of the neutron, which is denied by experiments. Their proposal was then to include a new field in the Lagrangian, with pseudoscalar symmetry, in order eliminate this incongruence. Quantisation of this new field, then implies the existence of a new particle, the axion [4, 5]. According to more recent views, this hypothetical elementary particle would have a very small mass (possibly in the meV or micro-eV range) and couples very weakly with quarks, leptons and photons. For this reason axions could be very good candidates for the explanation of the large amount of dark matter observed in the Universe [6].

Dark matter is one of the enigmas of our present understanding of the cosmos, and an essential problem in modern cosmology (see an interesting historical account in [7]). The observation of clusters of galaxies, star rotation in spiral galaxies and

doi:10.1088/978-0-7503-2786-2ch14

galaxy stability models, led to an accumulation of evidence that a large amount of non-observed mass must exist in the Universe [8, 9]. Our present understanding, based on microwave survey maps [10], shows that nearly 23% of all matter is dark matter today, that is to say, matter of unknown origin, and was more than 63% some 13.7 billion years ago in the far past. To the fundamental question: what is dark matter, many different answers have been advanced, from weakly interacting massive particles (WIMPS) to weakly interacting slim particles (WISPS). But, to our knowledge, the axions and axion-like particles (ALP), seem to be a very good candidate for dark matter. Several experiments and projects of different sort are presently in operation, trying to observe this hypothetical particle. They include passive astrophysical observations and active laboratory experiments.

In this chapter, we review some of the basic properties of axions, and discuss the basic axion–photon coupling mechanisms. These mechanisms can be used in passive as well in active devices. First, they can be used as passive instruments to detect axions coming from solar or other astronomical sources. Second, as active instruments to create and excite the axion field in the laboratory. For these two purposes we can use, not only the static magnetic fields as initially conceived [11, 12], but also intense laser beams propagating in vacuum [13, 14].

The possible creation of axions in the laboratory, and their extremely weak interaction with matter, was at the origin of an interesting concept called *light shinning through wall experiment* [16]. Examples of passive experiments include the Cern Axion Solar Telescope (CAST), which is basically a canon-type optical telescope with a strong magnetic field pointing to the Sun [17], and the Axion Dark Matter Experiment at the University of Florida (ADMX), which is based on a magnetised microwave cavity [18]. Laboratory experiments are typically of the type *shinning through wall experiments*, such as the ALPS at DESY in Germany [19], where the axions are (supposedly) created by a laser beam in the active part of the device and then detected in the passive part, after crossing the wall chamber, which absorbs the laser photons but is transparent to axions. But other device configurations can be envisaged, as discussed here.

The axion–photon coupling mediated by a magnetic field, initially conceived in vacuum, can be extended to plasmas due to the existence of axion–plasmon decay interactions, first noted by [20], and to dynamical coupling with plasma oscillations. It has been shown recently, that this axion–plasmon coupling mechanism leads to the formation of a polariton state [21], similar to those considered in condensed matter. This would enhance the axion coupling with matter by several orders of magnitude. Intense laser-plasma interactions, and the formation of electrostatic wakefields, could be used as active sources of axions in the laboratory [22]. This would lead to the new concept of *plasma shinning through wall* experiment, proposed in [23].

14.1 Axion–photon coupling

The axions are the elementary excitations of a pseudoscalar field, a. They are coupled to the electromagnetic field by an interaction Lagrangian of the form

$$\mathcal{L}_{\text{int}} = -\frac{1}{4} g_{a\gamma} F_{\mu\nu} \tilde{F}^{\mu\nu} a = g_{a\gamma} (\mathbf{E} \cdot \mathbf{B}) a, \tag{14.1}$$

where $F^{\mu\nu}$ is the electromagnetic field tensor, $\tilde{F}^{\mu\nu}$ its dual, and $g_{a\gamma}$ is the coupling constant, to be determined. The Lagrangian $\mathcal{L}_{\text{int}}$ couples axions and photons, described by the fields a and $\mathbf{E}$, in the presence of a static magnetic field $\mathbf{B}$. This is a triangular process, known as *Primakov process*, which also couples other low mass bosons, such as pions or gravitons, with photons in an external field. The above Lagrangian remains constant under a charge-parity transformation, because $\mathbf{E}$ is a polar vector and $\mathbf{B}$ is an axial one. As a result the product $(\mathbf{E} \cdot \mathbf{B})$ is a pseudoscalar. Due to the existence of this Lagrangian term, the axion field is described by a Klein–Gordon equation with a source term, as

$$(\partial_t^2 - \nabla^2 + m_a^2) a = g_{a\gamma} (\mathbf{E} \cdot \mathbf{B}). \tag{14.2}$$

On the other hand, Maxwell's equations are modified, in such a way that the electric field $\mathbf{E}$ is replaced by $(\mathbf{E} + g_{a\gamma} a \mathbf{B})$, and the magnetic field $\mathbf{B}$ by $(\mathbf{B} - g_{a\gamma} a \mathbf{E})$, The resulting wave equation (using $c = 1$) is

$$(\partial_t^2 - \nabla^2) \mathbf{E} = -g_{a\gamma} \mathbf{B}\, \partial_t^2 a, \tag{14.3}$$

where m_a is the axion mass, still to be found. For field perturbations oscillating with frequency ω and propagating in a direction z perpendicular to the static magnetic field $\mathbf{B}_0$, we can describe the coupled axion–photon field with

$$(\omega^2 + \partial_z^2 + \bar{M}^2) \begin{bmatrix} E \\ a \end{bmatrix} = 0, \tag{14.4}$$

where we have used the mass matrix

$$\bar{M}^2 = \begin{bmatrix} -m_\gamma^2 & \Omega_{a\gamma}^2 \\ \Omega_{a\gamma}^2 & -m_a^2 \end{bmatrix}, \quad \Omega_{a\gamma} = \sqrt{g_{a\gamma} (\mathbf{e} \cdot \mathbf{B}_0) \omega}\,. \tag{14.5}$$

Here, the quantity $\Omega_{a\gamma}$ plays the role of a coupling frequency. Note that we have introduced an effective photon mass m_γ, representing the possible transverse dimensions of the photon mode, as $\nabla_\perp^2 E = -m_\gamma^2 E$, which can be due to waveguide propagation or to a plasma background. The first would be relevant to detection devices [12], the second to astrophysical environments [24, 25]. But plasma effects can also be used to improve detection [26]. The mass matrix can be diagonalised using a rotation angle θ, such that

$$\frac{1}{2} \tan(2\theta) = \frac{2\Omega_{a\gamma}^2}{(m_a^2 - m_\gamma^2)}. \tag{14.6}$$

This angle sets the scale for energy oscillations between the photon and the axion mode. The resulting frequency eigenvalues are

$$\omega_\pm^2 = \frac{1}{2}\left(m_a^2 + m_\gamma^2\right) \pm \frac{1}{2}\sqrt{\left(m_a^2 - m_\gamma^2\right)^2 + 4\Omega_{a\gamma}^4}\,, \tag{14.7}$$

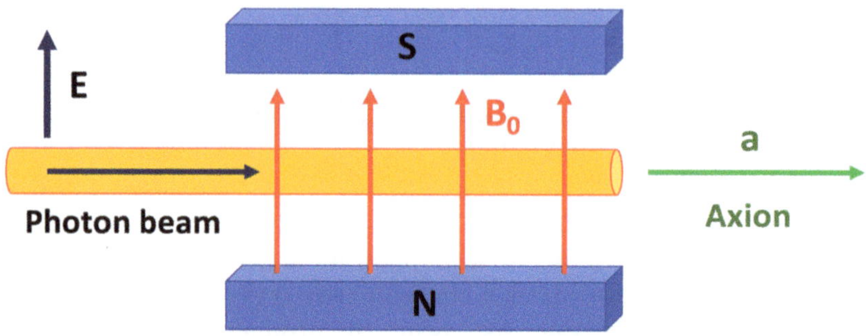

Figure 14.1. Basic geometry of axion–photon interactions: (i) a photon beam propagates in vacuum across a static magnetic field $\mathbf{B}_0$, with electric field $\mathbf{E}$ polarised along the static field; (ii) the resulting axions are emitted along the direction of the photon beam. The same geometry can be used for detection: axions emitted from a distance source (e.g. the Sun) are transformed into photons by the reverse process.

and the corresponding eigenfrequencies

$$\omega_\pm^2 = m_\pm^2 + k^2, \tag{14.8}$$

The probability for axion conversion into a photon, at a given energy (frequency) ω is determined by

$$P_{a\gamma}(z) = \sin^2(2\theta)\sin^2\left[(k_+ - k_-)\frac{z}{2}\right]. \tag{14.9}$$

The conversion rate, after a distance L, can then be estimated as (figure 14.1)

$$P_{a\gamma}(L) \simeq g_{a\gamma}^2 B_0^2 L^2, \tag{14.10}$$

The same theoretical description is valid if, instead of using the axion–photon coupling to detect axions arriving from the Sun and from distant astronomical sources, we consider active sources of axions, where an intense laser field $\mathbf{E}$ is the source of axions. In this case, the probability for axion excitation, will be also be proportional to the laser pulse energy, as discussed later.

14.2 Axion polariton

In the above description of axion–photon coupling the eventual existence of a plasma background leaded to the introduction of a photon mass, but did not affect the basic vacuum model. However, the axion–plasma processes are much more interesting, because of the existence of longitudinal photons, the plasmons, and the possible occurrence of axion instabilities. These new processes will be described in the subsequent sections of this chapter. First, we examine the axion–plasmon coupling, which is considerably different from the axion–photon coupling. The reason is the intersection of the axion and the plasmon dispersion curves, which leads to the formation of a kind of *axion polariton*. For that purpose, we go back to Maxwell's equations. When the axion coupling is included, they can be stated as

$$\nabla \cdot (\mathbf{B} - g_{a\gamma} a\mathbf{E}) = 0, \quad \nabla \times (\mathbf{B} - g_{a\gamma} a\mathbf{E}) = \frac{\partial}{\partial t}(\mathbf{E} + g_{a\gamma} a\mathbf{B}) + \mathbf{J}, \tag{14.11}$$

$$\nabla \cdot (\mathbf{E} + g_{a\gamma} a\mathbf{B}) = \frac{1}{\epsilon_0}\rho, \quad \nabla \times (\mathbf{E} + g_{a\gamma} a\mathbf{B}) = -\frac{\partial}{\partial t}(\mathbf{B} - g_{a\gamma} a\mathbf{E}) + \mu_0 \mathbf{J}, \tag{14.12}$$

where the axion field a is still described by equation (14.2), and the charge ρ and current $\mathbf{J}$ density distributions are determined by the plasma response. For simplicity, we focus on high frequency oscillations, with frequencies of the order of the electron-plasma frequency $\omega^2 \sim \omega_p^2$, and neglect the ion response. In that case, we have $\rho = e(n_0 - n_e)$ and $\mathbf{J} = -en_e\mathbf{v}$, where e is the electron charge (absolute value), n_0 is the ion density assumed constant, and the electron density and velocity are determined by the non-relativistic fluid equations

$$\frac{\partial n_e}{\partial t} + \nabla \cdot (n_e \mathbf{v} = 0, \quad \left(\frac{\partial n_e}{\partial t} + \nabla \cdot \mathbf{v}\right)\mathbf{v} = -\frac{e}{m_e}(\mathbf{E} + \mathbf{v} \times \mathbf{B}) - \frac{\nabla P_e}{n_e m_e}, \tag{14.13}$$

where P_e is the electron pressure. We assume the usual equation of state, $\nabla P_e = 3T_e \nabla n_e$, where T_e is the electron temperature. The ion motion is neglected, but could be important for low frequency electrostatic oscillations, such as the ion acoustic modes. However, these modes usually have a very small electric field and their coupling with axions is negligible. Low frequency electromagnetic modes, such as Alfven waves, could eventually be interesting for axion detection, but have not yet been considered. Relativistic effects, in contrast, are important and will be included later. We apply this simple model to a magnetised plasma, and assume wave propagation with frequency ω and wavevector $\mathbf{k}$ along the static magnetic field $\mathbf{B}_0 = B_0\mathbf{e}_z$. We therefore have $\mathbf{E}\|\mathbf{k}\|\mathbf{B}_0$. Linearising the above electron equations, we arrive at a pair of coupled equations for the electron density and axion perturbations, $\tilde{n} = n_e - n_0$ and $\tilde{a}$, of the form

$$\left(\omega^2 - \omega_p^2 - S_e^2 k^2\right)\tilde{n} = ig\frac{eB_0}{m_e} k n_0 \tilde{a}, \quad (\omega^2 - \tilde{m}_a^2 - k^2)\tilde{a} = -ig_{a\gamma}\frac{eB_0}{k}\tilde{n}, \tag{14.14}$$

where $S_e = \sqrt{3T_e/m_e}$ is the electron thermal velocity, and $\tilde{m}_a = \sqrt{m_a^2 + g_{a\gamma}^2 B_0^2}$ is an effective axion mass, modified by the static magnetic field. Non-trivial solutions of these equations imply that

$$\left(\omega^2 - \omega_{\rm pl}^2\right)\left(\omega^2 - \omega_a^2\right) - \Omega^4 = 0, \tag{14.15}$$

where we have defined the frequencies

$$\omega_{\rm pl} = \sqrt{\omega_p^2 + S_e^2 k^2}, \quad \omega_a = \sqrt{\tilde{m}_a^2 + k^2}, \quad \Omega = \sqrt{g_{a\gamma} B_0 \omega_p}. \tag{14.16}$$

Equation (14.15) clearly states that the two independent modes, with frequencies $\omega = \omega_{\rm pl}$ and $\omega = \omega_a$, become coupled by the existence of a Rabi frequency, Ω. This gives rise to two frequency branches, $\omega_p m$, which are determined by

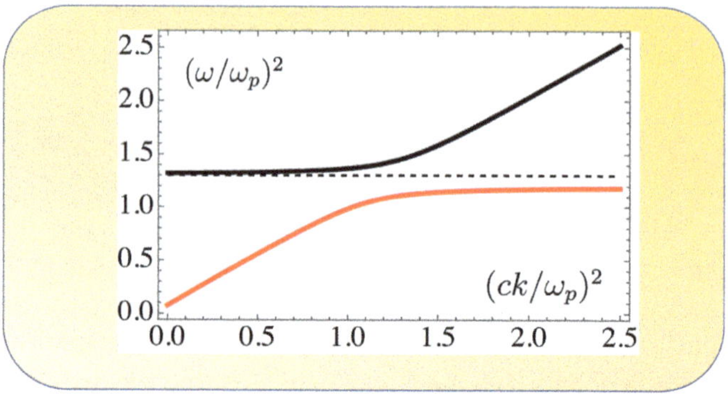

Figure 14.2. Dispersion curve of axion–plasmon polariton: frequency square $(\omega/\omega_p)^2$, versus momentum square $(ck/\omega_p)^2$, for $(\Omega/\omega_p)^2 = 0.5$, and $(m_a/\omega_p)^2 = 0.3$. Upper-plasmon branch (in black) and lower-axion branch (in red). The independent plasmon branch, is shown for comparison (dashed).

$$\omega_\pm^2 = \frac{1}{2}\left(\omega_a^2 + \omega_{\text{pl}}^2\right) \pm \frac{1}{2}\sqrt{\left(\omega_a^2 - \omega_{\text{pl}}^2\right)^2 + 4\Omega^4}. \tag{14.17}$$

These are mixed oscillations of the axion and plasmon (electrostatic) field, which are very similar to the polariton branches in condensed matter, and by that reason can be called *polariton modes*. These two polariton branches are illustrated in figure 14.2. We should also call the attention to the formal analogy between equation (14.17) valid for plasmons, and equation (14.7) valid for photons. But, despite this strong analogy, intersection of the two independent axion and plasmon branches is possible due to the low value of the electron thermal velocity $S_e/c \ll 1$, and never occurs for the two independent axion and photon branches. Therefore, the dispersion changes of the plasmon mode are much more dramatic than those of the photon mode, and can eventually be directly observed.

14.3 Axion beam instability

We now consider the mechanisms that can excite the above axion–plasmon modes from noise level to observable amplitudes. The unstable mechanisms needed to excite these modes are relativistic electron beams, and intense laser beams. Let us first consider the electron-beam case. In order to describe it, we assume a relativistic electron beam with density n_{b0}, moving along the applied magnetic field $B_0\mathbf{e}_z$. In order to describe the electron beam, we need to and beam fluid equations to the above axion and plasma equations. We then get, for the electron density perturbation

$$\left(\partial_t^2 - S_e^2\nabla^2 + \omega_p^2\right)\tilde{n}_e = -g_{a\gamma}\frac{en_{e0}}{m_e}\mathbf{B}_0 \cdot \nabla a - \omega_p^2\tilde{n}_b, \tag{14.18}$$

where $\tilde{n}_b$ is the perturbation induced in the beam density. A similar equation describes the evolution of $\tilde{n}_b$, of the form

$$\left[(\partial_t + \mathbf{u}_0 \cdot \nabla)^2 + \frac{\nu_b}{\gamma_0}\omega_p^2\right]\tilde{n}_b = -g_{a\gamma}\frac{en_{b0}}{m_e}\mathbf{B}_0 \cdot \nabla a - \frac{\nu_b}{\gamma_0^3}\omega_p^2\tilde{n}_e. \tag{14.19}$$

Here, we have neglected beam temperature effects, and assumed a relative beam density $\nu_b = n_{b0}/n_{e0}$, moving at relativistic speed $\mathbf{u}_0 = u_0\mathbf{e}_z$. The corresponding relativistic factor is $\gamma_0 = 1/\sqrt{1 - u_0^2}$. For perturbations with frequency ω, propagating in the z-direction, we obtain the dispersion relation

$$\left(\omega^2 - \omega_{\rm pl}^2 - \omega_b^2\right)\left(\omega^2 - \omega_a^2\right) = \omega_g^2\omega_c^2\left[1 + \frac{\omega_b^2}{\omega^2}\right], \tag{14.20}$$

where we have introduced the beam frequency ω_b, the electro-cyclotron frequency ω_c, and the axion-coupling frequency ω_g, defined by

$$\omega_b^2 = \frac{\nu}{\gamma_0^3}\frac{\omega^2\omega_p^2}{(\omega - ku_0)^2}, \quad \omega_c = \frac{eB_0}{m_e}, \quad \omega_g = g_{a\gamma}\sqrt{n_{e0}m_e}. \tag{14.21}$$

For $\nu_b = 0$, we are reduced to our previous equation (14.15). This dispersion relation is satisfied for complex values of the mode frequency $\omega = \omega_r + i\gamma_p$, with regions of positive growth rate $\Gamma > 0$. Instability occurs over a range of wavenumbers, such that $k \leqslant \omega_p/u_0$, where we approximately have $\omega_r \simeq ku_0$, and

$$\gamma_p \simeq \frac{\nu_b^{2/3}}{\sqrt{3}\gamma_0^2}\frac{k^2u_0^2}{\omega_p}. \tag{14.22}$$

The instability growth rate is therefore of order $\gamma_p \sim \nu_n^{2/3}\omega_p$. This shows that the axion field can extract energy from the relativistic electron beam through the beam-plasma interaction mechanism (figure 14.3).

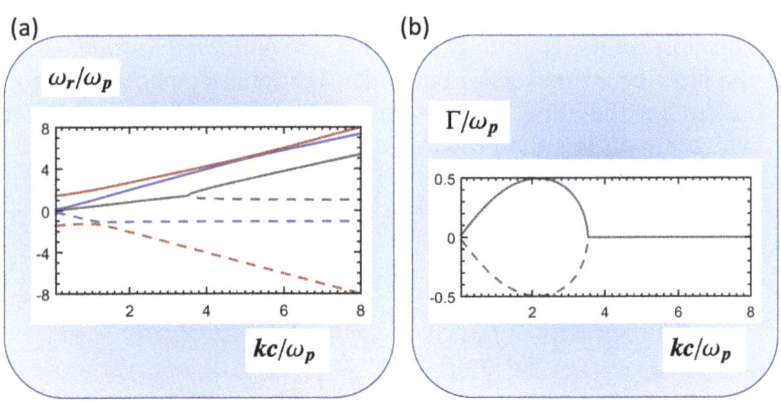

Figure 14.3. Modulation instability of an intense laser beam in a magnetised plasma: (a) real part of the frequency ω_r/ω_p, and (b) instability growth rates Γ/ω_p, as a function of the normalised wavenumber ck/ω_p. Positive and negative solutions are shown. (Courtesy J D Rodrigues.)

Similarly, an intense laser beam propagating in a magnetised plasma, along the magnetic field direction is also able to drive an instability. This is not surprising if we realise that the laser beam can be described as a beam of charged particles, physically analogous to an electron beam, because of the photons effective change in the medium. We should notice that, if the laser beam propagates along the static field $\mathbf{B}_0$, it cannot directly couple to the axions. Therefore, axion–photon coupling only occurs due to possible electrostatic fields in the magnetic field direction, and not due to the laser field itself. But this is not a problem, because we have experimental evidence that these two electric fields, the transverse laser field and the longitudinal field excited by laser instabilities, can be comparable.

In order to describe the laser beam instability, we use a modified version of the electron density equation (14.18), where the density perturbation $\tilde{n}_b$ by an equivalent perturbation due to the laser ponderomotive force, as

$$\tilde{n}_b \to \frac{2\hbar}{\gamma_0^2 m_e \omega_0} \nabla^2 \int \tilde{N}(\mathbf{k}) dk, \tag{14.23}$$

where ω_0 is the laser frequency and $\gamma_0 = \sqrt{1 + (eE_0/m\omega_0)^2}$ is the average relativistic factor of the electrons on the laser field E_0. Here, we have also used the spectral photon number density $N(\mathbf{k})$. On the other hand, the electron beam equation has to be replaced by a photon kinetic equation. where the ponderomotive coupling with the electron density perturbations $\tilde{n}_e$ is included [27]. Using a standard perturbative approach, we can then derive a dispersion equation, formally identical to the above equation (14.20), but where the beam frequency ω_b takes the form

$$\omega_b^2 = \frac{\nu}{\gamma_0} \frac{\omega^2 \nu_{\text{ph}}^2}{(\omega - kv_0)^2}, \quad \nu_{\text{ph}}^2 = \frac{k^2}{m_e n_0} \frac{\omega_p^2}{\omega} N_0, \tag{14.24}$$

where v_0 is the laser group velocity and N_0 is mean photon number density. As in the previous case of an electron beam, an instability of the coupled axion–plasmon mode can occur, as determined by the complex solutions, $\omega = \omega_r + i\Gamma$, of this equation. This instability is usually called a *modulational instability*, because it modulates the laser beam power, or equivalently, the laser photon number density. In order to estimate the expected growth rates, we consider the triple resonance condition, which maximises the growth rate, and is defined by

$$\omega_r = \omega_a = kv_0 = \frac{\omega_{\text{pl}}}{\gamma_0}. \tag{14.25}$$

In this case, equation (14.24) leads to the growth rate

$$\Gamma = \frac{\sqrt{3}}{2} \omega_p^{1/3} \left(\frac{keE_0}{4m_e\omega_0} \right)^{2/3}. \tag{14.26}$$

The instability increases with the cubic root of the laser beam power, as $\Gamma \propto E_0^{2/3}$. This shows the formal analogy between the electron beam and laser beam driven instability of the axion modes in a magnetised plasma. Both configurations can be used for active experiments aiming to demonstrate the existence of axions.

As a final comment on these instability mechanisms, we should notice that in both cases the beams interact resonantly with a narrow frequency range of possible axion modes, which depends on the coupling constant $g_{a\gamma}$, as well as on the axion bare mass m_a. Given the lack of knowledge on these constants, this property could eventually be used to determine their value. For that purpose, we go back to the axion wave equation and to the polariton dispersion relation, to establish a relation between the axion field $\tilde{a}$ and the electron density perturbation $\tilde{n}_e$. We get

$$(g_{a\gamma}\tilde{a}) = -i\frac{\omega_c}{kc}\frac{\omega_g^2}{(\omega^2 - \omega_a^2)}\frac{\tilde{n}_e}{n_0}, \tag{14.27}$$

where, for dimensionality reasons, we have reintroduced the speed of light c. We can see that, over the entire k range of unstable modes, only the axion modes satisfying the condition $\omega_p \sim \omega_a = \sqrt{\tilde{m}_a^2 + k^2c^2}$ will be resonantly excited.

14.4 Axion wakes

The above discussion on laser beam instabilities is only valid for very long pulses, with durations much larger than the electron-plasma wave period, $\Delta t \gg 1/\omega_p$. For shorter pulses, the modulation instability has no time to develop. Fortunately, other unstable mechanisms exist for very short pulses, such that $\Delta t \ll 1/\omega_p$. In this case, the laser pulses are able to produce electrostatic wakefields. Assuming the hypothetical coupling between axions and photons, axion wakefields could eventually be excited. We focus on the one-dimensional problem, which contains the main features of wakefield formation, and assume propagation along the z-direction. Retaining relativistic corrections, the electron density equation becomes

$$\left(\gamma_0\partial_t^2 + \omega_p^2\right)\tilde{n}_e = -g_{a\gamma}\frac{en_{e0}}{m_e}\mathbf{B}_0 \cdot \partial_z a + \frac{n_0}{2\gamma_0}\partial_z^2 I_0(z - v_0t), \tag{14.28}$$

where $I_0(z - v_0t)$ is the intensity of the driving laser pulse. Going into the pulse frame, using the variable $\xi = (z - v_0t)$, and noting that the axion field satisfies the equation

$$\partial_z a = -2eg_{a\gamma}\frac{B_0}{m_a^2}\tilde{n}_e, \tag{14.29}$$

we obtain

$$\left[\partial_\xi^2 + k^2(\xi)\right]\tilde{n}_e = \frac{n_0}{2\gamma_0(\xi)}\partial_\xi^2 I_0(\xi), \tag{14.30}$$

where we have

$$k^2(\xi) = \frac{\omega_p^2}{\gamma_0(\xi)v_0^2}. \tag{14.31}$$

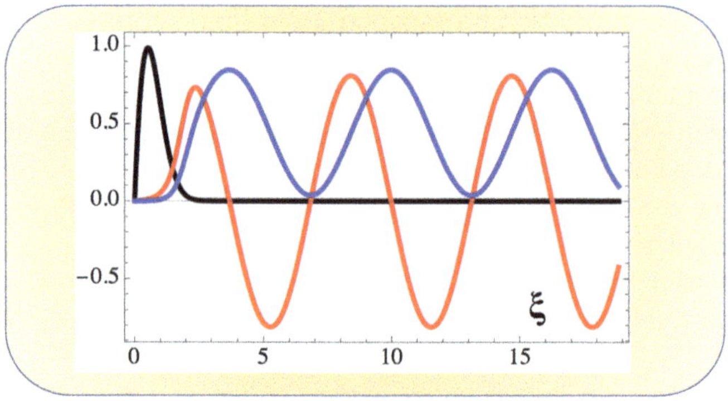

Figure 14.4. Laser pulse wakes: electron-plasma density, $\tilde{n}_e(\xi)/n(0)$, in red; normalised axion field, $a(\xi)g_{a\gamma}(\omega_p^2/\omega_g^2)$, in blue, and laser pulse shape, $I_0(\xi)/I_0(0)$, in black. High laser intensities, such that $\gamma_a = 10$ are assumed.

This shows that the wavenumber of the expected density wake varies inside the pulse, due to relativistic corrections, but is equal to (ω_p/v_0) in the wake region behind the pulse.

For relevant initial conditions, the density perturbations $\tilde{n}_e(\xi)$ describing the laser wake are determined by an integral over the pulse shape, of the form

$$\tilde{n}_e(\xi) = \frac{n_0}{2}\int_\infty^\xi \frac{1}{\gamma_0^2}(\partial_{\xi'}I_0)\cos[k(\xi - \xi')]d\xi'. \tag{14.32}$$

On the other hand, equation (14.29) implies the co-existence of an axion wake, determined by

$$\varphi(\xi) = -2e\frac{g\hbar B_0}{m_\varphi^2}\int_\infty^\xi \tilde{n}(\xi')dx'. \tag{14.33}$$

These double-wake solutions are illustrated in figure 14.4. From this analysis we conclude that, if the axion field exists, and is part of our physical world, an axion wake will be excited in intense laser-plasma interactions using very short pulses. These experimental configurations are currently being used to explore electron acceleration by laser wakefields [28]. It means that, in these experiments, a tiny axion wakefield is always present. The problem is, how to detect it.

14.5 Shinning through wall

We have seen that the axion–photon interaction process, in vacuum or in plasmas, could be used to detect or to excite this hypothetical particle. The same process has been used in passive devices, which are detectors of axion-like particles coming from the Sun or from other astronomical objects, as well as in active devices, which intend to create axions in the laboratory. These various approaches have be shortly mentioned above, but our main focus here is the concept of shinning-through wall experiment, and its more recent version based on plasma instabilities. No sign of axions has been reported yet, but an eventual detection would significantly change our understanding of the physical world.

One of the main problems related with the axion detection strategy is our ignorance of the axion mass m_a, and of the coupling constant $g_{a\gamma}$. First of all, the value of the mass depends on some unknown symmetry scale, f_a, and is currently defined as

$$m_a = 6\,\mathrm{eV}\left(10^6\mathrm{GeV}\frac{1}{f_a}\right). \tag{14.34}$$

In the original Peccei–Quinn model [3–5], usually called the *standard axion model*, the symmetry scale is of the order of electroweak scale, $f_a \sim g_{\mathrm{weak}} \simeq 250$ GeV, which would lead to very heavy axion particles, $m_a \sim$ 100KeV. Such high values of the mass are now discarded by accelerator experiments. Information due to various negative detection approaches, has been pushing the expected value of m_a down to the meV and μeV regions. On the other hand, the coupling constant $g_{a\gamma}$ is also assumed to be inversely proportional to the symmetry scale f_a. This can be stated as

$$g_{a\gamma} = \alpha\chi_\gamma\frac{1}{\pi f_a}, \tag{14.35}$$

where α is the fine-structure constant and χ_γ a dimensionless quantity of order one. If we take the two most relevant theoretical models, the *KSVZ model* due to [29, 30], gives $\chi_\gamma = 0.96$, while the *DFSZ model* [31, 32] points to a somewhat lower value, $\chi_\gamma \simeq 0.36$. But these are debatable predictions, and we need to rely on astrophysical evidence and laboratory experiments [33].

The existing experiments aiming to detect axions are mainly based on the axion decay into photons. This can occur in the presence of a static magnetic field, as described in section 15.1. The reverse case of photons decaying into axions in a similar magnetised vacuum have also been considered, using a laser beam. This led to the interesting concept of *light shinning-through wall* experiment, where an active section (the emitter) is separated by a thick wall from a passive section (the detector), both based on the same photon–axion decay process. In this configuration, the axions are created by an intense laser beam propagating in vacuum, perpendicularly to a static magnetic field. The hypothetical axions are emitted along the laser beam direction and enter the detection area after crossing a thick wall, and decay into photons [16].

But, the interesting axion–plasma properties, described in the previous sections, open the way to other experimental strategies. First, the axion-polariton state described in section 15.2 shows that the dispersion curve of the plasmon mode is modified by the axion field. Study of the dispersion properties of electron-plasma waves in the polariton range of wavenumbers could eventually lead to axion field detection. On the other hand, unstable oscillations of the axion-polariton field could be actively excited by relativistic electron beams and by intense laser pulses, as described in the previous two sections. These unstable configurations could be used

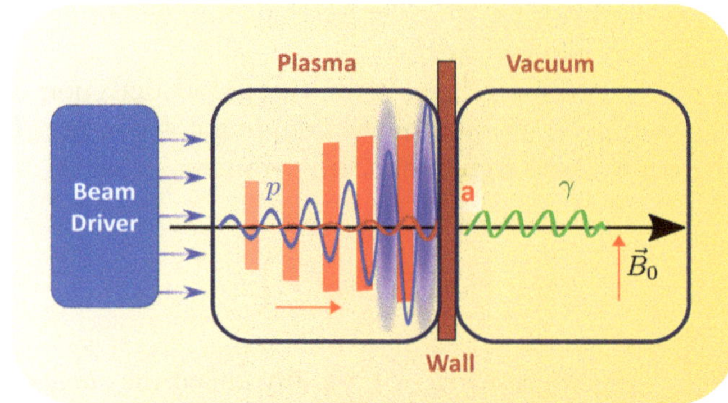

Figure 14.5. Plasma shinning through wall concept: (i) an intense beam driver (electron or laser beam) propagates in a magnetised plasma, along the static magnetic field $\mathbf{B}_0$; (ii) plasmon oscillations are excited, coupled with axions; (iii) the axions cross a thick wall and decay into photons in a vacuum magnetised chamber. The magnetic field directions (red arrows) are different on the two sides of the wall.

as the active part of a new device, the *plasma shinning through wall* device. See figure 14.5 for an illustration of the concept. The active section is based on axion–plasma beam instabilities, while the passive section (or detector) is that of the axion–photon vacuum decay process.

Although similar to the original light shinning through wall device, the magnetic field in the active part of this new configuration is parallel and not perpendicular to the beam direction of propagation. Most importantly, it has the advantage of using a beam instability, which could eventually enhance the number of produced axions by several orders of magnitude. But, when extrapolating from simple theoretical models to actual experiments, we should take into account other important factors that could limit the exponential growth of the axion instability. The first is the conversion of axions into plasmons, as a result of the same polariton coupling. The other is the saturation of the beam instability due to the nonlinear plasma response. As a result, the probability for axion creation by an electron beam instability is determined by an expression which deviates from a purely exponential law, and is given by [23]

$$P_a = \frac{g_{a\gamma}^2 B_0^2}{4\left[g_{a\gamma}^2 B_0^2 - (\omega_r - \omega_a)^2 - \gamma_p^2\right]} F(t), \tag{14.36}$$

where the temporal evolution is determined by

$$F(t) = \sin^2\left[\sqrt{g_{a\gamma}^2 B_0^2 - (\omega_r - \omega_a)}\, t/2\right] \exp(2\gamma_p t), \tag{14.37}$$

where an oscillating factor is superposed to the exponential factor, and γ_p is the growth rate determined by equation (14.22). If, instead of using an electron beam as the instability driver, we used an intense laser beam, we would need to replace γ_p by the growth rate Γ given by equation (14.26). In this expression, the exponential time factor stays is valid as long as the length of the plasma column L stays smaller than the saturation length $L_{sat} = t_{sat}/v_a$, where v_a is determined by the axion time of flight, and the saturation time is determined by

$$t_{sat} \simeq \frac{\omega_p}{\nu_b^{1/3}\gamma_p^2}. \tag{14.38}$$

For $L \geqslant L_{sat}$, we have to replace t by t_{sat} in the exponential. On the other hand, the resonant condition for axion emission, stated in equation (14.27), seems to suggest that a large range of possible axion masses m_a can be continuously explored with this kind of device, just by tuning the plasma frequency ω_p.

References

[1] Kim J E and Carosi G 2010 Axions and the strong CP problem *Rev. Mod. Phys.* **82** 557–602

[2] Sanders R H 2010 *The Dark Matter Problem* (Cambridge: Cambridge University Press)

[3] Peccei R D and Quinn H R 1977 CP conservation in the presence of instantons *Phys. Rev. Lett.* **38** 1440–3

[4] Weinberg S 1978 A new light boson? *Phys. Rev. Lett.* **40** 223–6

[5] Wilczek F 1978 Problem of strong P and T invariance in the presence of instantons *Phys. Rev. Lett.* **40** 279–82

[6] Raffelt G G 1996 *Stars as Laboratories for Fundamental Physics* (Chicago: University of Chicago Press)

[7] Bertone G and Hooper D 2018 History of dark matter *Rev. Mod. Phys.* **90** 045002

[8] Zwicky F 1933 Die Rotverschiebung von extragalaktischen *Nebeln Helv. Phys. Acta* **6** 110–27

1937 On the Masses of Nebulae and of Clusters of Nebulae *Astrophys. J.* **86** 217–46

[9] Smith S 1936 The mass of the virgo cluster *Astrophys. J.* **83** 23–30

[10] Spergel D N *et al* 2007 The-year Wilkinson Microwave Anisotropy Probe (WMAP) observations: implications for cosmology *Astrophys. J. Suppl.* **170** 377–408

[11] Raffelt G and Stodolsky L 1988 Mixing of the photon with low-mass particles *Phys. Rev.* D **37** 1237–49

[12] Sikivie P 1983 Experimental tests of the “Invisible” axion *Phys. Rev. Lett.* **51** 1415–7

[13] Mendonça J T 2007 Axion excitation by intense laser fields *Europhys. Lett.* **79** 21001

[14] Dödrich B and Gies H 2010 Axion-like-particle search with high-intensity lasers *J. High Energy Phys.* **10** 1–27

[15] Villalba-Chávez S, Podszus T and Müller C 2017 Polarization-operator approach to optical signatures of axion-like particles in strong laser pulses *Phys. Lett.* B **769** 233–41

[16] van Bibber K, Dagdeviren N R, Koonin S E, Kerman A K and Nelson H N 1987 Proposed experiment to produce and detect light pseudoscalars *Phys. Rev. Lett.* **59** 759–62

[17] CAST Collaboration 2017 New CAST limit on the axion-photon interaction *Nat. Phys.* **13** 584–90

[18] ADMX Collaboration 2010 A SQUID-based microwave cavity search for Dark-Matte axions *Phys. Rev. Lett.* **104** 041301

[19] Isleif K-S, ALPS Collaboration 2022 The any light particle search experiment at DESY (arXiv:2202.07306v1)

[20] Mikheev N V, Raffelt G and Vassilevskaya L A 1998 Axion emission by magnetic-field induced conversion of longitudinal plasmons *Phys. Rev.* D **58** 055008

[21] Terças H, Rodrigues J D and Mendonça J T 2018 Axion-plasmon polaritons in strongly magnetized plasmas *Phys. Rev. Lett.* **120** 181803

[22] Burton D A and Noble A 2018 Plasma-based wakefield accelerators as sources of axion-like particles *New J. Phys.* **20** 033022

[23] Mendonça J T, Rodrigues J D and Terças H 2020 Axion-plasmon polaritons in strongly magnetized plasmas *Phys. Rev.* D **101** 051701

[24] Huang F P, Kadota K, Sekiguchi T and Tashiro H 2018 Radio telescope search for the resonant conversion of cold dark matter axions from the magnetized astrophysical sources *Phys. Rev.* D **97** 123001

[25] Hook A, Kahn Y, Safdi B R and Sun Z 2018 Radio signals from axion dark matter conversion in neutron star magnetospheres *Phys. Rev. Lett.* **121** 241102

[26] Lawson M, Millar A J, Pancaldi M, Vitagliano E and Wilczek F 2019 Tunable axion plasma haloscopes *Phys. Rev. Lett.* **123** 141802

[27] Mendonça J T, Terças H and Rodrigues J D 2020 Axion excitation by intense laser fields in a plasma *Phys. Scr.* **95** 045601

[28] Tajima T and Malka V 2020 Laser plasma accelerators *Plasma Phys. Control. Fusion* **62** 034004

[29] Kim J E 1979 Weak-interaction singlet and strong CP invariance *Phys. Rev. Lett.* **43** 103–7

[30] Shifman M A, Vainshtein A and Zakharov V I 1980 Instantons in non-perturbative QCD vacuum *Nucl. Phys.* B **165** 45–54

[31] Dine M, Fischler W and Srednicki M 1981 A simple solution to the strong CP problem with a harmless axion *Phys. Lett.* B **104** 199–202

[32] Zhitnitskii A P 1980 Possible suppression of axion-hadron interactions *Sov. J. Nucl. Phys.* **31** 260

[33] Graham P W, Irastorza I G, Lamoreaux S K, Lindner A and van Bibber K A 2015 Experimental searches for the axion and axion-like particles *Annu. Rev. Nucl. Part. Sci.* **65** 485–515

IOP Publishing

The Quantum Nature of Light
From photon states to quantum fluids of light
J T Mendonça

Appendix A

Elementary quantum

A.1 Schrödinger equation

An isolated quantum system, or a quantum particle, is described by a ket-vector $\mid \psi\rangle$ in the space of states, the *Hilbert space*. And the evolution of its state in time is determined by the *wave equation*

$$i\hbar\frac{\partial}{\partial t}\mid \psi\rangle = H\mid \psi\rangle, \tag{A.1}$$

where $\hbar$ is the Planck's constant divided by 2π, and H is the *Hamiltonian operator*. This operator is hermitian, $H = H^\dagger$, and its explicit form will be dictated by the configuration of the system. For a non-relativistic particle with mass m, moving with momentum p along the coordinate q it can be written as

$$H = \frac{p^2}{2m} + V(q, t), \tag{A.2}$$

where the first term represents the kinetic energy operator and $V(q, t)$ is the potential. In this case, the wave equation is usually called the *Schrödinger equation*.

Given a function $F(q, p)$, the following commutation relations should be satisfied

$$[q, F(q, p)] = i\hbar\frac{\partial F}{\partial p}, \quad [p, F(q, p)] = -i\hbar\frac{\partial F}{\partial q}. \tag{A.3}$$

These relations generalise equations (2.3) of chapter 2. Other useful commutation relation involving these quantities are

$$(a)\ [q^n, p] = i\hbar n q^{n-1}, \quad [q, p^n] = i\hbar n p^{n-1}; \tag{A.4}$$

$$(b)\ [f(q, p), p] = i\hbar\frac{\partial f}{\partial q}, \quad [q, f(q, p)] = i\hbar\frac{\partial f}{\partial p}; \tag{A.5}$$

$$(c)\ [q, qp] = i\hbar(pq + qp). \tag{A.6}$$

doi:10.1088/978-0-7503-2786-2ch15

The quantity $\langle \psi | \psi \rangle$ is real and positive, and can be normalised, such that $\langle \psi | \psi \rangle = 1$. Any *observable* A, such as the position, the momentum, or the energy is represented by a hermitian operator, such that $A = A^\dagger$. For a given state of the system, its expectation value is real, and defined by

$$\langle A \rangle = \langle \psi | A | \psi \rangle. \tag{A.7}$$

Given two different state vectors $| \psi \rangle$ and $| \phi \rangle$, and a hermitian operator A, we can write

$$\langle \psi | A^\dagger | \phi \rangle = \langle \phi | A | \psi \rangle^*. \tag{A.8}$$

It is also possible to show that, given two normalised state vector $| \psi_1 \rangle$ and $| \psi_2 \rangle$, the following inequality holds

$$| \langle \psi_1 | \psi_2 \rangle |^2 \leqslant 1. \tag{A.9}$$

This is known as the *Schwarz inequality*.

A.2 Representations

In quantum mechanics we can use different representations. They correspond to different choices of reference frames to represent the state vectors. These reference frames should be made of an ensemble $| u_i \rangle$ of state vectors satisfying the orthonormality condition $\langle u_i | u_j \rangle = \delta_{ij}$. This allows us to write

$$| \psi \rangle = \sum_i c_i | u_i \rangle, \quad c_i = \langle u_i | \psi \rangle, \tag{A.10}$$

where the complex numbers c_i are the coefficients defining the state vector on that frame of reference. This representation should be complete, in the sense that the operator identity I can be defined by

$$\sum_i | u_i \rangle \langle u_i | = I. \tag{A.11}$$

This can be generalised to a continuous reference frame $| u_\alpha \rangle$, where α is not a discrete but a real variable, satisfying the new orthonormality condition $\langle u_\alpha | u_{\alpha'} \rangle = \delta(\alpha - \alpha')$. In this case we have

$$| \psi \rangle = \int c(\alpha) | u_\alpha \rangle d\alpha, \quad c(\alpha) = \langle u_\alpha | \psi \rangle, \tag{A.12}$$

and the identity operator I is defined as

$$\int | u_\alpha \rangle \langle u_\alpha | \, d\alpha = I. \tag{A.13}$$

The internal product between two state vectors $|\psi\rangle$ and $|\phi\rangle$ can then be defined as

$$\langle\psi|\phi\rangle = \langle\psi|I|\phi\rangle = \sum_i \langle\psi|u_i\rangle\langle u_i|\phi\rangle = \sum_i c_i^* b_i, \tag{A.14}$$

where c_i and b_i are the coefficients of $|\psi\rangle$ and $|\phi\rangle$ respectively. A particular example of such a discrete representation is given in section 2.2, dealing with the simple harmonic oscillator, where the vectors $|u_i\rangle$ are energy state vectors. For a continuous reference frame, we have

$$\langle\psi|\phi\rangle = \langle\psi|I|\phi\rangle = \int \langle\psi|u_\alpha\rangle\langle u_\alpha|\phi\rangle d\alpha = \int c_\alpha^* b_\alpha \, d\alpha. \tag{A.15}$$

Important examples of a continuous reference frame are the position and momentum representations, where the real variable α is replaced by the real position q, or by the momentum p. In this case, the basic vectors $|u_\alpha\rangle$ can be written as $|u_\alpha\rangle$ and $|u_\alpha\rangle$, respectively, and the coefficients c_α are nothing but the wavefunction $\psi(q, t)$ and its Fourier transform $\psi(p, t)$. We get

$$\psi(q) = \langle q|\psi\rangle, \tag{A.16}$$

and

$$\psi(p) = \langle p|\psi\rangle = \int \psi(q) e^{-iqp/\hbar} \frac{dp}{2\pi\hbar}. \tag{A.17}$$

Generalisation to three dimensions is straightforward. In the position representation, the operator p is represented by $-i\hbar\partial/\partial q$, and the Schrödinger equation (A.1)–(A.2) takes its usual form

$$i\hbar\frac{\partial}{\partial t}\psi(q) = \left[-\frac{\hbar^2}{2m}\frac{\partial^2}{\partial q^2} + V(q, t)\right]\psi(q). \tag{A.18}$$

The wavefunction $\psi(q)$ is particularly useful to describe the state of non-relativistic massive particles, and its Fourier transform $\psi(p)$ is well adapted to describe the state of massless particles such as the photon, as shown later. Due to their non-commutativity, the position and momentum variables satisfy an *uncertainty relation* such that

$$\Delta q \Delta p \geqslant \frac{\hbar}{2}, \tag{A.19}$$

where the uncertainty of an operator A can be defined by the deviations with respect to the expectation value $\langle A\rangle$, as

$$\Delta A = [(A - \langle A\rangle)^2]^{1/2}. \tag{A.20}$$

It can indeed be shown, for two non-commuting operators A and B, such that $[A, B] = i\alpha$, where α is a constant, we have

$$(\Delta A)^2(\Delta B)^2 \geqslant \frac{1}{4}|\alpha|^2. \tag{A.21}$$

The position-momentum uncertainty (A.19) is a particular case of this inequality. We can also derive it directly, using the properties of the Fourier transformation (A.17). Another useful result concerning two different operators, A and B is the *Glauber formula*. It states that, if $[A, [A, B]] = [B, [A, B]] = 0$, the following equality is true

$$e^A e^B = e^{A+B} e^{\frac{1}{2}[A, B]}. \tag{A.22}$$

Furthermore, if A and B are two non-commuting operators, and α is a constant, it can be shown that

$$e^{-\alpha A} B e^{\alpha A} = B - \alpha[A, B] + \frac{\alpha^2}{2!}[A, [A, B]] + \cdots \tag{A.23}$$

These identities are demonstrated at the end of the appendix.

A.3 Evolution operator

Let us now turn to the evolution of a quantum system, as described by the wave equation (2.1). In order to solve this equation, we can introduce an *evolution operator* $U(t, t_0)$, such that

$$|\psi(t)\rangle = U(t, t_0)|\psi(t_0)\rangle. \tag{A.24}$$

The initial condition is used to define an identity operator

$$U(t_0, t_0) = I. \tag{A.25}$$

Applying twice the definition of $U(t, t_0)$, we can derive the group property

$$U(t, t_0) = U(t, t_1)U(t_1, t_0), \tag{A.26}$$

and the inverse operator $U^{-1}(t, t_0) = U(t_0, t)$. Using the conservation of probability $\langle\psi | \psi\rangle$, we can write

$$\langle\psi(t)| \psi(t)\rangle = \langle\psi(t_0)| U^\dagger(t, t_0)U(t, t_0)\psi(t_0)\rangle = \langle\psi(t_0)| \psi(t_0)\rangle. \tag{A.27}$$

From where we conclude that $U^\dagger(t, t_0)U(t, t_0) = I$. Similarly, we have $U(t, t_0)U^\dagger(t, t_0) = I$. This shows that the evolution operator is *unitary*. Let us now find an explicit expression for $U(t, t_0)$. Replacing the definition (A.24) in the wave equation (A.1), we obtain an evolution equation of the form

$$i\hbar\frac{\partial U}{\partial t} = HU. \tag{A.28}$$

This differential equation, complemented with the initial condition (A.25), can be replaced by an integral equation

$$U(t, t_0) = I - \frac{i}{\hbar}\int_{t_0}^{t} HU(t', t_0)dt'. \tag{A.29}$$

From here we get, for an infinitesimal time interval δt

$$U(t_0 + \delta, t_0) = I - \frac{i}{\hbar} H\delta t. \tag{A.30}$$

This is valid to the first order in δt, and shows that the Hamiltonian H is the generator of an infinitesimal *time translation*. Of particular interest is the case of a time-independent Hamiltonian. In this case, a solution of equation (A.28) satisfying the initial condition (A.25) is

$$U(t, t_0) = \exp\left[-\frac{i}{\hbar} H(t - t_0)\right]. \tag{A.31}$$

This is the case of a particle moving in a static potential $V(q)$. When the state vector $|\,\psi\rangle$ is an energy *eigenvector* satisfying the equation $H\,|\,\psi\rangle = E\,|\,\psi\rangle$, for a given energy eigenstate $E = \langle\psi\,|\,H\,|\,\psi\rangle$, we obtain the time-dependent solution

$$|\,\psi(t)\rangle = \exp[-i\omega(t - t_0)]\,|\,\psi(t_0)\rangle, \tag{A.32}$$

where we have introduced the proper frequency $\omega = E/\hbar$.

A.4 Quantum pictures

Different pictures can be used to describe quantum mechanics. The most commonly used are the Schrödinger and the Heisenberg pictures, but other useful pictures, such as the interaction picture, will be introduced later. The picture we have been using so far is the *Schrödinger picture*, where the operators position and momentum q and p are time-independent, and the temporal evolution of the system is contained in the state vector $|\,\psi(t)\rangle$. For convenience, we will call them here q_S, p_S and $\left|\,\psi_S(t)\right\rangle$. A different approach is that of Heisenberg, where the operators become time-dependent and the state vector becomes constant. Let us denote these new quantities as q_H, p_H and $|\,\psi_H\rangle$. A state vectors of the two pictures can be related by equation (A.24), we can be rewritten here as

$$|\,\psi_S(t)\rangle = U(t, t_0)|\,\psi_H(t_0)\rangle. \tag{A.33}$$

The Schrödinger state vector is evolving in time, while the Heisenberg state vector is fixed. They only coincide at $t = t_0$. Similarly, we can write

$$|\,\psi_H\rangle = U^\dagger(t, t_0)|\,\psi_S(t)\rangle = |\,\psi_S(t_0)\rangle. \tag{A.34}$$

The expectation value of an operator A_S, is determined by

$$\langle A\rangle = \langle\psi_S(t)|\,A_S\,|\,\psi_S(t)\rangle = \langle\psi_H(t_0)|\,U^\dagger A_S U\,|\,\psi\rangle. \tag{A.35}$$

This shows that an operator A_S, defined in the Schrödinger picture, we can obtain the corresponding operator in the Heisenberg pictures, using the same unitary transformation

$$A_H(t) = U^\dagger(t, t_0)A_S U(t, t_0). \tag{A.36}$$

This new operator will depend on time, even if A_S is time-independent. This definition of A_H will guarantee that the expectation value $\langle A \rangle$ is the same in the two pictures. Let us now take the time derivative of this equation, and using (A.28), we get

$$\frac{d}{dt}A_H = \frac{1}{i\hbar}(U^\dagger A_S H U - U^\dagger H A_S U) + U^\dagger \frac{\partial A_S}{\partial t} U. \tag{A.37}$$

Defining the Heisenberg Hamiltonian operator $H_H = UH_SU^\dagger$ and the unitary operator $I = U^\dagger U = UU^\dagger$, we obtain

$$\frac{d}{dt}A_H = \frac{1}{i\hbar}[A_H,\, H_H] + U^\dagger \frac{\partial A_S}{\partial t} U. \tag{A.38}$$

This is the *Heisenberg equation* for the observable A. An important example is that of the Hamiltonian, $A = H$. For a conservative system, the last term in this equation is equal to zero, $\partial H_S / \partial T = 0$. We then have $H_H = H_S$ and $[H,\, U] = 0$. In the general case, if A_S is not explicitly dependent on time, this equation reduces to

$$\frac{d}{dt}A_H = \frac{1}{i\hbar}[A_H,\, H_H], \tag{A.39}$$

which shows that A_H is a constant of motion if it commutes with the Hamiltonian. Important examples are those of q and p, the position and momentum operators. Because they don't depend explicitly on time, they satisfy the following Heisenberg equation

$$\frac{d}{dt}q_H = \frac{1}{i\hbar}\left[q_H,\, H_H\right], \quad \frac{d}{dt}p_H = \frac{1}{i\hbar}\left[p_H,\, H_H\right]. \tag{A.40}$$

Furthermore, using equation (A.3), we can rewrite them as

$$\frac{d}{dt}q_H = \frac{\partial H_H}{\partial p_H}, \quad \frac{d}{dt}p_H = -\frac{\partial H_H}{\partial q_H}. \tag{A.41}$$

The Heisenberg equations for the operators therefore show remarkable formal similarities with the Hamiltonian equations of Classical Mechanics.

A.5 Density operator

For a quantum system in a state $|\,\psi\rangle$, we can define the density operator ρ as

$$\rho = |\,\psi\rangle\langle\psi\,|. \tag{A.42}$$

This definition shows that the density operator is hermitian, $\rho^\dagger = \rho$. If the state vector is represented in a discrete vector basis $|\,u_i\rangle$, the identity operator will be represented on that basis by

$$\rho = \sum_{ij} c_i c_j^* \,|\, u_i\rangle\langle u_j\,|. \tag{A.43}$$

We can therefore say that ρ will be represented on that basis by a matrix, usually called the *density matrix*, with elements

$$\rho_{ij} = \langle u_i \mid \rho \mid u_j \rangle = c_i c_j^*. \tag{A.44}$$

For a normalised state vector, $\langle \psi \mid \psi \rangle = 1$, we can see that the trace of the density matrix (the sum of its diagonal elements) is equal to 1

$$Tr.(\rho) = \sum_i \rho_{ii} = \sum_i \mid c_i \mid^2 = 1. \tag{A.45}$$

It is also useful to consider the trace of the operator $\mid \psi \rangle \langle \phi \mid$, as

$$Tr.(\mid \psi \rangle \langle \phi \mid) = \sum_i \langle u_i \mid \psi \rangle \langle \phi \mid \rangle = \sum_i \langle \phi \mid u_i \rangle \langle u_i \mid \psi \rangle. \tag{A.46}$$

From the completeness condition (2.7), we can then conclude that

$$Tr.(\mid \psi \rangle \langle \phi \mid) = \langle \phi \mid \psi \rangle. \tag{A.47}$$

This result allows us to consider the trace of the expectation value of a generic observable A, as defined by (2.4). If we replace in this definition $A \mid \psi \rangle$ with $\mid \phi \rangle$, we can easily conclude that

$$\langle A \rangle = Tr.(A \mid \psi \rangle \langle \psi \mid) = Tr.(A\rho). \tag{A.48}$$

Similarly, we could also find that $Tr.(\rho A) = Tr.(A\rho)$. This is a very useful way to calculate expectation values. Another property of the density operator is that

$$\rho^2 = \mid \psi \rangle \langle \psi \mid \psi \rangle \langle \psi \mid = \rho, \tag{A.49}$$

from where we get $Tr.(\rho^2) = 1$. Let us now consider the evolution equation for the ρ. Taking the time derivative of its definition (A.42), and using the wave equation (A.1), we can easily get

$$\frac{\partial \rho}{\partial t} = \left(\frac{\partial \mid \psi \rangle}{\partial t} \right) \langle \psi \mid + \mid \psi \rangle \left(\frac{\partial \langle \psi \mid}{\partial t} \right) = -\frac{1}{i\hbar}[\rho, H]. \tag{A.50}$$

This is the *von Neumann equation*. Another important application of the concept of density operator is for statistical problems, where the state of the system is not completely known. In such cases, we cannot claim that the system is described by a single state vector $\mid \psi \rangle$, but instead that the system has a certain probability p_k of being described by a state vector $\mid \psi_k \rangle$, such that

$$\sum_k p_k = 1, \quad p_k \leqslant 1 \quad (k = 1, 2, 3, \ldots). \tag{A.51}$$

In this case, we can reformulate the definition (A.42), and use

$$\rho = \sum_k p_k \rho_k, \quad \rho_k = \mid \psi_k \rangle \langle \psi_k \mid. \tag{A.52}$$

Most of the above properties of the density operator remain valid, in what concerns the trace

$$Tr.(\rho) = \sum_k p_k Tr.(\rho_k) = \sum_k p_k = 1, \tag{A.53}$$

and the expectation values $\langle A \rangle = Tr.(\rho A)$. The von Neumann equation (2.25) stays also valid. However, we now have, in general

$$\rho^2 \neq \rho, \quad Tr.(\rho^2) \leqslant 1. \tag{A.54}$$

We complete our short survey of some basic properties of quantum mechanics, with the demonstration of the Glauber formula, equation (A.22).

A.6 Glauber formula

We consider the Glauber formula, which is a special case of the *Baker–Hausdorff theorem* of group theory. It states that, if A and B are two non-commuting operators satisfying the conditions

$$[A, [A, B]] = [B, [A, B]] = 0, \tag{A.55}$$

then, the following equalities apply

$$e^{A+B} = e^A e^B e^{-[A,\, B]/2} = e^B e^A e^{+[A,\, B]/2}. \tag{A.56}$$

In order to demonstrate this important result, we introduce an auxiliary operator of the form

$$f(\xi) = e^{\xi A} e^{\xi B}, \tag{A.57}$$

where ξ is a c-number, not an operator. Taking the derivative with respect to ξ, we get

$$\frac{df}{d\xi} = Af(\xi) + e^{\xi A} B e^{\xi B}. \tag{A.58}$$

Using the identity operator

$$I = e^{-\xi A} e^{\xi A}, \tag{A.59}$$

we rewrite the last term of equation (A.58) as

$$e^{\xi A} B e^{\xi B} = e^{\xi A} B e^{-\xi A} e^{\xi A} e^{\xi B} = e^{\xi A} B e^{-\xi A} f(\xi). \tag{A.60}$$

This allows us to write equation (A.58) as

$$\frac{df}{d\xi} = (A + e^{\xi A} B e^{-\xi A}) f(\xi). \tag{A.61}$$

In order to proceed further, we consider a new operator, defined by

$$g(\xi) = e^{\xi A} B e^{-\xi A}. \tag{A.62}$$

We note that $g(0) = B$. Expanding $g(\xi)$ as a power series in ξ, we get

$$g(\xi) = g(0) + \xi\left(\frac{dg}{d\xi}\right)_0 + \frac{\xi^2}{2}\left(\frac{d^2g}{d\xi^2}\right)_0 + \cdots \tag{A.63}$$

We then have

$$\frac{dg}{d\xi} = Ae^{\xi A}Be^{-\xi A} - e^{\xi A}Be^{-\xi A}A = [A, g(\xi)]. \tag{A.64}$$

And, consequently

$$\left(\frac{dg}{d\xi}\right)_0 = [A, B]. \tag{A.65}$$

Similarly, we can obtain

$$\left(\frac{d^2g}{d\xi^2}\right)_0 = \left[A, \frac{dg}{d\xi}\right]_0 = [A, [A, g(\xi)]]_0 = [A, [A, B]], \tag{A.66}$$

and so on. Replacing this in the expansion (A.63), we arrive at the following equality

$$e^{\xi A}Be^{-\xi A} = B + \xi[A, B] + \frac{\xi^2}{2}[A, [A, B]] + \cdots \tag{A.67}$$

But, if we assume that the operators satisfy the conditions (A.55), this will reduce to

$$e^{\xi A}Be^{-\xi A} = B + \xi[A, B]. \tag{A.68}$$

Replacing this in equation (A.61), we get

$$\frac{df}{d\xi} = (A + B)f(\xi) + \xi[A, B]f(\xi). \tag{A.69}$$

But, from the same conditions (A.55), we notice that $(A + B)$ commutes with the commutator $[A, B]$. This means that the differential equation (A.69) can be treated as a normal equation involving commuting variables. Integrating this equation, we get the obvious solution

$$f(\xi) = f(0)\exp\left\{(A + B)\xi + [A, B]\frac{\xi^2}{2}\right\}. \tag{A.70}$$

Noting that $f(0) \equiv I$ is the identity operator, this solution can also be written as

$$f(\xi) = \exp\left\{(A + B)\xi + [A, B]\frac{\xi^2}{2}\right\}. \tag{A.71}$$

If we take $\xi = 1$, and multiply from the left both sides of this equation by $\exp(-[A, B]/2)$, we finally obtain

$$e^A e^B e^{-[A,\, B]/2} = e^{A+B}. \tag{A.72}$$

This demonstrates the first identity of the theorem (A.56). Obviously, because the operators A and B commute with their commutator $[A, B]$, this identity can also be written as

$$e^{A+B} = e^{-[A,\, B]/2} e^A e^B. \tag{A.73}$$

The second identity in equation (A.56) could be demonstrated in the same way.

IOP Publishing

The Quantum Nature of Light
From photon states to quantum fluids of light
J T Mendonça

Appendix B

Lagrangians

The Lagrange function, or simply *Lagrangian*, plays a central role in our theoretical description of the physical world, because with the help of a variational principle, they allow us to describe the basic equations of particles and fields. Here we assemble some of the basic Lagrangians that are relevant to the physical processes described in this work.

B.1 Particle in a potential

The Lagrangian $L(x, v)$ of a classical particle with velocity $v = \dot{x}$ and mass m moving in an external potential, is given by the difference between its kinetic and potential energies, as

$$L(x, v) = T(\dot{x}) - V(x, t) = \frac{1}{2}mv^2 + xF(t), \tag{B.1}$$

where $F(t)$ is the force.

B.2 Relativistic particle

The Lagrangian of a relativistic particle in the absence of a potential is

$$L(x, v) = -mc^2\sqrt{1 - \frac{v^2}{c^2}}. \tag{B.2}$$

In the weakly relativistic limit, for $v^2 \ll c^2$, it reduces to

$$L = -mc^2 + \frac{1}{2}mv^2. \tag{B.3}$$

B.3 Charged particle

Lagrangian of a particle with charge e, mass m, and in the presence of an electromagnetic field,

$$L_{\rm ch}(x, v) = -mc^2\sqrt{1 - \frac{v^2}{c^2}} - eV + \frac{e}{c}\mathbf{A}\cdot\mathbf{v}, \tag{B.4}$$

where V and $\mathbf{A}$ are the scalar and vector potentials.

B.4 System of charged particles

The Lagrangian of a system of particles with charges e_i and masses m_i, in the weakly relativistic limit, $|v_i^2 \ll c^2$ is

$$L = \frac{1}{2}\sum_i \left(1 + \frac{v_i^2}{4c^2}\right) m_i v_i^2 - \sum_{j>i} \frac{e_i e_j}{R_{ij}} \left\{1 + \frac{1}{2c^2}\left[\mathbf{v}_i \cdot \mathbf{v}_j + (\mathbf{v}_i \cdot \mathbf{e}_{ij})(\mathbf{v}_i \cdot \mathbf{e}_{ij})\right]\right\}, \quad (B.5)$$

where $R_{ij} = |\mathbf{r}_i - \mathbf{r}_j|$ is the inter-particle distance, and $\mathbf{e}_{ij} = \mathbf{R}_{ij}/R_{ij}$ is the unit vector pointing from particle i to particle j.

B.5 Scalar field

The Lagrangian of a massive scalar field $\varphi(x)$, where x is the 4-vector position, is

$$L(t) = \int d^3x \mathcal{L}(x), \quad (B.6)$$

where the Lagrangian density is

$$\mathcal{L}(x) = -\frac{1}{2}\partial^\mu \varphi(x) \partial_\mu \varphi(x) - \frac{1}{2} m^2 \varphi^2(x). \quad (B.7)$$

Using the action principle, for a variation $\delta\varphi$, as

$$\delta S = \int d^4x \mathcal{L}(x) \delta\varphi = 0, \quad (B.8)$$

we get

$$\int d^4x \left[\partial^\mu \partial_\mu \varphi(x) - m^2 \varphi(x)\right] \delta\varphi = 0. \quad (B.9)$$

This leads to the Klein–Gordon equation, as

$$(\partial^2 - m^2)\varphi(x) = 0. \quad (B.10)$$

Here, and in the following, we use units $c = 1$, and $\hbar = 1$.

B.6 Electromagnetic field

The Lagrangian density for the electromagnetic field is written in terms of the 4-potential A^μ, and takes the form

$$\mathcal{L}_{\text{em}}(x) = -\frac{1}{4} F_{\mu\nu} F^{\mu\nu} - J_\mu A^\mu, \quad (B.11)$$

where $F^{\mu\nu}$ is the electromagnetic tensor and J^μ the 4-current, defined as

$$F^{\mu\nu} = \partial^\mu A^\nu - \partial^\nu A^\mu, \quad J^\mu = (\rho, \mathbf{J}). \quad (B.12)$$

Another useful form of the Lagrangian density is

$$\mathcal{L}_{\text{em}}(x) = \frac{1}{2}(\mathbf{E}^2 - \mathbf{B}^2) - J_\mu A^\mu, \tag{B.13}$$

where **E** and **B** are the electric and the magnetic fields.

B.7 Dirac field

For the Dirac field $\psi(x)$ discussed in chapter 11, the Lagrangian density is

$$\mathcal{L}_D(x) = \bar{\psi}(x)\left(i\gamma^\mu\partial_\mu - m\right)\psi, \tag{B.14}$$

where $\bar{\psi} = \psi^\dagger\gamma^0$, and γ^μ are the 4-matrices defined in section 11.2.

B.8 Axion–photon field

The axion pseudo-scalar field a, coupled with the electromagnetic field A^μ, are described by the total Lagrangian density

$$L(x) = \mathcal{L}_{\text{em}}(x) + \mathcal{L}_a(x) + \mathcal{L}_{\text{int}}(x), \tag{B.15}$$

where the first term represents the electromagnetic field, and is defined in vacuum by equation (B.11) or (B.13) with $J^\mu = 0$, and the axion term is determined by

$$\mathcal{L}_a = \frac{1}{2}\partial^\mu a^*\partial_\mu a - \frac{1}{2}m_a^2|a|^2. \tag{B.16}$$

Finally, the interaction term is

$$\mathcal{L}_{\text{int}}(x) = \frac{1}{4}g_{a\gamma}aF_{\mu\nu}\tilde{F}^{\mu\nu} = -g_{a\gamma}a(\mathbf{E}\cdot\mathbf{B}), \tag{B.17}$$

where $g_{a\gamma}$ is the coupling constant.

Appendix C

Photon kinetic equation

We start from the wave equation in a generic optical medium.

$$\left[\nabla \times (\nabla \times\ ) + \frac{1}{c^2}\frac{\partial^2}{\partial t^2}\right]\mathbf{E} = -\mu_0 \frac{\partial \mathbf{J}}{\partial t}, \tag{C.1}$$

where $\mathbf{E}$ is the electric field. The conductive current $\mathbf{J}$ can be defined in a non-homogeneous and non-stationary medium, by

$$\mathbf{J}(\mathbf{r}, t) = \sum_\alpha \int dt' \int d\mathbf{r}' \bar{\bar{\sigma}}_\alpha(\mathbf{r}, \mathbf{r}', t, t') \cdot \mathbf{E}(\mathbf{r}', t'), \tag{C.2}$$

where $\alpha = i, e$ refers to the relevant particle species in the medium (e.g. atoms or electrons), and $\bar{\bar{\sigma}}_\alpha$ is the conductivity tensor of each species. For a given field mode $\mathbf{E} \propto \exp(i\mathbf{k} \cdot \mathbf{r} - i\omega t)$, when the scales of inhomogeneity and non-stationarity are much larger than $1/k$ and $1/\omega$, these equations lead to a dispersion relation of the form

$$\mathcal{D}(\omega, \mathbf{k}) \equiv \mathbf{e}_\omega \cdot \bar{\bar{\mathcal{D}}}(\omega, \mathbf{k}) \cdot \mathbf{e}_\omega = 0, \tag{C.3}$$

for a field eigenmode with unit polarisation vector $\mathbf{e}_\omega$. Here, $\bar{\bar{\mathcal{D}}}(\omega, \mathbf{k})$ is the dispersion tensor defined by

$$\bar{\bar{\mathcal{D}}}(\omega, \mathbf{k}) = (\mathbf{NN} - N^2\bar{\bar{\mathrm{I}}} + \bar{\bar{\kappa}}), \quad \bar{\bar{\kappa}} = \bar{\bar{\mathrm{I}}} + \frac{i}{\omega\epsilon_0}\sum_\alpha \bar{\bar{\sigma}}_\alpha, \tag{C.4}$$

where we have used the normalised wave vector $\mathbf{N} = (c/\omega)\mathbf{k}$, and the linear conductivity tensors $\bar{\bar{\sigma}}_\alpha$. This is only valid for a linear response of the medium, and for local homogeneity. In order to generalise this result we can assume linear or nonlinear deviations from such an ideal situation, due for instance to the existence of space and time dependent density perturbations of the different species $\delta n_\alpha(\mathbf{r}, t)$, which are added to the equilibrium particle density n_0. Fluctuations in other external parameters $\delta\Lambda$ could also be included. In this case, the unperturbed result (C.3) is replaced by

$$\left[\mathcal{D}\left(\omega + i\frac{\partial}{\partial t}, \mathbf{k} + i\nabla\right) + \delta\mathcal{D}(\mathbf{r}, t)\right]E = 0, \tag{C.5}$$

where E is the field amplitude, and

$$\delta\mathcal{D}(\mathbf{r}, t) = \sum_{\alpha}\left(\frac{\partial\mathcal{D}}{\partial n_\alpha}\right)\delta n_\alpha(\mathbf{r}, t) + \frac{\partial\mathcal{D}}{\partial\Lambda}\cdot\delta\Lambda. \tag{C.6}$$

Expanding around the local dispersion (C.3), we obtain an evolution equation for the wave amplitude as

$$\left(\frac{\partial}{\partial t} + \mathbf{v}_k\cdot\nabla\right)E = iV(\mathbf{r}, t)E, \tag{C.7}$$

where the mode group velocity $\mathbf{v}_k$, and the perturbation potential $V(\mathbf{r}, t)$, are defined by

$$\mathbf{v}_k = -\frac{(\partial\mathcal{D}/\partial\mathbf{k})}{(\partial\mathcal{D}/\partial\omega)}, \qquad V(\mathbf{r}, t) = \frac{\delta\mathcal{D}(\mathbf{r}, t)}{(\partial\mathcal{D}/\partial\omega)}. \tag{C.8}$$

This is valid for a local mode approach, where the wave mode is usually described by plane waves, but an extension to global modes such as those of a cavity is also possible. We now introduce the field auto-correlation function. Using to the Wigner function $W(\mathbf{r}, t, \omega, \mathbf{k})$, as the double Fourier transform of this auto-correlation function, we obtain

$$\left(\frac{\partial}{\partial t} + \mathbf{v}_k\cdot\nabla\right)W = \int\frac{d\mathbf{q}}{(2\pi)^3}\int\frac{d\Omega}{2\pi}V(\Omega, \mathbf{q})[W_- - W_+]\exp(i\mathbf{q}\cdot\mathbf{r} - i\Omega t), \tag{C.9}$$

where $W_\pm = W(\omega \pm \Omega/2, \mathbf{k} \pm \mathbf{q}/2)$, and $V(\Omega, \mathbf{q})$ is the frequency and wave vector spectrum of the perturbation potential associated with the deviations from the equilibrium state of the medium. It is useful to consider the reduced Wigner function $W_k = W(\omega, \mathbf{k})\delta(\omega - \omega_k)$, where ω_k is a solution of the linear dispersion (C.3). The evolution equation of this reduced function will be formally identical with the above equation, but with W replaced by W_k.

As in our previous discussion of photon kinetics, It is now useful to take the limit of geometric optics expansion, valid when the scale length of the background density is much larger than the wavelength of the wave modes described by the quasi-distribution function W_k, or $|\mathbf{k}| \gg |\mathbf{q}|$. We then have

$$W_k^\pm \simeq W_k \pm \frac{\mathbf{q}}{2}\cdot\frac{\partial W_k}{\partial\mathbf{k}} + \frac{1}{2}\left(\frac{\mathbf{q}}{2}\cdot\frac{\partial}{\partial\mathbf{k}}\right)^2 W_k \pm \frac{1}{3!}\left(\frac{\mathbf{q}}{2}\cdot\frac{\partial}{\partial\mathbf{k}}\right)^3 W_k. \tag{C.10}$$

Replacing this in the Wigner–Moyal equation for quasi-particles, we obtain an approximate kinetic equation for rays with distribution function W_k, such that

$$\left(\frac{\partial}{\partial t} + \mathbf{v}_k \cdot \nabla - \nabla V(\mathbf{r}, t) \cdot \frac{\partial}{\partial \mathbf{k}} + \frac{1}{3 \cdot 2^3} \nabla^3 V \cdot \frac{\partial^3}{\partial \mathbf{k}^3}\right) W_k = 0. \tag{C.11}$$

This is an improved form of the kinetic equation for ray trajectories, where the diffraction effects associated with the wave character of these physical objects is taken into account (last term, with a third order derivative). When diffraction effects are neglected, this equation reduces to a Vlasov equation for quasi-particle, as given by

$$\frac{d}{dt} W_k(\mathbf{r}, \mathbf{k}, t) = 0, \tag{C.12}$$

with the total time derivative defined as

$$\frac{d}{dt} \equiv \frac{\partial}{\partial t} + \frac{d\mathbf{r}}{dt} \cdot \nabla + \frac{d\mathbf{k}}{dt} \cdot \frac{\partial}{\partial \mathbf{k}}. \tag{C.13}$$

Notice that the quasi-particle ray equations are now defined by

$$\frac{d\mathbf{r}}{dt} = \mathbf{v}_k, \quad \frac{d\mathbf{k}}{dt} = -\nabla V(\mathbf{r}, t). \tag{C.14}$$

As noted before, an additional ray equation should be added, which is usually ignored in ray tracing codes. It describes the frequency shift along the ray trajectories, due to the temporal variation of the medium, and can be stated as

$$\frac{d\omega}{dt} = \frac{\partial}{\partial t} V(\mathbf{r}, t). \tag{C.15}$$

This determines the frequency shift of an arbitrary quasi-particle described by the dispersion relation (C.3). Such frequency shifts can be significant in optics, but are usually very small and ignored in the description of Bose–Einstein condensation of photons.

IOP Publishing

The Quantum Nature of Light
From photon states to quantum fluids of light
J T Mendonça

Appendix D

Curved spacetime

We briefly discuss the electromagnetic wave equation in a curved spacetime. The interval ds, defined as the infinitesimal displacement in spacetime, is defined by

$$ds^2 = dx_\mu dx_\mu = g_{\nu\mu} dx^\nu dx_\mu, \tag{D.1}$$

where the 4-position x^μ, and $\mu = 0, 1, 2, 3$. In Cartesian coordinates, we could write $x^\mu = (ct, \mathbf{r})$. We also have used the metric tensor $g_{\nu\mu}$. In the flat spacetime of special relativity, we are reduced to the Minkowski metric tensor

$$g_{\nu\mu} = \eta_{\nu\mu} \equiv diag.\,(1, -1, -1, -1). \tag{D.2}$$

In the general case, we have $g \equiv det.(g_{\nu\mu}) < 0$. We can describe photons, and electromagnetic fields in general, using Maxwell's equations where conventional derivatives $\partial_\alpha = \partial/\partial x^\alpha$ are replaced by *covariant derivatives*, D_α. This allows us to write

$$D_\nu F^{\mu\nu} = -\mu_0 J^\mu, \tag{D.3}$$

where μ_0 is the vacuum permeability, as before, and J^μ is the 4-current

$$J^\mu = \rho \frac{d}{ds} x^\mu = \rho u^\mu, \tag{D.4}$$

where $d\tau = ds/c$ is the proper time element, ρ the charge density and u^μ the 4-velocity. The electromagnetic field tensor $F^{\mu\nu}$ is defined in terms of the 4-vector potential A^μ by the expression

$$F_{\mu\nu} = D_\mu A_\nu - D_\nu A_\mu. \tag{D.5}$$

This is an anti-symmetric tensor, and for that reason, this expression can be reduced to

$$F_{\mu\nu} = \partial_\mu A_\nu - \partial_\nu A_\mu. \tag{D.6}$$

doi:10.1088/978-0-7503-2786-2ch18

Replacing equation (D.5) in equation (D.3), we get

$$-D_\nu D^\nu A^\mu + D_\nu D^\mu A^\nu = -\mu_0 J^\mu, \tag{D.7}$$

Here, we can use the following identity

$$D_\nu D^\mu A^\nu = D^\mu D_\nu A^\nu + R_\nu^\mu A^\nu, \tag{D.8}$$

where R_ν^μ is the second order curvature tensor. This allows us to interchange the covariant derivatives in (D.7), leading to

$$-D_\nu D^\nu A^\mu + D^\mu D_\nu A^\nu + R_\nu^\mu A^\nu = -\mu_0 J^\mu. \tag{D.9}$$

Now, using the Lorentz condition

$$D_\nu A^\nu = 0, \tag{D.10}$$

we can finally write the wave equation in the form

$$\Box A^\mu = \mu_0 J^\mu + R_\nu^\mu A^\nu, \tag{D.11}$$

where the covariant *d'Alembert operator* is determined by

$$\Box = D_\nu D^\nu = \frac{1}{\sqrt{-g}} \partial_\nu \sqrt{-g}\, \partial^\nu. \tag{D.12}$$

Assuming that the current J^μ is only determined by free electrons, we can use

$$u^\mu = \frac{e}{m_e} A^\mu, \tag{D.13}$$

which leads to

$$\frac{1}{\sqrt{-g}} \partial_\nu \left(\sqrt{-g}\, \partial^\nu A^\mu \right) + k_p^2 A^\mu = R_\nu^\mu A^\nu, \tag{D.14}$$

where $k_p^2 = \omega_p^2 / c$ is the effective photon mass associated with an eventually existing background plasma. In order to illustrate this equation, we consider the simple case of a flat spacetime. In this case, we have $\partial_\nu \sqrt{-g} = 0$, and $R_\nu^\mu = 0$. We are then reduced to $(\partial_\nu \partial^\nu + k_p^2) A^\mu = 0$. Let us now define $A^\mu = e^\mu A$, and use a simpler description of the photon behaviour in curved spacetime. The evolution of the wave amplitude A is that of a scalar field, with an equation of motion

$$\left(\Box + k_p^2 + \xi R \right) A = 0, \tag{D.15}$$

where $R = R_\nu^\nu$ is the Ricci scalar, and ξ is the minimal coupling constant. Such an equation can be formally derived from the action

$$S = \int \sqrt{-g}\, d^4x \left[g^{\alpha\beta} (\partial_\alpha A)(\partial_\beta A) - \left(k_p^2 + \xi R \right) A^2 \right], \tag{D.16}$$

following the usual procedure. For $\xi = 0$, we have the so-called *minimal coupling*. Notice that retaining the plasma contribution k_p^2, is equivalent to the assumption of a finite photon mass. The value $\xi = 1/6$ corresponds to the case of *conformal coupling*.

www.ingramcontent.com/pod-product-compliance
Ingram Content Group UK Ltd.
Pitfield, Milton Keynes, MK11 3LW, UK
UKHW051904100926
8556IPUK00005B/757

9 780750 327848